Communications and Control Engineering

Communications and Control Engineering is a high-level academic monograph series publishing research in control and systems theory, control engineering and communications. It has worldwide distribution to engineers, researchers, educators (several of the titles in this series find use as advanced textbooks although that is not their primary purpose), and libraries.

The series reflects the major technological and mathematical advances that have a great impact in the fields of communication and control. The range of areas to which control and systems theory is applied is broadening rapidly with particular growth being noticeable in the fields of finance and biologically inspired control. Books in this series generally pull together many related research threads in more mature areas of the subject than the highly specialised volumes of *Lecture Notes in Control and Information Sciences*. This series's mathematical and control-theoretic emphasis is complemented by *Advances in Industrial Control* which provides a much more applied, engineering-oriented outlook.

Indexed by SCOPUS and Engineering Index.

Publishing Ethics: Researchers should conduct their research from research proposal to publication in line with best practices and codes of conduct of relevant professional bodies and/or national and international regulatory bodies. For more details on individual ethics matters please see: https://www.springer.com/gp/authors-editors/journal-author/journal-author-helpdesk/publishing-ethics/14214.

Juan Martinez-Piazuelo ·
Carlos Ocampo-Martinez · Nicanor Quijano

Generalized Nash Equilibrium Seeking in Population Games

Applications in the Control of Large-Scale Multi-agent Systems

Springer

Juan Martinez-Piazuelo
Automatic Control Department (ESAII)
Universitat Politècnica de Catalunya
Barcelona, Spain

Carlos Ocampo-Martinez
Automatic Control Department (ESAII)
Universitat Politècnica de Catalunya
Barcelona, Spain

Nicanor Quijano
Department of Electrical and Electronics Engineering
Universidad de los Andes
Bogotá, Colombia

ISSN 0178-5354 ISSN 2197-7119 (electronic)
Communications and Control Engineering
ISBN 978-3-032-06080-8 ISBN 978-3-032-06081-5 (eBook)
https://doi.org/10.1007/978-3-032-06081-5

Mathematics Subject Classification: 91A22, 91A80, 90C29, 58E25, 93B45, 93C35, 90C06, 93A15, 93A16

This Springer imprint is published by the registered company Springer Nature Switzerland AG
The registered company address is: Gewerbestrasse 11, 6330 Cham, Switzerland

To my family

—Juan Martinez-Piazuelo

To Nahia and Lissa

—Carlos Ocampo-Martinez

To Oriana, Carlos, Lilia, and Catalina

—Nicanor Quijano

Foreword

The proliferation of systems at large or societal scale, and the ability to measure and influence their behavior, has motivated research into scalable methods for analysis and design that remain tractable despite the complexity of these systems. This effort has been driven in part by the challenge of predicting unexpected instabilities, as seen in interconnected power grids, epidemics, and even congestion pricing. For instance, in 2024 in Virginia (US), congestion pricing went unstable and charged drivers hundreds of dollars for a short commute.

A promising approach to the analysis and design of such systems leverages the formalism of population games, in which participants are strategic agents. These agents are grouped into populations with similar characteristics and preferences, and may be artificial, human, or animal.Within each population, agents select strategies from a predefined set and can repeatedly revise their strategy choices to improve their outcomes, as measured by the strategies' payoffs. A payoff mechanism, either designed by a coordinator or abstractly representing the strategic environment, ascribes the payoffs. Typically, the agents are indistinguishable, which allows the prediction of the population's strategic profile through finite-dimensional evolutionary systems that are both insightful and tractable.

This formulation draws inspiration from earlier work in Evolutionary Game Theory, where strategies were phenotypes and agents were organisms evolving in an environment over generations to adopt phenotypes yielding higher fitness. In an analogy with population games, strategies and their payoffs correspond to phenotypes and fitness, respectively. Evolutionary biologists and economists, notably Joseph Hofbauer and William Sandholm, popularized and substantially generalized these ideas to study societal-scale systems, such as traffic congestion in transportation networks, in which the agents are commuters, the strategies are the available routes, and the possible reductions in commuting durations across the routes are the payoffs. Besides the now-common term population games, popularized by Sandholm, one could also adopt John M. Smith's view and refer to the framework as learning in large populations. This term reflects the fact that evolutionary dynamics is used to model not only the agents' preferences but also their learning behavior. An important learning objective is Nash equilibrium seeking, which describes the ability of the populations to reach and maintain a Nash equilibrium of the payoff mechanism. Since Nash equilibria often yield performance guarantees, other desirable properties or even optimality in some cases, establishing Nash equilibrium seeking is a central theme of this book.

As evolutionary biologists and economists employed stability concepts and system-theoretic methods to study Nash equilibrium seeking, the control systems community realized that evolutionary dynamics, which describe the populations' learning behavior, together with the payoff mechanism, form a nonlinear feedback system. This insight opened the possibility of applying not only advanced system-theoretic tools but also of developing new ones to study Nash equilibrium seeking in broader contexts, thus expanding the range of applications. Over the last 15 years, a new wave of methods has emerged from control systems researchers, among whom the authors of this book are leaders. These methods address two key challenges that are central to this book. The first is the presence of constraints on the agents' learning behavior, leading to generalized Nash equilibria. The second is the presence of dynamics in payoff mechanisms, which in evolutionary biology and economics were mostly considered memoryless. These advances have greatly expanded the applicability of population games, as evidenced by pioneering work by the authors on decentralized constrained optimization, optimal dispatch in microgrid control, water distribution systems, building temperature control, decentralized control of urban drainage systems, and decentralized control more broadly.

The book offers a concise, accessible, and thorough presentation of the foundations of population games, with an emphasis on generalized Nash equilibrium seeking and recent methods grounded in new and promising notions of system-theoretic passivity theory. It is the first book to comprehensively integrate design methodologies for population games, where steering the populations' strategic profile is viewed as a control problem.

Several examples drawn from the authors' extensive work and the available literature are discussed throughout. The techniques and methods presented will be relevant to many future applications, including the design of payoff mechanisms to regulate the behavior of strategic populations involving agentic AI. In this setting, the speed and efficiency of AI agents may, on the one hand, render the population's learning behavior more predictable, but, on the other hand, may introduce unforeseen dynamic effects. The tractable approach proposed in the book provides systematic tools to study and mitigate such effects.

This book is ideal for researchers and students seeking to learn the foundations and recent results in generalized Nash equilibrium seeking and system-theoretic methods for the analysis and design of population games, along with their applications. It is equally valuable for those already familiar with the classic theory of population games who wish to learn these new methods. Researchers in the social sciences will find that this book significantly expands the analysis and design techniques at their disposal, while evolutionary biologists and control systems researchers will encounter concepts and methods that allow them to apply their expertise to new problems involving learning in large populations.

As population games continue to grow in relevance as a framework for modeling and designing societal-scale systems in the age of AI, this book will stand as an indispensable reference for all who aim to advance the principled analysis and design of such systems.

Maryland, USA
September 2025

Nuno C. Martins

Preface

The growing complexity of modern infrastructures, cyber-physical systems, and networked multi-agent systems has introduced new challenges in modeling, analysis, and control. Traditional centralized approaches often become infeasible in large-scale scenarios, where the collective behavior of many interacting decision-makers must be understood and guided. In recent years, population games and evolutionary dynamics have emerged as a powerful paradigm for modeling such large-scale multi-agent decision-making processes. At the same time, generalized Nash equilibrium (GNE) problems and their control-oriented formulations have become central to the study of distributed and networked systems from an optimization-based perspective.

This book stems from the authors' research efforts to unify these two viewpoints, namely population games and optimization-based control of large-scale complex systems (LSCSs), into a common framework. It aims to provide a structured and self-contained treatment of how the theory of population games can be leveraged to analyze, design, and implement decentralized control strategies for large-scale multi-agent systems. In doing so, it bridges the gap between theoretical game-theoretic foundations, dynamical systems analysis, and practical controller design for LSCSs.

The book is intended for graduate students, researchers, and practitioners working in control theory, game theory, optimization, and networked systems. It assumes familiarity with basic concepts in dynamical systems and optimization but includes optional appendices reviewing the Karush–Kuhn–Tucker optimality conditions and Lyapunov stability theory to keep the material as self-contained as possible.

The content is organized into three main parts. Part I introduces the modeling framework, including the formulation of population games, payoff mechanisms, and revision processes, as well as their connection to GNE problems from an optimization-based perspective. Part II develops the necessary theoretical tools, most notably δ-dissipativity theory, to establish convergence guarantees in a broad class of large-scale multi-agent settings. Finally, Part III focuses on controller design, presenting both prescriptive and non-prescriptive approaches for steering the collective behavior of large populations and extending the discussion to LSCSs beyond classical population-game models.

Throughout the book, several illustrative numerical examples are provided, together with open-source *Python* implementations using the `popgames` package. Readers are encouraged to replicate and adapt these examples to gain hands-on experience with the theoretical ideas discussed in the text.

It is the authors' hope that this book will serve both as an accessible introduction and as a valuable reference for those interested in the intersection of population games, evolutionary dynamics, and distributed control of large-scale systems.

Barcelona, Spain — Juan Martinez-Piazuelo
Barcelona, Spain — Carlos Ocampo-Martinez
Bogotá, Colombia — Nicanor Quijano
July 2025

Acknowledgements

This work has been partially supported by the following projects:

- Predoctoral grant FPI-UPC from the Universitat Politècnica de Catalunya and Banco Santander.
- The Spanish project entitled *Sustainable learning-based Management of Multi-resource Large-scale Systems* (SEAMLESS), with reference number PID2023-148840OB-I00, funded by Ministerio de Ciencia, Innovación y Universidades and the *Agencia Estatal de Investigación* (10.13039/501100011033/FEDER, UE).
- The Colombian project *Desarrollo de un modelo transaccional de energía no convencional de múltiples agentes para el departamento de Nariño* funded by the CTeI, SGR and Ministerio de Ciencia Tecnología e Innovación (Colombia), project with reference number BPIN 2021000100499.

We would like to acknowledge all the colleagues who have contributed in some sense to the research underpinning this book. In particular, we are deeply grateful to Prof. Eduardo F. Camacho, Prof. José M. Maestre and Dr. Ana Sánchez-Amores from Universidad de Sevilla, as well as Prof. Mariagrazia Dotoli, Prof. Raffaele Carli and Dr. Nicola Mignoni from Politecnico di Bari, for generously sharing and discussing several novel research ideas that have significantly enriched the work culminating in this book. Likewise, we wish to express our sincere appreciation to Prof. Sergio Grammatico from Delft University of Technology, Prof. Germán Obando from Universidad de Nariño, and Dr. Wicak Ananduta for their valuable discussions and insightful contributions.

Beyond these financial supports and academic collaborations, each of us would like to express personal gratitude to those who have provided unwavering support throughout the preparation of this book.

I wish to express my deepest gratitude to my parents, Juan Pablo and María Isabel, to my brother Felipe, and to my girlfriend Lorena, for their unconditional support, encouragement and love throughout the preparation of this book. Your presence has been a constant source of strength and inspiration.

I am equally grateful to my coauthors and dear friends, Carlos and Nicanor. Working alongside you on this journey has been both a privilege and a joy. Your insights and relentless dedication have shaped not only the content of this book but also my own growth as a researcher.

—Juan Martinez-Piazuelo

I want to express my deepest gratitude to my coauthors and friends, Juan and Nicanor, for the inspiring conversations and fruitful discussions we've shared throughout this journey. Working together on such a fascinating topic not only deepened our understanding (at least mine!), but also connected us through a shared passion for research and science.

To my close family circle, thank you for your unwavering support. A special thank you to my twin daughters—for their patience, for waiting, and for missing me while I was immersed in the work behind this book. Your love kept me grounded and gave meaning to every effort.

... and to that presence that no longer walks beside me, but still lives in the deepest part of my convictions and aspirations—thank you for being the silent strength that drives me forward, always!

—Carlos Ocampo-Martinez

First of all, I would like to express my deepest gratitude to my co-authors, Carlos and Juan. With Carlos, I have shared nearly 14 years of insightful and enriching discussions. He has supported many of my research ideas in ways I cannot fully express, including providing essential funding that helped Juan begin his Ph.D. journey. With Juan, I have shared almost five years of intense and rewarding collaboration. Throughout this time, he has consistently demonstrated both his exceptional talent as a researcher and his remarkable qualities as a person. His creativity and ability to articulate ideas scientifically have been instrumental in shaping the direction of this book. Together, these collaborations laid the foundation for the work presented here. We sincerely hope that this book will serve as a valuable resource for all those interested in this exciting area of research.

Second, I would like to thank all those who have supported me throughout my research journey. Had Kevin Passino not introduced me to the fascinating world of population dynamics and evolutionary games, I might never have discovered this remarkable area with applications across so many fields. I am also grateful to Andrea Serrani, who showed me how nonlinear analysis and passivity theory form some of the most fundamental tools in systems theory—insights that are reflected throughout this book. I also want to acknowledge the many students who have worked alongside us on topics related to population games and evolutionary dynamics. Their dedication and contributions have enriched our work and helped shape the research that forms the backbone of this text.

Last but not least, I would like to thank my parents, Carlos Enrique and Lilia, my sister Catalina, and especially my partner in "love and crime," Oriana. John von

Neumann once said that “happiness is an eminently empyrical (sic) proposition”—and that is precisely what Oriana has given me. Happiness is not a theory; it is you. It is the moments with you that prove to me, empirically, undeniably, that joy exists.

—Nicanor Quijano

Contents

Abbreviations and Notation

Abbreviations

BNN	Brown–von Neumann–Nash
CV	Constraint Violation
EDM	Evolutionary Dynamics Model
GNE	Generalized Nash Equilibrium
IPC	Impartial Pairwise Comparison
KKT	Karush–Kuhn–Tucker
KPI	Key Performance Indicator
LSCS	Large-Scale Complex System
NE	Nash Equilibrium
ODE	Ordinary Differential Equation
PBR	Perturbed Best-Response
PDM	Payoff Dynamics Model
PEV	Plug-in Electric Vehicle
SEPT	Separable Excess Payoff Target

Notation

$\{\cdots\}$	Set
$x \in \boldsymbol{\mathcal{X}}$	x is an element of the set $\boldsymbol{\mathcal{X}}$
$x \notin \boldsymbol{\mathcal{X}}$	x is not an element of the set $\boldsymbol{\mathcal{X}}$
$\emptyset$	Empty set
$\mathbb{R}$	Set of real numbers
$\mathbb{R}_{\geq c}$	$\mathbb{R}_{\geq c} = \{x \in \mathbb{R} : x \geq c\}$, for some $c \in \mathbb{R}$
$\mathbb{R}_{>c}$	$\mathbb{R}_{>c} = \{x \in \mathbb{R} : x > c\}$, for some $c \in \mathbb{R}$
$\mathbb{R}^n$	Space of n-dimensional (column) vectors with real entries

$\mathbb{R}^{m\times n}$	Space of $m \times n$ matrices with real entries
$\mathbb{Z}$	Set of integers
$\mathbb{Z}_{\geq c}$	$\mathbb{Z}_{\geq c} = \{x \in \mathbb{Z} : x \geq c\}$, for some $c \in \mathbb{Z}$
$\mathbb{Z}_{>c}$	$\mathbb{Z}_{>c} = \{x \in \mathbb{Z} : x > c\}$, for some $c \in \mathbb{Z}$
$\mathbb{Z}_{\gg c}$	Set of integers much greater than $c \in \mathbb{Z}$
$\mathbb{N}$	Set of natural numbers, i.e., $\mathbb{N} = \mathbb{Z}_{>0}$
$\mathbb{B}$	Set of binary numbers, i.e., $\mathbb{B} = \{0, 1\}$
$\mathcal{X}(\subset) \subseteq \mathcal{Y}$	Set $\mathcal{X}$ is a (strict) subset of $\mathcal{Y}$
$\mathcal{X} \cap \mathcal{Y}$	Intersection of sets $\mathcal{X}$ and $\mathcal{Y}$
$\mathcal{X} \cup \mathcal{Y}$	Union of sets $\mathcal{X}$ and $\mathcal{Y}$
$\mathcal{X} \backslash \mathcal{Y}$	Set of elements in $\mathcal{X}$ but not in $\mathcal{Y}$
$\mathcal{X} \times \mathcal{Y}$	Cartesian product of sets $\mathcal{X}$ and $\mathcal{Y}$
$\mathcal{X}^n$	n-dimensional Cartesian product $\prod_{i=1}^{n} \mathcal{X} = \mathcal{X} \times \cdots \times \mathcal{X}$
$\mathrm{int}(\mathcal{X})$	Relative interior of set $\mathcal{X}$, i.e., the interior of $\mathcal{X}$ relative to its affine hull
$T\mathcal{X}(\mathbf{x})$	Tangent cone of the (closed convex) set $\mathcal{X} \subseteq \mathbb{R}^n$ at $\mathbf{x} \in \mathcal{X}$, i.e., $T\mathcal{X}(\mathbf{x}) = \mathrm{cl}\,\{\mathbf{z} \in \mathbb{R}^n : \mathbf{z} = \alpha(\mathbf{y} - \mathbf{x})$, for some $(\mathbf{y}, \alpha) \in \mathcal{X} \times \mathbb{R}_{\geq 0}\}$ (here $\mathrm{cl}(\cdot)$ denotes the closure operator)
$\mathbf{x}^\top (\mathbf{X}^\top)$	Transpose of vector $\mathbf{x}$ (matrix $\mathbf{X}$)
$\mathbf{X}^{-1}$	Inverse of matrix $\mathbf{X}$
$\mathrm{col}(\cdot)$	Column vector concatenation operator
$\mathrm{diag}(\cdot)$	Block diagonal matrix concatenation operator
$\lvert\alpha\rvert$	Absolute value of the scalar $\alpha \in \mathbb{R}$
$\lvert\mathcal{X}\rvert$	Cardinality of the set $\mathcal{X}$
$\lVert\mathbf{x}\rVert$, $\lVert\mathbf{x}\rVert_2$	Euclidean norm of vector $\mathbf{x}$
$\lVert\mathbf{x}\rVert_{\mathbf{M}}^2$	If $\mathbf{x} \in \mathbb{R}^n$ and $\mathbf{M} \in \mathbb{R}^{n\times n}$, then $\lVert\mathbf{x}\rVert_{\mathbf{M}}^2 = \mathbf{x}^\top \mathbf{M}\mathbf{x}$
$\lVert\mathbf{X}\rVert$, $\lVert\mathbf{X}\rVert_2$	Spectral norm of matrix $\mathbf{X}$
$\lVert\mathbf{x}\rVert_\infty$	Infinity norm of vector $\mathbf{x}$
$\lambda_{\max}(\mathbf{X})$	Maximum eigenvalue of the square matrix $\mathbf{X}$
$\lambda_{\min}(\mathbf{X})$	Minimum eigenvalue of the square matrix $\mathbf{X}$
$\mathbf{1}_n(\mathbf{0}_n)$	n-dimensional vector of all ones (zeros)
$\mathbf{1}_{m\times n}(\mathbf{0}_{m\times n})$	$m \times n$ matrix of all ones (zeros)
$\mathbf{I}_n$	Identity matrix of dimension n
$\mathbf{X} \succ 0\,(\prec 0)$	$\mathbf{X}$ is a positive (negative) definite matrix
$\mathbf{X} \succeq 0\,(\preceq 0)$	$\mathbf{X}$ is a positive (negative) semi-definite matrix
$\mathbf{X} \succ \mathbf{Y}$	$\mathbf{X} - \mathbf{Y} \succ 0$ (similar notation hold for $\succeq$, $\prec$ and $\preceq$)
$\mathbf{x} \succ \mathbf{0}_n (\succeq \mathbf{0}_n)$	Every element of the vector $\mathbf{x} \in \mathbb{R}^n$ is positive (non-negative)
$\mathbf{x} \prec \mathbf{0}_n (\preceq \mathbf{0}_n)$	Every element of the vector $\mathbf{x} \in \mathbb{R}^n$ is negative (non-positive)
$\odot$	Hadamard/element-wise product
$\otimes$	Kronecker product
$\nabla f(\cdot)$	(Column vector) gradient of scalar-valued function $f(\cdot)$
$D\mathbf{f}(\cdot)$	Jacobian matrix of vector-valued function $\mathbf{f}(\cdot)$
$\dot{x}(t)$	Time derivative of signal $x(t)$, i.e., $\dot{x}(t) = dx(t)/dt$
$[\cdot]_+$	$[\cdot]_+ = \max\{\cdot, 0\}$

$[\cdot]_c$	$[\cdot]_c = \max\{\cdot, c\}$, for some $c \in \mathbb{R}$
$[\cdot]_a^b$	$[\cdot]_a^b = \min\{\max\{\cdot, a\}, b\}$, for some $a, b \in \mathbb{R}$
$[\cdot]_+^c$	$[\cdot]_+^c = \min\{\max\{\cdot, 0\}, c\}$, for some $c \in \mathbb{R}$
$\mathrm{supp}(\mathbf{x})$	$\mathrm{supp}(\mathbf{x}) = \{i \in \{1, 2, \ldots, n\} : x_i > 0\}$, with $\mathbf{x} = \mathrm{col}(x_i)_{i=1}^n$
$\mathrm{fix}(T)$	If $T : \mathcal{D} \to \mathcal{D}$, with $\mathcal{D} \subseteq \mathbb{R}^n$, then $\mathrm{fix}(T) = \{\mathbf{x} \in \mathcal{D} : \mathbf{x} = T(\mathbf{x})\}$

Chapter 1
Introduction

This book explores the strategic behavior that emerges in large populations of decision-making agents when engaged in repeated interactions. The framework under consideration consists of three fundamental elements:

E1 The first element, the *population game*, defines the underlying strategic environment by specifying:

  i the populations of decision-making agents;
  ii their available choices and coupling constraints (if any); and
  iii the desirability of each available choice as a function of the distribution of choices across the populations, i.e., the *fitness functions*.

E2 The second element, the *payoff mechanism*, determines the payoffs or rewards that agents perceive according to their strategic choices. In other words, the payoff mechanism establishes how agents "sense" the strategic environment in which they are involved, in terms of the performance of their selected actions.

E3 The final element, the *revision process*, governs when and how agents update their choices in response to their perceived payoffs. Therefore, the revision process is the learning rule that allows agents to autonomously modify their choices according to their performance.

Within this framework, the population game establishes the strategic setting by specifying its variables, constraints, and objectives; the revision process defines the rules that guide adaptation over time; and the payoff mechanism is the interface between the population game and the revision process, linking strategic choices to perceived rewards (e.g., economic incentives, performance indicators).

The considered framework is particularly relevant in contexts where decision-making is decentralized, and agents operate under limited information. Unlike centralized control approaches, where a single authority dictates the actions of individual agents, the considered framework allows for self-organizing behavior to emerge through repeated interactions. Its applications span biological, social, and economic

J. Martinez-Piazuelo et al., *Generalized Nash Equilibrium Seeking in Population Games*, Communications and Control Engineering,
https://doi.org/10.1007/978-3-032-06081-5_1

contexts, as well as technological domains where many autonomous agents make strategically interdependent decisions. Although precise technical conditions will be progressively introduced throughout this book, there are three fundamental properties that an application context must satisfy to be well posed for the considered framework:

P1 *Large populations:* Each population consists of a *large* number of agents.
P2 *Finite roles:* Each agent belongs to exactly one of a finite number of populations. All agents within the same population share the same finite set of available choices and perceive the same payoff as any other agent making the same choice, regardless of their individual identity.
P3 *Negligible individual influence:* An individual agent's choice has little to no impact on the problem's coupling constraints, the fitness functions, and the payoffs perceived by the remainder agents from a population. These elements depend on the aggregate choice distribution of the populations rather than on the decisions of specific agents.

Unlike classical game theory, which primarily models the logical reasoning and strategic choices of individual players (often assuming perfect rationality), the considered framework focuses on the aggregate behavior of large populations of autonomous decision-makers. Instead of tracking individual agents or enforcing specific choices, the goal is to characterize how the distribution of choices within a set of populations evolves over time. This shift in perspective enables the study of emergent, collective behaviors in large-scale decentralized (yet interdependent) decision-making processes.

Within the considered framework, two fundamental questions arise:

Q1 *Does an equilibrium state exist?* An equilibrium is a choice distribution in which no agent has an incentive to unilaterally deviate once it is reached. That is, given the current state, no single agent can improve their payoff by switching choices, assuming others remain unchanged. If such a state exists, it can be interpreted as a self-enforcing agreement among the populations, providing a natural solution concept for the underlying (potentially non-cooperative) multi-agent decision-making problem.
Q2 *Will the populations converge to an equilibrium state, provided one exists?* Given a predefined set of population games, payoff mechanisms, and revision processes, can it be ensured (at least in a probabilistic sense) that the system will settle into a stable, self-enforcing agreement? Since agents' choices are interdependent, this question is not merely about individual adaptation but about the collective dynamics of the entire system. If convergence occurs, the equilibrium state acts as a reliable long-term predictor of the populations' choice distribution.

Answering these fundamental questions not only provides a theoretical understanding of equilibrium behavior in large populations, but also offers key insights for designing mechanisms to control such systems. For instance, if the context allows it, one may employ prescriptive approaches where the revision processes of the individual decision-making agents are designed to guarantee the populations' convergence toward a desired equilibrium. Alternatively, non-prescriptive approaches may leverage incentive design such that, under the assumption that agents operate by their own fixed revision processes, appropriate payoff mechanisms steer the populations toward an intended outcome. In both cases, accurately modeling and understanding the coupled dynamics of large-scale multi-agent interactions are essential aspects for achieving effective coordination and control.

The goal of this introductory chapter is to provide an initial overview of the modeling framework considered throughout this book and to illustrate the scope of its potential applications. To this end, Sect. 1.1 briefly introduces the three fundamental elements of the framework: Elements E1–E3, and contextualizes the specific equilibrium notion studied in this book: the generalized Nash equilibrium. Such a section is intended as a high-level introduction rather than a formal treatment of these topics, seeking to familiarize the reader with key concepts before they are rigorously developed in Chap. 2. Section 1.2 presents and discusses several application examples that align with the proposed framework, motivating different control perspectives and their relevance for large-scale multi-agent systems. Finally, Sect. 1.3 outlines the book's structure and provides instructions for replicating the various numerical examples that will be presented throughout the book.

1.1 Initial Overview of the Modeling Framework

As mentioned above, the modeling framework considered throughout this book consists of three fundamental elements: the population game, the payoff mechanism, and the revision process.

The Population Game

Consider a finite set of $P \in \mathbb{Z}_{\geq 1}$ populations, denoted by

$$\mathcal{P} = \{1, 2, \ldots, P\}.$$

Each population $k \in \mathcal{P}$ consists of a large but finite number of agents. To simplify the formulation at this introductory stage, the agents of each population k are modeled as a continuum of mass $m^k \in \mathbb{R}_{>0}$. Note that the superscript k in m^k is an index representing an arbitrary population, not an exponent. Hence, $m^k \neq \prod_{i=1}^{k} m$. This notational convention is adopted consistently throughout the book.

All agents within the same population share a finite set of available choices, referred to as *strategies*. The set of strategies available to the agents of population k is given by

$$\mathcal{S}^k = \{1, 2, \ldots, n^k\}.$$

Here, n^k denotes the number of strategies available to agents in population k, and it is assumed that $n^k \in \mathbb{Z}_{\geq 2}$, ensuring that each agent has at least two options to choose from. The total number of strategies across all populations is thus $n = \sum_{k \in \mathcal{P}} n^k$.

During game play, each agent in population k selects a (pure) strategy from $\mathcal{S}^k$. The aggregate choice distribution of population k is then described by the vector

$$\mathbf{x}^k = \mathrm{col}\left(x_i^k\right)_{i\in\mathcal{S}^k} = \left[x_1^k,\, x_2^k,\, \cdots,\, x_{n^k}^k\right]^\top,$$

where $x_i^k \in [0, m^k]$ represents the mass of agents in population k choosing strategy i. Similarly, the strategic distribution across all populations is given by

$$\mathbf{x} = \mathrm{col}\left(\mathbf{x}^k\right)_{k\in\mathcal{P}} = \left[\mathbf{x}^{1\top},\, \mathbf{x}^{2\top},\, \cdots,\, \mathbf{x}^{P\top}\right]^\top.$$

Throughout, $\mathbf{x}^k$ and $\mathbf{x}$ are referred to as the *strategic distribution* (or simply the *state*) of population k and the entire society, respectively. It follows that the set of all population states for population k is denoted by

$$\Delta^k = \left\{\mathbf{x}^k \in \mathbb{R}_{\geq 0}^{n^k} : \sum_{i\in\mathcal{S}^k} x_i^k = m^k\right\},$$

and the set of all society states is given by $\Delta = \prod_{k\in\mathcal{P}} \Delta^k$.

The population game may be subject to coupling constraints imposed on the strategic distributions. While Δ defines the set of all possible strategic distributions of the society, depending on the problem one may be interested in a particular subset of Δ that satisfies additional constraints. More precisely, let $\Omega \subseteq \mathbb{R}^n$ represent the set of constraints, and define

$$\mathcal{X} = \Delta \cap \Omega.$$

Thus, $\mathcal{X}$ is the set of *feasible society states* with respect to Ω. That is, a society state $\tilde{\mathbf{x}} \in \Delta$ is said to be feasible if and only if $\tilde{\mathbf{x}} \in \mathcal{X}$. Since society states capture aggregate choice distributions, the constraint set Ω is agnostic to the individual identities of the agents.

For each population $k \in \mathcal{P}$, every strategy $i \in \mathcal{S}^k$ is characterized by a so-called *fitness function* $f_i^k : \Delta \to \mathbb{R}$, which assigns a *fitness value* $f_i^k(\mathbf{x})$ to strategy i in population k for every society state $\mathbf{x} \in \Delta$. The fitness functions determine the desirability of each strategy as a function of the society's strategic distribution and independent of the constraints given by Ω. In other words, the fitness value of a given strategy at a particular society state represents how desirable that strategy is at that state, regardless of the constraint set Ω. That is, fitness functions account only for the underlying objective and not the constraints. Establishing a parallel with mathematical optimization helps illustrate these concepts.

Example 1.1 Consider a population game consisting of a single population. In this case, $\mathcal{P} = \{1\}$, which allows the superscript k to be omitted from all notation. Suppose that each agent seeks to autonomously choose its own strategy so that the resulting strategic distribution of the population, namely $\mathbf{x}^* \in \mathcal{X}$, solves the following optimization problem:

$$\max_{\mathbf{x}\in\Delta} \quad \varphi(\mathbf{x}) \quad \text{s.t.} \quad \mathbf{x} \in \Omega, \tag{1.1}$$

where $\varphi : \mathbb{R}^n_{\geq 0} \to \mathbb{R}$ is a concave and differentiable function. In this context, $\varphi(\cdot)$ is interpreted as a collective objective function that the population seeks to maximize. Disregarding the constraint set Ω, the desirability of a strategy depends solely on its marginal contribution to the increase of $\varphi(\cdot)$ at a given population state. This concept corresponds precisely to the interpretation of a fitness function. Specifically, in this example, the fitness function for each strategy $i \in \mathcal{S}$ is given by

$$f_i(\mathbf{x}) = \frac{\partial \varphi(\mathbf{x})}{\partial x_i}.$$

This formulation aligns with the principle that strategies with higher marginal benefits tend to be more desirable.

It is important to note that, for the population to effectively solve the given constrained optimization problem (1.1), the constraint set Ω should indeed be taken into account in the decision-making process of the agents. However, as will be discussed in the forthcoming sections, this goal is achieved through payoff mechanisms and revision processes rather than through the fitness functions.

Furthermore, the analogy with mathematical optimization is intended purely for illustrative purposes. In general, population games do not necessarily involve the maximization of a collective objective function. ♢

As with population and society states, the fitness functions are concatenated into vector-valued functions as

$$\mathbf{f}^k(\mathbf{x}) = \text{col}\left(f_i^k(\mathbf{x})\right)_{i\in\mathcal{S}^k}, \quad \forall k \in \mathcal{P},$$
$$\mathbf{f}(\mathbf{x}) = \text{col}\left(\mathbf{f}^k(\mathbf{x})\right)_{k\in\mathcal{P}}.$$

Throughout, $\mathbf{f}^k(\cdot)$ is referred to as the *fitness function of population k*, while $\mathbf{f}(\cdot)$ is denoted as the *fitness function of the entire society*. Similar to Ω, since fitness functions take aggregate choice distributions as inputs, they are agnostic to the individual identities of the agents.

Under the previous definitions, a population game is defined in normal form as

$$G = (\mathcal{P}, \mathcal{X}, \mathbf{f}(\cdot)).$$

This representation encapsulates the set of populations ($\mathcal{P}$), the feasible strategic distributions ($\mathcal{X} = \Delta \cap \Omega$), and the fitness function of the entire society ($\mathbf{f}(\cdot)$).

The Payoff Mechanism

The population game G is played repeatedly over time. Therefore, it is essential to formally define how agents perceive the evolving strategic environment in which they operate.

As previously discussed, fitness functions characterize the desirability of each strategy based on the society state, independent of the problem's constraints. Payoffs, on the other hand, determine the rewards that agents perceive for each of their available strategies. While fitness functions define the underlying objectives of the population game, payoffs represent how agents actually experience the strategic environment.

Let $t \in \mathbb{R}_{\geq 0}$ denote the continuous-time index, and let $\mathbf{x}(t)$ be the society state at time t. In the simplest case, the payoff associated to strategy i at time t could be given by

$$p_i^k(t) = f_i^k(\mathbf{x}(t)), \qquad \forall i \in \mathcal{S}^k, \ \forall k \in \mathcal{P},$$

indicating that the payoffs perceived by agents correspond exactly to the fitness values of the respective strategies at the current society state. This model, however, neglects the possibility that agents' perception of the environment may involve dynamic effects not captured by the fitness functions. For instance, phenomena such as anticipation, delay, estimation, or inertia cannot be accounted for within such a memoryless payoff definition. Moreover, this simple model excludes the possibility for payoffs to depend on additional state variables other than $\mathbf{x}(t)$, which could represent for instance dynamical systems external to (yet interacting with) the society of agents.

To allow for greater generality, payoffs are instead defined as the output of a dynamical system representing the so-called payoff mechanism. Specifically, in this book the focus is on continuous-time dynamical systems of the form

$$\begin{aligned} \dot{\mathbf{q}}(t) &= \mathcal{W}(\mathbf{q}(t), \mathbf{x}(t)), \\ \mathbf{p}(t) &= \mathcal{H}(\mathbf{q}(t), \mathbf{x}(t)), \end{aligned}$$

where[1] $\mathcal{W} : \mathbb{R}^d \times \Delta \to \mathbb{R}^d$ governs the evolution of additional state variables $\mathbf{q}(t)$, and $\mathcal{H} : \mathbb{R}^d \times \Delta \to \mathbb{R}^n$ defines how these additional state variables, together with the society state, influence the *overall payoff vector*

$$\mathbf{p}(t) = \operatorname{col}\left(\operatorname{col}\left(p_i^k(t)\right)_{i \in \mathcal{S}^k}\right)_{k \in \mathcal{P}}.$$

Notably, setting $d = 0$ and defining $\mathcal{H}(\cdot) = \mathbf{f}(\cdot)$ recovers the memoryless case, where payoffs match the fitness functions as $\mathbf{p}(t) = \mathbf{f}(\mathbf{x}(t))$. Nonetheless, when $d > 0$, the dynamic model may capture richer dynamic effects, allowing the payoff mechanism to depend not only on the current society state $\mathbf{x}(t)$ but also on additional state variables $\mathbf{q}(t)$.

To illustrate these concepts, it is convenient to consider a primal-dual optimization perspective under the light of Example 1.1.

Example 1.2 Recall the scenario of Example 1.1 and, for simplicity, let the constraint set Ω be given by

[1] Both $\mathcal{W}(\cdot,\cdot)$ and $\mathcal{H}(\cdot,\cdot)$ are assumed to be locally Lipschitz continuous.

$$\Omega = \{\mathbf{x} \in \mathbb{R}^n : \mathbf{A}\mathbf{x} = \mathbf{b}\},$$

where $\mathbf{A} \in \mathbb{R}^{d \times n}$ and $\mathbf{b} \in \mathbb{R}^d$ define the constraints. To incentivize agents to solve the constrained optimization problem (1.1), the payoff mechanism can be potentially defined as

$$\begin{aligned}\dot{\mathbf{q}}(t) &= \mathbf{A}\mathbf{x}(t) - \mathbf{b} \\ \mathbf{p}(t) &= \mathbf{f}\left(\mathbf{x}(t)\right) - \mathbf{A}^\top \mathbf{q}(t).\end{aligned} \tag{1.2}$$

By setting $\mathbf{f}(\mathbf{x}) = \nabla_{\mathbf{x}} \varphi(\mathbf{x})$ and interpreting the additional state variables $\mathbf{q}(t)$ as the dual-Lagrange multipliers associated with the equality constraints in Ω, it follows that $\mathbf{p}(t)$ corresponds to the gradient (with respect to $\mathbf{x}$, evaluated at time t) of the dual-Lagrangian function

$$L(\mathbf{x}, \mathbf{q}) = \varphi(\mathbf{x}) - \mathbf{q}^\top \left(\mathbf{A}\mathbf{x} - \mathbf{b}\right).$$

That is, $\mathbf{p}(t) = \nabla_{\mathbf{x}} L\left(\mathbf{x}(t), \mathbf{q}(t)\right)$. Under this formulation, the payoff associated with each strategy depends not only on its underlying desirability, as determined by the fitness function, but also on the influence of the constraints at the current society state. In fact, the proposed payoff mechanism increases the elements of $\mathbf{q}(t)$ corresponding to constraint violations in Ω, thereby adjusting the perceived desirability of the related strategies through their payoffs. ◊

Who Computes the Payoffs?

The modeling framework discussed so far is motivated by large-scale decentralized decision-making scenarios. Thus, a natural question that may arise is: *Who computes the payoffs?* Or equivalently, *how is the payoff mechanism implemented in practice?* The answer depends on the specific application and the interpretation of the payoff mechanism.

In the literature on population games and evolutionary dynamics, the process by which payoffs are computed is often modeled implicitly rather than explicitly. This fact is because, in many applications, particularly in biological and sociological contexts, payoffs emerge from complex dynamics involving how intelligent agents perceive and interpret the outcomes of their strategic behaviors. As such, high-level models are often used to implicitly capture the agents' learning processes, incorporating factors such as anticipation, delay, estimation, or inertia. In these cases, payoffs are computed *internally* by the agents, and this process is implicitly represented through the payoff mechanism.

Conversely, in many practical engineering applications, payoffs have a concrete meaning. For instance, they may correspond to economic incentives, performance indicators, or constraint violations, as in Example 1.2. In such contexts, a well-defined higher-level entity is responsible for computing payoffs and broadcasting them to the decision-making agents. This entity, which may be a central controller, a regulatory authority, or a distributed computational system, collects relevant information from the system, processes it, and then produces payoff signals to guide the agents' strategic choices. In these scenarios, payoffs are computed *externally* to the agents and subsequently communicated to them. The mechanisms through which these payoffs are communicated may vary, ranging from direct signals in automatic control systems to publicly available information in economic and social systems.

To maintain a flexible framework, later in this book the notion of *payoff providers* is introduced. A payoff provider may correspond to a tangible entity, such as a physical system or institution that explicitly computes and disseminates payoff signals, an intangible entity that represents an underlying cognitive process within the agents, or a combination of both. The payoff mechanism, in turn, serves as the mathematical abstraction that models how payoffs are computed, regardless of whether they arise from external structures, internal cognition, or a mixture of both. This distinction allows the framework to accommodate a wide range of applications, from biological and socioeconomic systems to decision-making environments in engineering.

The Revision Process

The final element of the modeling framework is the revision process. While the population game defines the problem under consideration and the payoff mechanism determines how agents perceive the environment, the revision process governs when and how agents update their decisions. In this sense, the revision process is the core of the decision-making dynamics of the society.

To determine when and how to update their selected strategy, each agent is equipped with an *alarm clock* and a *revision protocol*. The alarm clock provides sporadic and random opportunities for each individual agent to revise their strategy, thereby specifying the *when*. In contrast, the revision protocol is the rule that a revising agent follows to select a new strategy, defining the *how*. A detailed explanation of the mechanics behind these components is addressed later in Chap. 2. However, some key observations are highlighted below:

- Revision times are independent and identically distributed across all agents, enabling random and asynchronous decision-making processes.
- Revision protocols are mappings from the society state and payoff vector to the probability distribution over each agent's available strategies, allowing for uncertainty and the possibility of mistakes in the agents' decisions.

Example 1.3 illustrates these concepts.

Example 1.3 Consider a population game with a single population ($P = 1$), where each agent makes decisions repeatedly over time at a constant average rate. Namely, the time interval between two consecutive revision opportunities for each agent is random but occurs on average every $1/R$ time units, where $R \in \mathbb{R}_{>0}$. This scenario can be modeled by equipping agents with *Poisson alarm clocks*, where each individual revision opportunity is triggered according to an independent exponential distribution with rate R.

On the other hand, suppose that agents are self-interested and always seek to choose the strategy leading to the highest payoff. However, agents are imperfect and prone to making mistakes. This scenario can be modeled by equipping agents with revision protocols characterized by the *softmax* function given by

$$\mathbb{P}(i) = \frac{\exp\left(\frac{p_i(\tilde{t})}{\eta}\right)}{\sum_{j \in \mathcal{S}} \exp\left(\frac{p_j(\tilde{t})}{\eta}\right)}, \quad \forall i \in \mathcal{S}, \tag{1.3}$$

where $\mathbb{P}(i)$ denotes the probability that an agent, revising their decision at time t, chooses i as their new strategy; $\tilde{t} < t$ is an arbitrary time instant between the previous revision time of any agent in the population and time t; and $\eta \in \mathbb{R}_{>0}$ represents the noise level in the decision-making process.[2] Namely, the parameter η accounts for the possibility that agents may make mistakes and do not always select the strategy with the highest payoff.

To further clarify the role of the revision process and its interplay with the other fundamental elements of the modeling framework, recall Examples 1.1 and 1.2. From such an optimization perspective, the population game corresponds to the problem definition, specifying the variables, constraints, and objective function (captured through the fitness functions). The payoff mechanism (1.2) characterizes the dynamics of the dual-Lagrange multipliers, enabling agents to account for constraint violations through the perceived payoffs. The revision process, in turn, corresponds to the *primal dynamics* that govern the evolution of the primal variables ($\mathbf{x}$) over time. Once again, this optimization interpretation serves only as an illustration, yet it provides intuition into the role of each fundamental element.

To illustrate the interplay between the three fundamental elements under the considered optimization perspective, a numerical simulation is depicted in Fig. 1.1. For the considered instance of (1.1), the set of strategies is taken as $\mathcal{S} = \{1, 2, 3\}$ (thus, $n = 3$); the objective function is defined as

$$\varphi(\mathbf{x}) = \frac{1}{2}\mathbf{x}^\top \mathbf{Q}\mathbf{x} + \mathbf{r}^\top \mathbf{x},$$

where $\mathbf{Q} = \operatorname{diag}(-0.6, -0.7, -0.8)$ and $\mathbf{r} = \begin{bmatrix}0, 0.1, 0.1\end{bmatrix}^\top$; and the coupling constraint is given by $\mathbf{A} = \begin{bmatrix}0.5, 0.5, 0\end{bmatrix}$ and $b = 0.2$. Furthermore, the payoff mechanism is set according to (1.2), and a population of 1000 agents is considered, represented as a unitary mass ($m = 1$). The Poisson alarm clocks are set with rate $R = 1$, and the noise level of the revision protocols (1.3) fixed at $\eta = 10^{-2}$. Without loss of generality, it is assumed that the game begins with all agents choosing the first strategy, i.e., $\mathbf{x}(0) = \begin{bmatrix}1, 0, 0\end{bmatrix}^\top$, and the initial condition of the payoff mechanism is set to $q(0) = 0$.

As depicted in Fig. 1.1, the agents initially move toward the (unconstrained) optimizer of $\varphi(\mathbf{x})$ within Δ. However, over time, they adjust their strategies in response

[2] Taking the limit $\eta \to \infty$ results in a uniform distribution over $\mathcal{S}$, where $\mathbb{P}(i) = \frac{1}{n}$ for all $i \in \mathcal{S}$, while $\eta \to 0$ results in a (possibly degenerate) distribution where $\mathbb{P}(i) = \frac{1}{c(t)}$ for every $i \in \arg\max_{j \in \mathcal{S}} p_j(t)$ and $\mathbb{P}(i) = 0$ otherwise. Here, $c(t) = \left|\{i : i \in \arg\max_{j \in \mathcal{S}} p_j(t)\}\right|$.

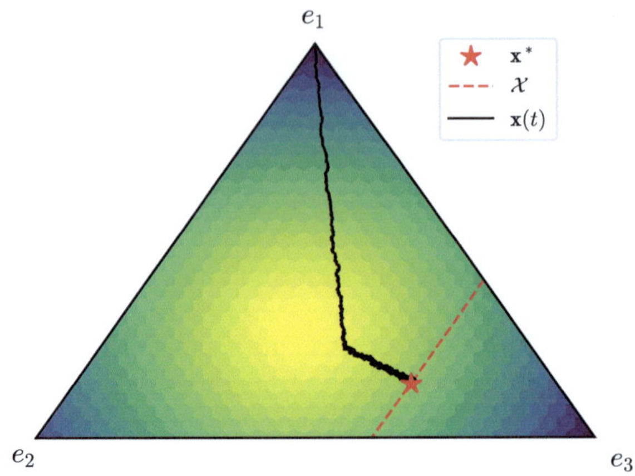

Fig. 1.1 A numerical simulation of Example 1.3. The heatmap depicts the projection of the objective function $\varphi(\cdot)$ over the simplex Δ, and $\mathbf{x}^* = \arg\max_{\mathbf{x}\in\mathcal{X}} \varphi(\mathbf{x})$

to their perceived payoffs and converge toward the state $\mathbf{x}^* = \arg\max_{\mathbf{x}\in\mathcal{X}} \varphi(\mathbf{x})$, successfully satisfying the coupling constraint of Ω and solving the constrained optimization problem (1.1). Consequently, under the light of the considered optimization perspective, the interplay between the population game, the payoff mechanism, and the revision process, corresponds to a primal-dual optimization algorithm, where the agents collectively solve (1.1) in a decentralized fashion. $\diamondsuit$

Remark 1.1 (*A Note on Alarm Clocks*) Example 1.3 introduces the notion of Poisson alarm clocks, which model the scenario in which each agent's strategy revision times occur independently and repeatedly over time, following a Poisson process with a constant average rate. While other random processes could be used to model agents' alarm clocks, this book places particular emphasis on Poisson clocks due to their suitability for various practical scenarios and their analytical tractability. In particular, the mathematical properties of Poisson clocks enable the derivation of key theoretical results that will play a fundamental role later in this book. $\diamondsuit$

The Generalized Nash Equilibrium Notion

As discussed before, two key questions naturally arise within the modeling framework considered in this book. The first, Question Q1, concerns the existence of an equilibrium state. Specifically, does there exist a society state $\mathbf{x}^* \in \mathcal{X}$ in which no agent has an incentive to unilaterally deviate once it is reached? The second, Question Q2, pertains to the convergence of the system's dynamics: given an equilibrium $\mathbf{x}^*$, will the agents' decision-making processes drive the system toward it over time?

These questions are fundamental to understanding the viability and predictive power of the modeling framework. The existence of an equilibrium provides a well-defined solution concept, capturing a state of self-enforcing stability in a multi-agent decision-making process. However, existence alone is insufficient if the system fails to reach such a state in practice. This is where the second question becomes critical:

when an equilibrium exists, the interplay between population games, payoff mechanisms, and revision processes must be examined (and potentially designed) to assert whether agents' strategic adjustments will ultimately lead to such an equilibrium.

The equilibrium state described in Question Q1 is known in the game theory literature as a Nash equilibrium (NE) or a generalized Nash equilibrium (GNE) when coupling constraints exist (i.e., if $\Omega \subset \mathbb{R}^n$), and the existence of such equilibria can be well established under mild regularity conditions. The key distinction between an NE and a GNE is that the latter explicitly accounts for coupling constraints among agents' decisions. While formal definitions are provided in Chap. 2, intuitively, a GNE is a strategic distribution $\mathbf{x}^* \in \mathcal{X}$ in which no agent has an incentive to unilaterally deviate from their chosen strategy. If no explicit coupling constraints exist over the strategic distributions, then the notion of GNE simplifies to that of an NE. For the sake of generality, this book focuses on the GNE notion.

It turns out, that in the context of Examples 1.1–1.3, the unique GNE of the game G is precisely $\mathbf{x}^* = \arg\max_{\mathbf{x}\in\mathcal{X}} \varphi(\mathbf{x})$. A formal proof of this claim is provided later in Chap. 2.

1.2 Practical Application Examples

Large-scale multi-agent decision-making problems are ubiquitous in real-world scenarios, many of which can be effectively modeled within the framework presented in this book. To illustrate its versatility and applicability, this section introduces six practical application examples. The goal is to establish a foundational understanding of how different problem structures can be mapped onto the framework.

The first example focuses on modeling rather than control, serving as an initial illustration of the framework's capabilities. It demonstrates how its three fundamental components (the population game, the payoff mechanism, and the revision process) can be structured within a practical context.

The next two examples illustrate scenarios where prescriptive control approaches are required. Specifically, they present cases in which the revision processes of decision-making agents must be designed to achieve a desired system performance, such as the emergence of an equilibrium outcome.

In contrast, the final three examples focus on non-prescriptive control approaches, where agents' revision processes are assumed to be fixed, and control is instead exerted through the design of the payoff mechanism. These examples highlight how indirect interventions, such as incentives or monetary penalties, play a role in shaping system-wide behavior without directly enforcing individual decision-making policies.

These six examples will serve as recurring references throughout this book, providing practical application contexts that motivate the exploration of subsequent theoretical developments and analytical tools.

1.2.1 Scientific Research Topic Selection

Researchers choose a specific topic for their upcoming scientific endeavors. This decision is influenced by factors such as available funding, prevailing trends in the scientific community, and the potential for making novel contributions. Researchers tend to favor topics that are not overly saturated, where securing funding and producing original work are more feasible. Yet, they seek areas with abundant collaboration opportunities, avoiding both overrepresented and underpopulated topics.

This complex dynamic process of how researchers choose their topics can be well modeled under the considered framework by defining each of its fundamental elements as outlined below.

- **Population Game:**
  - Multiple populations represent different research domains (e.g., machine learning, automatic control).
  - The decision-making agents of each population are the researchers (or research groups) specializing in the corresponding domain.
  - The available strategies within each domain correspond to specific research topics. For example, in machine learning, strategies may include natural language processing or computer vision, while in automatic control, they may include model predictive control or optimal control. Notably, some topics may overlap across multiple domains (e.g., reinforcement learning).
  - For simplicity, explicit coupling constraints in this context may be neglected, leading to $\Omega = \mathbb{R}^n$.
  - The fitness functions should capture the fact that both overrepresented and underpopulated topics are less desirable for researchers. To this end, a suitable formulation is given by

$$f_i^k(\mathbf{x}) = \sigma_i^k \left(x_i^k + \sum_{\ell \in \mathcal{P} \setminus \{k\}} \sum_{j \in \mathcal{S}^\ell} \omega_{ij}^{k\ell} x_j^\ell \right), \quad \forall i \in \mathcal{S}^k, \ \forall k \in \mathcal{P},$$

    where

    - $\omega_{ij}^{k\ell} = 1$ if $i \in \mathcal{S}^k$ and $j \in \mathcal{S}^\ell$ correspond to the same (overlapping) topic in different populations $k \neq \ell$, and $\omega_{ij}^{k\ell} = 0$ otherwise.
    - $\sigma_i^k : \mathbb{R}_{\geq 0} \to \mathbb{R}_{\geq 0}$ is a *bell-shaped* function modeling the fact that topics with either too few or too many researchers are less desirable. A canonical example is the Gaussian function:

$$\sigma_i^k(z) = a_i^k \exp\left(-\left(\frac{z - b_i^k}{c_i^k} \right)^2 \right),$$

      with parameters $a_i^k, b_i^k, c_i^k \in \mathbb{R}_{>0}$.

- **Payoff Mechanism:**
  Depending on the modeling goal, the payoff mechanism can be formulated as

$$\begin{aligned} \dot{\mathbf{q}}(t) &= \tau\left(\mathbf{f}\left(\mathbf{x}(t)\right) - \mathbf{q}(t)\right), \\ \mathbf{p}(t) &= \alpha_0 \mathbf{f}\left(\mathbf{x}(t)\right) + \alpha_1 \mathbf{q}(t) + \alpha_2 \dot{\mathbf{q}}(t), \end{aligned} \tag{1.4}$$

  which can capture combinations of *memoryless* ($\alpha_0 = 1, \alpha_1 = \alpha_2 = 0$), *smoothing* ($\alpha_0 = 0$, $\alpha_1 = 1$, $\alpha_2 = 0$), and *anticipatory* ($\alpha_0 = 1$, $\alpha_1 = 0$, $\alpha_2 > 0$) behaviors based on their parameters (in all cases with $\tau \in \mathbb{R}_{>0}$). This payoff mechanism allows modeling dynamic effects such as delay, anticipation, and inertia in how researchers perceive payoffs.
- **Revision Process:**
  Given that researchers make decisions based on personal preferences and experiences, their specific revision processes cannot be accurately modeled or designed. However, meaningful analysis can still be conducted by considering broad families of intuitive decision-making policies.
  As an example, since revision times correspond to moments when researchers (or research groups) reassess their research priorities, they may be modeled using Poisson clocks (as in Example 1.3) with a time scale on the order of years. The revision protocols, in turn, can be assumed to follow probabilistic rules. For instance, one may consider that a researcher currently specializing in topic $i \in \mathcal{S}^k$ switches to topic $j \in \mathcal{S}^k$ with a probability proportional to the difference in their associated payoffs. Specifically, the decision-making process may be characterized by

$$\begin{aligned} \mathbb{P}(j|i) &\propto \left[p_j^k(\tilde{t}) - p_i^k(\tilde{t})\right]_+, \quad \forall j \neq i, \\ \mathbb{P}(i|i) &= 1 - \sum_{j \in \mathcal{S}^k \setminus \{i\}} \mathbb{P}(j|i), \end{aligned}$$

  where $[\cdot]_+ := \max\{\cdot, 0\}$, $\mathbb{P}(j|i)$ denotes the probability that a researcher revising their strategy at time t will switch from topic i to j; and $\tilde{t} < t$ is an arbitrary time instant between the previous revision time of any researcher and time t.
  By analyzing broad families of decision-making policies, valuable insights can be gained into the factors influencing researchers' decisions and the long-term equilibrium states that may emerge.

1.2.2 Load-Balancing in a Computing Center

Load-balancing schedulers assign incoming jobs to a set of available servers in a computing center. Specifically, each time a scheduler receives a job, it must select a server for allocation. The processing speed of a server degrades as its workload increases. Therefore, schedulers aim to optimize overall system performance by dynamically selecting the most efficient server for each arriving job.

This problem can be formulated within the modeling framework as follows:

- **Population Game:**
  - A single population represents the computing center (thus, $P = 1$).
  - The decision-making agents are the incoming job schedulers.
  - The strategies correspond to the servers, which are shared among schedulers.
  - No explicit constraints are considered, implying that $\Omega = \mathbb{R}^n$.
  - The fitness function associated with each server is given by

$$f_i(x_i) = s_i - \alpha_i x_i, \quad \forall i \in \mathcal{S},$$

  where $s_i \in \mathbb{R}_{>0}$ denotes the nominal processing speed of server i, and $\alpha_i \in \mathbb{R}_{\geq 0}$ is a weighting parameter. This formulation reflects a preference for servers with higher nominal computing speeds and promotes an equitable distribution of the load in proportion to the servers' associated weighting parameters.

- **Payoff Mechanism:**
  The payoff mechanism is modeled as a memoryless map fully determined by the fitness functions. Specifically,

$$p_i(t) = f_i(x_i(t)),$$

  for all $i \in \mathcal{S}$. As such, schedulers are encouraged to assign tasks to the most efficient servers, naturally balancing the load across the computing center.
- **Revision Process:**
  In this context, revision times correspond to the arrival of incoming jobs at each scheduler. Specifically, when a scheduler receives a new job, it must make a decision and select a server for allocation. Consequently, modeling revision times using Poisson clocks is appropriate, as it accurately captures the stochastic nature of job arrivals.
  On the other hand, since schedulers operate as automated programs, their revision protocols must be explicitly designed by the system's control engineer. For instance, the engineer may equip the schedulers with strategy updating rules such as (1.3). However, it is crucial to analytically ensure that the resulting dynamical process remains stable and achieves the desired system performance.

1.2.3 Multi-robot Distributed Garbage Collection

A fleet of mobile robots allocates itself to different zones of the city for distributed garbage collection. Each robot autonomously selects the zone in which it will operate. The rate at which garbage accumulates in a given zone decreases as more robots simultaneously work in that zone, while garbage accumulates more rapidly in less populated zones. The robots' goal is to collectively minimize the overall accumulated garbage across all zones, with heavier penalties given to larger accumulations. Occasionally, each robot may choose to work in a different zone to help reduce garbage in that area.

This application can be formulated within the considered framework as follows.

- **Population Game:**
  - A single population represents the fleet of robots ($P = 1$).
  - The robots act as the decision-making agents.
  - The available strategies correspond to the garbage collection zones.
  - To prevent zones from becoming overcrowded, the robots must coordinate to respect capacity limits in each zone. Accordingly, the coupling constraints are defined as $\Omega = \{\mathbf{x} \in \mathbb{R}^n : \mathbf{x} \preceq \overline{\mathbf{x}}\}$, where $\overline{\mathbf{x}} \in \mathbb{R}^n_{\geq 0}$ specifies the maximum allowable mass of robots in each zone.
  - Although robots do not have intrinsic preferences for any specific zone, they are programmed to promote an equitable distribution across all zones. Accordingly, the fitness functions are defined as

$$f_i(x_i) = -\alpha_i x_i, \quad \forall i \in \mathcal{S},$$

    where $\alpha_i \in \mathbb{R}_{\geq 0}$ is a weighting parameter. When $\alpha_i > 0$ for all i, this formulation encourages the fleet toward a balanced allocation of effort.

- **Payoff Mechanism:**
  The payoff mechanism must capture the garbage accumulation dynamics, thereby informing robots about the zones that require greater intervention. This goal can be achieved using a payoff mechanism of the form

$$\begin{aligned} \dot{q}_i(t) &= \psi_i\left(q_i(t), \gamma_i - x_i(t)\right) \\ p_i(t) &= f_i\left(x_i(t)\right) + q_i(t), \end{aligned}$$

  for all $i \in \mathcal{S}$. Here, $\gamma_i \in \mathbb{R}_{>0}$ is the nominal *garbage accumulation rate* at zone i, the state $q_i(t) \in \mathbb{R}_{\geq 0}$ represents the amount of garbage in zone i at time t, and $\psi_i : \mathbb{R} \times \mathbb{R} \to \mathbb{R}$ is locally Lipschitz continuous and satisfies

$$\begin{cases} \alpha \leq 0 \Rightarrow \psi_i(\alpha, \beta) \geq 0, \ \forall \beta \in \mathbb{R} \\ \beta > 0 \Rightarrow \psi_i(\alpha, \beta) > 0, \ \forall \alpha \in \mathbb{R}_{\geq 0} \\ \beta < 0 \Rightarrow \psi_i(\alpha, \beta) < 0, \ \forall \alpha \in \mathbb{R}_{>0} \\ \beta = 0 \Rightarrow \psi_i(\alpha, \beta) = 0, \ \forall \alpha \in \mathbb{R}_{>0}. \end{cases}$$

  Notably, the properties of $\psi_i(\cdot, \cdot)$ imply that

  - $q_i(t)$ is always non-negative (provided that $q_i(0) \geq 0$).
  - $q_i(t)$ increases when $x_i(t)$ is less than the nominal accumulation rate γ_i.
  - $q_i(t)$ decreases (bounded within $\mathbb{R}_{\geq 0}$) when $x_i(t)$ exceeds γ_i.
  - $q_i(t)$ remains constant when $x_i(t) = \gamma_i$.

Under this model, the payoff $p_i(t)$ captures both the accumulated garbage in zone i, represented by $q_i(t)$, and the baseline incentive to maintain a balanced allocation across zones, given by $-\alpha_i x_i(t)$. Consequently, a robot's incentive to operate in a specific zone dynamically depends on both the current level of accumulated garbage and the distribution of robots across zones, making it more likely for robots to move toward zones that are more polluted and less populated.

- **Revision Process:**
  In this application, revision times correspond to the moments when a robot completes a garbage collection task in its current zone and must decide where to operate next. Assuming that task completion times follow an exponential distribution, Poisson alarm clocks are well suited to model revision times.
  Similar to the load-balancing example in Sect. 1.2.2, the system's control engineer is responsible for designing the revision protocols to ensure system stability and achieve the desired performance. Additionally, these protocols must adhere to the coupling constraints defined by Ω, ensuring that no zone is left with fewer robots than required.
  It is possible to design probabilistic revision protocols that enforce the constraints in Ω. For instance, the system's designer may implement protocols such that

$$\begin{aligned}
\mathbb{P}(j|i) &\propto \left[\overline{x}_i - x_i(\tilde{t})\right]_+ \left[p_j(\tilde{t}) - p_i(\tilde{t})\right]_+, \quad \forall j \neq i, \\
\mathbb{P}(i|i) &= 1 - \sum_{j \in \mathcal{S}} \mathbb{P}(j|i).
\end{aligned}$$

  As a result, any transition that would lead to $x_i(t) > \overline{x}_i$ is assigned zero probability, thereby ensuring that each zone satisfies the maximum allowed presence of robots.

1.2.4 Competition and Coordination in Ride-Hailing Apps

In a city, a population of users and a population of drivers choose among a finite set of ride-hailing mobile apps. Specifically, each time a user requires transportation, they select an app to request a ride, while each time a driver becomes available, they choose an app to offer their service. Users aim to minimize their waiting time and fare, whereas drivers seek to reduce their idle time and maximize their earnings. The waiting time and fare for users increase when more users and fewer drivers operate on the same app, while drivers experience higher earnings and lower idle times when more users and fewer competing drivers use their selected app. Consequently, a competitive dynamic emerges within each population: users compete for driver availability, while drivers compete for ride requests. Simultaneously, a coordination dynamic arises between the two populations, as both users and drivers benefit from aligning their choices to the same platforms.

This problem can be modeled within the considered framework as follows.

- **Population Game:**
  - Two populations are considered ($P = 2$): one representing users and the other representing drivers.
  - The decision-making agents are the users and drivers, respectively.
  - In both cases, the available strategies are the ride-hailing apps operating in the city.
  - To ease the exposition, no explicit coupling constraints are considered in this application (i.e., $\Omega = \mathbb{R}^n$).
  - The fitness functions should capture both the competition within the same population and the incentive for coordination with the opposing population. Letting $k = 1$ and $k = 2$ denote the populations of users and drivers, respectively, and defining $\mathcal{S} = \mathcal{S}^1 = \mathcal{S}^2 = \{1, 2, \ldots, \hat{n}\}$ as the set of ride-hailing apps operating in the city (with $\hat{n}$ denoting the total number of apps, and so $n = 2\hat{n}$), a suitable yet simple formulation for the fitness functions is

$$f_i^1(\mathbf{x}) = \epsilon_i^1 x_i^2 - (1 - \epsilon_i^1) x_i^1 - \gamma_i^1$$
$$f_i^2(\mathbf{x}) = \epsilon_i^2 x_i^1 - (1 - \epsilon_i^2) x_i^2 + \gamma_i^2,$$

  for all $i \in \mathcal{S}$. Here, the parameters $\epsilon_i^k \in (0, 1)$ define a convex combination weighting the importance of competition and coordination, and $\gamma_i^k \in \mathbb{R}_{>0}$ represents the *nominal fare and incentive* offered by app i to its users ($k = 1$) and drivers ($k = 2$), respectively.[3] Clearly, for app i to be economically sustainable, it must hold that $\gamma_i^1 > \gamma_i^2$, i.e., the operational margin, denoted $\bar{\gamma}_i = \gamma_i^1 - \gamma_i^2$, is positive. Therefore, this formulation reflects the intended trade-off: apps with more drivers and fewer users become more attractive to users, whereas apps with more users and fewer drivers are generally preferred by drivers.[4] In addition, apps that are nominally cheaper (smaller γ_i^1) are preferred by users, while apps that offer better nominal incentives (bigger γ_i^2) are more attractive to drivers.

- **Payoff Mechanism:**
  A profit-driven ride-hailing app company may seek to dynamically increase both its fares and incentives to attract more drivers, at the expense of increased costs for the users. Specifically, such a company might implement a payoff mechanism of the form

$$\dot{q}_i(t) = \left[\overline{q}_i - q_i(t)\right]_+ \left[\underline{x}_i + x_i^1(t) - x_i^2(t)\right]_+ - q_i(t)\left[-\underline{x}_i - x_i^1(t) + x_i^2(t)\right]_+$$
$$p_i^1(t) = f_i^1(\mathbf{x}(t)) - q_i(t)$$
$$p_i^2(t) = f_i^2(\mathbf{x}(t)) + q_i(t),$$

---

[3] The parameters γ_i^k are assumed to be scaled by appropriate proportionality constants.

[4] Recall that x_i^k represents the mass of agents choosing strategy i in population k. Thus, x_i^1 and x_i^2 do not simply denote numbers of users and drivers, but rather the proportion of total users and drivers selecting app i, scaled by their population masses m^1 and m^2, respectively.

where $q_i(t) \in \mathbb{R}_{\geq 0}$ represents the dynamic increment of both fares and incentives; the parameter $\underline{x}_i \in \mathbb{R}$ specifies the minimum desired *mass of drivers minus users*, i.e., the maximum allowed *deficit of drivers with respect to users*; and $\overline{q}_i \in \mathbb{R}_{\geq 0}$ is an upper bound on the dynamic adjustment of fares and incentives (set, for instance, by a regulatory entity). The considered payoff mechanism behaves as follows:

- $q_i(t)$ is always non-negative (provided that $q_i(0) \geq 0$).
- $q_i(t)$ is never greater than $\overline{q}_i$ (provided that $q_i(0) \leq \overline{q}_i$).
- $q_i(t)$ increases (bounded within $[0, \overline{q}_i]$) when $x_i^2(t) - x_i^1(t) < \underline{x}_i$.
- $q_i(t)$ decreases (bounded within $[0, \overline{q}_i]$) when $x_i^2(t) - x_i^1(t) > \underline{x}_i$.

Thus, when the mass of drivers minus the mass of users falls below $\underline{x}_i$, fares and incentives increase, potentially attracting more drivers at the expense of increased fares for the users. Conversely, when the mass of drivers minus the mass of users exceeds $\underline{x}_i$, fares and incentives decrease to balance the system. Note that, under this setup, the effective fare charged by app i to its users at time t is $\gamma_i^1 + q_i(t)$, while the effective incentive offered by app i to its drivers at time t is $\gamma_i^2 + q_i(t)$. As such, regardless of the dynamic adjustment q_i, app i's operational margin remains constant and equal to $\bar{\gamma}_i = \gamma_i^1 - \gamma_i^2$.
On the other hand, a more user-centric ride-hailing app company may choose to dynamically lower fares and increase incentives to attract both more users and drivers. While reduced fares can stimulate user demand and higher incentives can encourage driver participation, these adjustments come at the expense of the company's operational margins. As a result, dynamic incentives should comply with financial constraints of the app's company. One possibility to achieve this objective is through a payoff mechanism of the form

$$\begin{aligned}
\dot{q}_i(t) &= \left[\overline{q}_i - q_i(t)\right]_+ \left[\underline{x}_i - x_i^1(t) - x_i^2(t)\right]_+ - q_i(t)\left[-\underline{x}_i + x_i^1(t) + x_i^2(t)\right]_+ \\
p_i^1(t) &= f_i^1(\mathbf{x}(t)) + q_i(t) \\
p_i^2(t) &= f_i^2(\mathbf{x}(t)) + q_i(t),
\end{aligned}$$

where $q_i(t) \in \mathbb{R}_{\geq 0}$ represents the dynamic adjustment of both fares and incentives, the parameter $\underline{x}_i \in \mathbb{R}_{\geq 0}$ specifies the minimum desired *mass of users plus drivers*, and

$$\overline{q}_i = \frac{1}{2}\left(\gamma_i^1 - \gamma_i^2 - \bar{\gamma}_i^*\right), \tag{1.5}$$

where $\bar{\gamma}_i^* \in \mathbb{R}_{>0}$ specifies the minimum operational margin required by the company managing app i. Similar to the previous example, the next properties hold:

- $q_i(t)$ is always non-negative (provided that $q_i(0) \geq 0$).
- $q_i(t)$ is never greater than $\overline{q}_i$ (provided that $q_i(0) \leq \overline{q}_i$).
- $q_i(t)$ increases (bounded within $[0, \overline{q}_i]$) when $x_i^2(t) + x_i^1(t) < \underline{x}_i$.
- $q_i(t)$ decreases (bounded within $[0, \overline{q}_i]$) when $x_i^2(t) + x_i^1(t) > \underline{x}_i$.

Hence, when the mass of users plus drivers falls below $\underline{x}_i$, the app lowers fares and increases incentives. Conversely, when the mass of users plus drivers exceeds $\underline{x}_i$, fares are raised and incentives are reduced. Under this setup, the effective fare charged by app i to its users at time t is thus $\gamma_i^1 - q_i(t)$, while the effective incentive offered to app i's drivers at time t is $\gamma_i^2 + q_i(t)$. Consequently, the effective operational margin of app i at time t is given by

$$\bar{\gamma}_i(t) = \gamma_i^1 - \gamma_i^2 - 2q_i(t),$$

which, by (1.5), is guaranteed to be no lower than $\bar{\gamma}_i^*$.

- **Revision Process:**

  Given that both users and drivers make decisions autonomously based on personal preferences and experiences, their specific revision processes cannot be explicitly designed. Instead, they must be assumed to belong to certain classes of revision processes, maintaining as much generality as possible while still allowing meaningful analysis.

  The revision times can be effectively modeled by using Poisson clocks, with different time scales for users and drivers. For instance, drivers working on ride-hailing apps typically complete multiple rides per day, frequently switching apps based on demand and fare fluctuations. In contrast, users generally request only one or two rides per day on average. As a result, the strategy-switching rate is significantly higher for drivers than for users. However, this imbalance is offset by the fact that the total number of users is generally much larger than the total number of drivers, ensuring a sustained dynamic interaction between the two populations.

  The revision protocols should retain a probabilistic nature and may differ for users and drivers. For instance, users may select their apps following a rule of the form

$$\begin{aligned}\mathbb{P}(j|i) &\propto \left[p_j^1(\tilde{t}) - p_i^1(\tilde{t})\right]_+, \quad \forall j \neq i\\ \mathbb{P}(i|i) &= 1 - \sum_{j\in\mathcal{S}^1} \mathbb{P}(j|i),\end{aligned}$$

  which models the fact that each user may have a preferred app i, and would switch their preference to app j based on a pairwise comparison of their payoffs. On the other hand, drivers currently operating over app i may update their choices via rules of the form

$$\begin{aligned}\mathbb{P}(j|i) &\propto \left[p_j^2(\tilde{t}) - \frac{1}{m^2}\sum_{\ell\in\mathcal{S}^2} x_\ell^2(\tilde{t}) p_\ell^2(\tilde{t})\right]_+, \quad \forall j \neq i\\ \mathbb{P}(i|i) &= 1 - \sum_{j\in\mathcal{S}^2} \mathbb{P}(j|i),\end{aligned}$$

which captures the tendency of drivers to migrate toward apps offering higher-than-average payoffs (with the average weighted by how popular an app is in the population).
Effectively analyzing the interplay between user and driver decision-making can provide valuable insights for ride-hailing apps to optimize their fare and incentive policies while enhancing their competitiveness.

1.2.5 Traffic Routing in Urban Networks

Drivers have a preferred route for their daily commute from home to work. Occasionally, a driver may reconsider their preference and switch to a different route in an attempt to minimize their average travel time. The average travel time experienced by each driver depends on the congestion level of their preferred route, which, in turn, is influenced by the route preferences of all other drivers.

This problem can be formulated as follows:

- **Population Game:**
  - Multiple populations represent distinct groups of drivers commuting between different home and work locations.
  - Each driver acts as a decision-making agent.
  - The available strategies correspond to the set of accessible routes for each group of drivers.
  - Due to environmental concerns, a city planner seeks to limit the number of drivers allowed in certain areas of the city. These restrictions can be encouraged by implementing measures such as transit fees or tolls, which influence driver behavior and regulate traffic congestion. From a modeling perspective, the city planner aims to enforce constraints of the form

$$\Omega = \left\{ \mathbf{x} \in \mathbb{R}^n : \mathbf{g}(\mathbf{x}) \preceq \mathbf{0} \right\},$$

    where $\mathbf{g} : \mathbb{R}^n_{\geq 0} \to \mathbb{R}^d$ (convex and continuously differentiable) defines the restrictions imposed on aggregate traffic flows.
  - The fitness functions for this problem are defined as

$$\mathbf{f}^k(\mathbf{x}) = -\boldsymbol{\alpha}^k - \boldsymbol{\beta}^k(\mathbf{x}), \quad \forall k \in \mathcal{P},$$

    where $\boldsymbol{\alpha}^k \in \mathbb{R}^{n^k}_{\geq 0}$ represents the nominal travel times for the routes available to population k, and $\boldsymbol{\beta}^k : \mathbb{R}^n_{\geq 0} \to \mathbb{R}^{n^k}_{\geq 0}$ captures the congestion-induced delays on these routes, which depend on the strategic choices of all populations. That is, routes with shorter traveling times and lower congestion levels are more desirable.

- **Payoff Mechanism:**
  In addition to reflecting the fitness values, the payoff mechanism should account for the transit fees or tolls imposed by the city planner as the means to incentivize constraint satisfaction. One such possible payoff mechanism is

$$\begin{aligned}\dot{\mathbf{q}}(t) &= \left[\mathbf{g}\left(\mathbf{x}(t)\right)\right]_+ - \mathbf{q}(t) \odot \left[-\mathbf{g}\left(\mathbf{x}(t)\right)\right]_+ \\ \mathbf{p}(t) &= \mathbf{f}\left(\mathbf{x}(t)\right) - \mathrm{D}\,\mathbf{g}\left(\mathbf{x}(t)\right)^\top \mathbf{q}(t),\end{aligned}$$

  where $\odot$ denotes the Hadamard product and $\mathrm{D}\,\mathbf{g}(\cdot)$ is the Jacobian matrix of $\mathbf{g}(\cdot)$. Here, $\mathbf{q}(t) \in \mathbb{R}^d_{\geq 0}$ represents the dynamically computed costs associated with transit fees or tolls, which act as shadow prices to enforce constraint satisfaction.
- **Revision Process:**
  In this application, the revision times determine how frequently drivers evaluate and potentially change their preferred routes. Hence, these times may be in the order of weeks or months and can be indeed modeled by using Poisson clocks. Unlike engineering systems, driver behavior cannot be directly designed, as individuals make decisions based on personal preferences, habits, and perceptions of congestion. Consequently, in this application, revision protocols are not prescribed but rather assumed and analyzed. One possible assumption is that a driver whose current preferred route is $i \in \mathcal{S}^k$ will switch to route $j \in \mathcal{S}^k$ with a probability proportional to the perceived increase in payoff. Specifically, the switching behavior can be modeled as

$$\begin{aligned}\mathbb{P}(j|i) &\propto \left[p_j^k(\tilde{t}) - p_i^k(\tilde{t})\right]_+, \quad \forall i \neq j \\ \mathbb{P}(i|i) &= 1 - \sum_{j \in \mathcal{S}^k} \mathbb{P}(j|i).\end{aligned}$$

  By considering a family of intuitive revision protocols, a city planner can evaluate whether equilibrium behaviors naturally emerge from drivers' choices. This analysis can then inform the design of appropriate payoff mechanisms, such as transit fees and tolls, to guide the system toward a desired traffic distribution.

1.2.6 Distributed Energy Management in Microgrids

A network of interconnected microgrids operates within a city, each responsible for managing its local energy generation and consumption. These microgrids interact to coordinate power exchanges, seeking to ensure coupled constraint satisfaction across the system.

Each microgrid consists of multiple populations of energy producers as well as multiple populations of energy consumers, all with flexible generation and demand capabilities. Each producer (e.g., a renewable energy generator with battery storage) and each consumer (e.g., a plug-in electric vehicle) selects an energy management policy from a set of available options. For instance:

- A renewable energy source with battery storage selects a pair (*time slot*, *power*) from a predefined finite set of possible combinations, specifying the fixed power to be supplied during the corresponding time period.
- A plug-in electric vehicle selects a pair (*time slot*, *charging rate*) from a predefined finite set of possible combinations, specifying the constant rate at which it will charge during the chosen time window.

The available energy management policies (e.g., the possible combinations of time slots and corresponding supply/demand levels) may vary from one population to another, reflecting differences in generation capacity, demand flexibility, and operational constraints. In all cases, both producers and consumers seek to optimize their individual costs based on dynamic market prices, which are established by the microgrid network. Namely, these market prices are influenced by supply-demand interactions as well as the coupling constraints between multiple microgrids.

This problem can be cast within the modeling framework as outlined next.

- **Population Game:**
  - The society consists of all populations of producers and consumers. For instance, if there are five populations in total, $\mathcal{P} = \{1, 2, 3, 4, 5\}$, then populations $\{1, 2\}$ may represent producers, while populations $\{3, 4, 5\}$ may correspond to consumers. In turn, each microgrid is represented by a disjoint subset of populations. For example, in a system with two microgrids, one microgrid may include populations $\{1, 3, 4\}$, while another microgrid may consist of populations $\{2, 5\}$.
  - The individual producers and consumers act as the decision-making agents.
  - The set of strategies for each population corresponds to its available energy management policies. While all agents within the same population share the same set of strategies, different populations may have distinct strategy sets, reflecting heterogeneity in available policies across the system.
  - The coupling constraints capture the required balance between generation and demand across the network of microgrids. Consequently, the set Ω takes the form
  $$\Omega = \left\{\mathbf{x} \in \mathbb{R}^n : \mathbf{A}\mathbf{x} = \mathbf{b}\right\},$$
  where $\mathbf{A} \in \mathbb{R}^{d \times n}$ and $\mathbf{b} \in \mathbb{R}^d$ characterize the constraints.
  - The fitness function represents the fixed prices offered to agents, independent of aggregate supply and demand levels. Specifically, it may capture *time-of-use* pricing schemes that act as a baseline to the dynamic pricing mechanism implemented by the microgrid network. Accordingly, the fitness function is assumed to take the form
  $$\mathbf{f}(\mathbf{x}) = -\boldsymbol{\alpha},$$
  where $\boldsymbol{\alpha} \in \mathbb{R}^n_{\geq 0}$ specifies the fixed baseline prices for all strategies.

- **Payoff Mechanism:**
  The payoff mechanism defines the dynamic pricing scheme computed across the network of microgrids, incentivizing agents to adjust their supply and demand levels to satisfy the coupling constraints of Ω. Specifically, the payoff mechanism is modeled as

$$\dot{\mathbf{q}}(t) = \mathcal{D}\left(\mathbf{q}(t), \mathbf{x}(t)\right)$$
$$\mathbf{p}(t) = \mathbf{f}\left(\mathbf{x}(t)\right) - \mathbf{M}\mathbf{q}(t),$$

  where $\mathbf{M}\mathbf{q}(t)$ represents the market prices (paid *by* consumers *to* producers) that incentivize supply and demand balance across the system, $\mathbf{M} \in \mathbb{R}^{n \times d}$, and $\mathcal{D} : \mathbb{R}^d \times \mathbb{R}^n_{\geq 0} \to \mathbb{R}^d$ is locally Lipschitz continuous. Besides, the payoff mechanism satisfies the following properties:

  - *Equilibrium compliance:* If $\mathcal{D}\left(\mathbf{q}(t), \mathbf{x}(t)\right) = \mathbf{0}$, then $\mathbf{A}\mathbf{x}(t) = \mathbf{b}$, for all $t \geq 0$. That is, every equilibrium state of the payoff mechanism certifies that the constraints of Ω are satisfied.
  - *Distributed computability:* The function $\mathcal{D}(\cdot, \cdot)$ and the price vector $\mathbf{M}\mathbf{q}(t)$ can be computed in a distributed manner over the microgrid network, ensuring compatibility with any system's communication-related constraints.

  In other words, the payoff mechanism must incentivize the satisfaction of the coupling constraints while respecting the system's decentralized nature, allowing multiple microgrids to negotiate market prices in a distributed manner.
- **Revision Process:**
  In this context, revision times correspond to the moments when producers and consumers reassess their energy management policies. The frequency of these revisions depends on the nature of the agents. For instance, if the agents correspond to automated energy management systems operating in a time-ahead market, then their revision times may be on the scale of minutes or hours. In contrast, if the agents represent individuals manually selecting their energy management policies, revision times may occur over days or even weeks. Consequently, a reasonable assumption is that revision times are stochastically triggered, following Poisson alarm clocks with different rates across populations, reflecting the diverse time scales at which different types of agents operate.
  On the other hand, revision protocols also depend on the nature of the agents. For automated energy management systems, it may be possible to explicitly design and enforce a specific decision-making policy. However, for individuals, decision-making is inherently subjective and cannot be directly imposed. To maintain generality, it is assumed that revision protocols vary across populations and follow probabilistic rules, as seen in previous examples.
  By appropriately modeling and analyzing the revision processes of the agents, control engineers can design suitable payoff mechanisms to ensure system stability and constraint satisfaction under decentralized decision-making and distributed computations over the microgrid network.

1.3 How to Read This Book?

This book presents a structured treatment of the considered framework for modeling, analyzing, and designing control methods for large-scale multi-agent decision-making systems. It is intended to be as self-contained as possible, allowing readers to navigate the material based on their background and interests.

Specifically, the book is structured into three parts:

- **Part I: Modeling (Chap. 2)**
  This part formally introduces the core concepts, mathematical models, and key assumptions of the modeling framework. It provides an in-depth exploration of its three fundamental components: the population game, the payoff mechanism, and the revision process; discussing their interplay and presenting deterministic models that serve as foundational tools for formal analysis. Additionally, this part defines the role of control within the framework and formally states the GNE problem, drawing a parallel with an optimization-based perspective. Since this part establishes the fundamental concepts that underpin the rest of the book, it is essential for all readers.
- **Part II: Analysis (Chaps. 3 and 4)**
  This second part focuses on the theoretical analysis of the modeling framework, establishing key properties related to GNE existence, characterization, and convergence guarantees.
  Chapter 3 introduces the notion of δ-dissipative dynamical systems, providing a self-contained introduction to the main analytical tools used throughout the book to study the considered framework.
  Chapter 4 then presents the main theoretical results concerning the existence and characterization of generalized Nash equilibria in the considered framework, along with sufficient conditions on its fundamental elements to ensure convergence of the society toward a GNE. This chapter offers a rigorous mathematical foundation for analyzing decentralized decision-making processes in large-scale multi-agent systems.
  This part is particularly relevant for readers interested in understanding the formal technical conditions that guarantee equilibrium emergence and stability in large-scale multi-agent systems.
- **Part III: Design (Chaps. 5 and 6)**
  The final part of the book is dedicated to the design of control methods for GNE seeking in large populations and large-scale complex systems (LSCSs). It builds upon the theoretical tools introduced in Part II and demonstrates how they can be leveraged for controller design in distributed and networked settings.
  Chapter 5 focuses on two complementary approaches:
  - *Prescriptive methods*, where revision protocols are explicitly designed to ensure constraint satisfaction and system stability.

– *Non-prescriptive methods*, where payoff mechanisms are strategically designed to influence the behavior of decision-making agents operating under fixed decision rules, guiding the system toward a desired equilibrium.

Chapter 6 extends the design perspective beyond traditional multi-agent populations by exploring the role of population games and evolutionary dynamics in the control of LSCSs. It shows how LSCSs can be formulated as non-cooperative N-player games, where the underlying optimization-based control tasks can be cast as GNE seeking problems. Furthermore, it highlights that population games can themselves be interpreted as LSCSs, thereby linking the modeling concepts of Chap. 2 with optimization-based control. An analogy-based viewpoint is then introduced, illustrating how evolutionary dynamics can be applied even when the system does not correspond to a genuine population of autonomous agents.
Taken together, this part is intended for readers interested in translating the analysis and modeling frameworks of earlier chapters into practical control methodologies for large-scale multi-agent systems and complex networked infrastructures.
In addition, the book includes two optional appendices provided for the reader's convenience, helping to keep the text as self-contained as possible. Appendix A reviews the Karush–Kuhn–Tucker (KKT) optimality conditions, while Appendix B introduces the basics of Lyapunov stability theory. These appendices offer fundamental background concepts that support, but are not required for, the main material of the book.

Based on previous discussions, it is recommended that readers follow the book's structure sequentially, progressing through Part I, Part II, and Part III in order. However, Part III has been written in a way that allows readers to understand its main ideas skipping directly from Part I, without necessarily delving into the theoretical analysis of Part II. Therefore, a reader primarily interested in *controller design* rather than the underlying mathematical analysis may skip Part II.

Replicating the Simulation Examples

Throughout the book, several numerical simulation examples are presented, referencing the application examples in Sect. 1.2. To maximize the learning experience, readers are encouraged to replicate the examples and experiment with the different components of each of them. To facilitate this process, the book provides code snippets demonstrating how to reproduce the presented simulations in *Python* using the `popgames` package. This package is open source, publicly available,[5] and it can be easily installed using

```
pip install popgames
```

For instance, to replicate the simulations from Example 1.3, readers simply need to execute the code shown in Listing 1.1.

[5] Refer to: https://github.com/Martinez-Piazuelo/popgames.

Engaging with the provided code not only reinforces the theoretical insights developed throughout the book but also allows for a deeper understanding of the modeling and control techniques through hands-on experimentation. Whether used for academic exploration, teaching, or applied research, the `popgames` package serves as a practical tool for fast implementation and simulation. Further documentation, including installation instructions, usage examples, and extended tutorials, is available in the package's online repository.

Notes

Comprehensive treatments of population games and revision processes can be found in [4, 12], with related biological perspectives discussed in [13, 15, 17]. These frameworks, however, consider only memoryless payoff mechanisms, which are static maps fully characterized by the fitness function. The introduction of general dynamic payoff mechanisms first appeared in [10, 11]. All of these works focus on the convergence of the society toward a Nash equilibrium (NE) [9] (or a perturbed variant), without explicitly incorporating coupling constraints on the agents' decisions.

In contrast, explicit considerations of generalized Nash equilibrium (GNE) problems [2] within population games have been explored in [1] for memoryless payoff mechanisms and in [7] for dynamic payoff mechanisms.

Additional practical applications not covered in this book include bandwidth sharing and user-base-station association in wireless networks [3, 16], decentralized residential demand response [6, 14], subsidy-based control policies implemented by governmental entities [8], and hassle-versus-price games in manufacturing systems [5], among others [12].

```
import numpy as np
import popgames as pg
np.random.seed(698) # fix the random seed

Q = np.diag([-0.6, -0.7, -0.8])
r = np.array([0.0, 0.1, 0.1]).reshape(-1, 1)
A, b = np.array([[0.5, 0.5, 0]]), np.array([0.2])

def f(x):  # Fitness function
    return np.dot(Q, x) + r

def w(q, x):  # RHS of payoff dynamics model
    return np.dot(A, x) - b

def h(q, x):  # Output of payoff mechanism
    return f(x) - np.dot(A.T, q)

population_game = pg.SinglePopulationGame(
    num_strategies=3, fitness_function=f,
    mass=1, A_eq=A, b_eq=b
)

payoff_mechanism = pg.PayoffMechanism(
    h_map=h, w_map=w, n=3, d=1
)

revision_process = pg.PoissonRevisionProcess(
    Poisson_clock_rate=1.0,
    revision_protocol=pg.protocol.Softmax(eta=0.01)
)

sim = pg.Simulator(
    population_game=population_game,
    payoff_mechanism=payoff_mechanism,
    revision_processes=revision_process,
    num_agents = 1000
)

x0, q0 = np.array([1, 0, 0]).reshape(-1, 1), np.
   array([[0]])
sim.reset(x0=x0, q0=q0)
sim.run(T_sim=30)

def varphi(x): # Potential function to plot
    return 0.5*np.dot(x.T, np.dot(Q, x)) + np.dot(r
        .T, x)

sim.plot(
    plot_type='ternary', potential_function=varphi,
    plot_deterministic_approximation=True
)
```

Listing 1.1 Python code to replicate Example 1.3

References

1. Barreiro-Gomez J, Tembine H (2018) Constrained evolutionary games by using a mixture of imitation dynamics. Automatica 97:254–262. https://doi.org/10.1016/j.automatica.2018.08.014
2. Facchinei F, Kanzow C (2010) Generalized Nash equilibrium problems. Ann Oper Res 175(1):177–211
3. Fletscher LA, Barreiro-Gomez J, Ocampo-Martinez C, Peroni CV, Maestre JM (2018) Atomicity and non-anonymity in population-like games for the energy efficiency of hybrid-power hetnets. IEEE Trans Netw Serv Manage 15(4):1600–1614. https://doi.org/10.1109/TNSM.2018.2879450
4. Hofbauer J, Sigmund K (1998) Evolutionary games and population dynamics. Cambridge University Press
5. Kara S, Martins NC (2023) Pairwise comparison evolutionary dynamics with strategy-dependent revision rates: Stability and δ-passivity. IEEE Trans Control Netw Syst 10(4):1656–1668. https://doi.org/10.1109/TCNS.2023.3237485
6. Martinez-Piazuelo J, Ananduta W, Ocampo-Martinez C, Grammatico S, Quijano N (2023) Population games with replicator dynamics under event-triggered payoff provider and a demand response application. IEEE Control Syst Lett 7:3417–3422. https://doi.org/10.1109/LCSYS.2023.3285532
7. Martinez-Piazuelo J, Quijano N, Ocampo-Martinez C (2022) A payoff dynamics model for generalized Nash equilibrium seeking in population games. Automatica 140:110227. https://doi.org/10.1016/j.automatica.2022.110227
8. Morimoto T, Kanazawa T, Ushio T (2016) Subsidy-based control of heterogeneous multiagent systems modeled by replicator dynamics. IEEE Trans Autom Control 61(10):3158–3163. https://doi.org/10.1109/TAC.2015.2504519
9. Nash J (1951) Non-cooperative games. Ann Math 54(2):286–295
10. Park S, Martins NC, Shamma JS (2019) From population games to payoff dynamics models: a passivity-based approach. In: Proceedings of the 58th IEEE Conference on Decision and Control (CDC), pp 6584–6601 (2019). https://doi.org/10.1109/CDC40024.2019.9029756
11. Park S, Martins NC, Shamma JS (2019) Payoff dynamics model and evolutionary dynamics model: Feedback and convergence to equilibria. https://doi.org/10.48550/ARXIV.1903.02018. arXiv:1903.02018
12. Sandholm WH (2010) Population games and evolutionary dynamics. MIT Press
13. Smith MJ (1984) The stability of a dynamic model of traffic assignment: an application of a method of Lyapunov. Transp Sci 18(3):245–252
14. Srikantha P, Kundur D (2017) Resilient distributed real-time demand response via population games. IEEE Trans Smart Grid 8(6):2532–2543. https://doi.org/10.1109/TSG.2016.2526651
15. Taylor PD, Jonker LB (1978) Evolutionary stable strategies and game dynamics. Math Biosci 40(1):145–156. https://doi.org/10.1016/0025-5564(78)90077-9
16. Tembine H, Altman E, El-Azouzi R, Hayel Y (2010) Evolutionary games in wireless networks. IEEE Trans Syst Man Cybern Part B (Cybern) 40(3):634–646. https://doi.org/10.1109/TSMCB.2009.2034631
17. Weibull JW (1997) Evolutionary game theory. MIT Press

Part I
Modeling

Chapter 2
Modeling Decision-Making in Large Populations

This chapter presents a detailed treatment of the modeling framework introduced in Chap. 1. The focus is on the analysis of large populations of decision-making agents whose behavior evolves through repeated interactions within a non-cooperative, payoff-driven environment. Unlike classical game theory, which centers on the logical reasoning of individual agents, the perspective adopted here emphasizes the collective behavior of populations. Moreover, agents are assumed to have bounded rationality and make decisions asynchronously and under uncertainty, both in the timing of their strategic revisions and in the choices they make.

To study the temporal evolution of strategic behavior in this setting, appropriate analytical tools are required. This chapter introduces the notion of *evolutionary dynamics model* (EDM), which serves as a deterministic approximation characterizing the expected evolution of the society's strategic distribution. In addition, it presents the *payoff dynamics model* (PDM), which governs the dynamics of the payoff mechanism, i.e., the dynamics ruling the agents' perceived payoffs. The EDM and PDM are interconnected in a positive feedback structure, and the properties of this closed-loop system are central to determining whether the population converges toward a *generalized Nash equilibrium* (GNE): **a stable, feasible configuration of decisions that persists over time in the absence of external disturbances**. Such convergence serves as a long-term predictor of the society's collective behavior.

The remainder of this chapter is structured as follows. Sections 2.1–2.3 introduce the three fundamental elements of the modeling framework: the *population game*, the *payoff mechanism*, and the *revision process*, along with illustrative examples. Section 2.4 develops the deterministic approximation models that enable analytical tractability throughout the rest of the book. Section 2.5 establishes the notion of control within the considered framework and formally defines the concept of a GNE while providing a parallelism with optimization. Finally, Sect. 2.6 presents numerical simulations to further illustrate the discussed ideas.

J. Martinez-Piazuelo et al., *Generalized Nash Equilibrium Seeking in Population Games*, Communications and Control Engineering,
https://doi.org/10.1007/978-3-032-06081-5_2

2.1 Population Games

Consider a society with $P \in \mathbb{Z}_{\geq 1}$ populations of decision-making agents, and let

$$\mathcal{P} = \{1, 2, \ldots, P\}$$

be the set indexing the populations. For every population $k \in \mathcal{P}$, its set of agents is indexed by

$$\mathcal{A}^k = \{1, 2, \ldots, N^k\},$$

where $N^k \in \mathbb{Z}_{\geq 1}$ denotes the (fixed and typically large) number of agents that belong to population k. Under this notation,[1] $a \in \mathcal{A}^k$ represents an arbitrary agent a that belongs to population k.

Each agent $a \in \mathcal{A}^k$, for all $k \in \mathcal{P}$, is fully described by the strategy they choose. That is, the individual identities of agents are not relevant, only their selected strategies. Accordingly, let

$$\mathcal{S}^k = \{1, 2, \ldots, n^k\}$$

be the set of strategies[2] available to the agents of population k, where $n^k \in \mathbb{Z}_{\geq 2}$ specifies the total number of available choices. The strategic decision of agent $a \in \mathcal{A}^k$ is then represented by the binary vector

$$\mathbf{s}_a^k \in \left\{ \begin{bmatrix} 1 \\ 0 \\ \vdots \\ 0 \end{bmatrix}, \begin{bmatrix} 0 \\ 1 \\ \vdots \\ 0 \end{bmatrix}, \ldots, \begin{bmatrix} 0 \\ 0 \\ \vdots \\ 1 \end{bmatrix} \right\} \subset \mathbb{R}^{n^k},$$

which encodes the agent's strategic choice in a *one-hot* format: if agent a selects strategy $i \in \mathcal{S}^k$, then $\mathbf{s}_a^k$ is the ith canonical basis vector in $\mathbb{R}^{n^k}$, i.e., the ith column of the $n^k \times n^k$ identity matrix.

The total number of agents in population k selecting each strategy in $\mathcal{S}^k$ is then captured by the vector

$$\boldsymbol{\xi}^k = \sum_{a \in \mathcal{A}^k} \mathbf{s}_a^k,$$

where each element ξ_i^k of $\boldsymbol{\xi}^k = \operatorname{col}\left(\xi_i^k\right)_{i \in \mathcal{S}^k} \in \mathbb{R}^{n^k}$ yields the count of agents choosing strategy $i \in \mathcal{S}^k$. Consequently, the proportion of agents in population k choosing strategy i is given by

$$x_i^k = \frac{\xi_i^k}{N^k}, \quad \forall i \in \mathcal{S}^k.$$

[1] When a single population is considered, i.e., $P = 1$, the upper index ^{k} is omitted from all notation.

[2] A set of strategies refers to a (finite) set of possible decisions/actions.

Furthermore, letting $m^k \in \mathbb{R}_{>0}$ represent the total *mass* of agents in population k, yields that

$$x_i^k = m^k \chi_i^k, \quad \forall i \in \mathcal{S}^k$$

is the mass of agents in population k choosing strategy i. Clearly, if $m^k = 1$, then it trivially holds that $x_i^k = \chi_i^k$.

Remark 2.1 (*The Role of Masses*) The introduction of population masses $\{m^k\}_{k\in\mathcal{P}}$ provides additional flexibility for modeling heterogeneous populations. By allowing each population to have a distinct total mass, the framework can account for differences in population size, influence, or relevance within the overall system. Moreover, in the limiting case as $N^k \to \infty$, the agents within population k can be interpreted as a continuum of mass m^k, allowing the framework to abstract away from individual-level modeling while still capturing aggregate decision-making behavior at the population level. ◊

Under the considered setup, the *strategic distribution* of population k, which describes how the agents of population k are distributed across the available strategies in $\mathcal{S}^k$, is fully captured by the vector

$$\begin{aligned}\mathbf{x}^k &= \frac{m^k}{N^k}\sum_{a\in\mathcal{A}^k} \mathbf{s}_a^k \\ &= \operatorname{col}\left(x_i^k\right)_{i\in\mathcal{S}^k} \in \Delta^k,\end{aligned}$$

where

$$\Delta^k = \left\{\mathbf{x}^k \in \mathbb{R}_{\geq 0}^{n^k} : \mathbf{1}_{n^k}^\top \mathbf{x}^k = m^k\right\}$$

denotes the set of possible strategic distributions for population k. Similarly, the strategic distribution of the entire society is captured by the vector

$$\mathbf{x} = \operatorname{col}\left(\mathbf{x}^k\right)_{k\in\mathcal{P}} \in \Delta,$$

where $\Delta = \prod_{k\in\mathcal{P}} \Delta^k = \Delta^1 \times \Delta^2 \times \cdots \times \Delta^P$, is the set of possible strategic distributions for the entire society. It is worth highlighting that Δ^k is defined as a subset of the real space $\mathbb{R}_{\geq 0}^{n^k}$, regardless of the mass m^k. In particular, even in the case where $m^k = 1$, so that each element $x_i^k = \chi_i^k$ corresponds to a rational number, the set Δ^k is still treated as a subset of $\mathbb{R}_{\geq 0}^{n^k}$, disregarding any non-convexities that would arise from restricting attention to the rationals. This abstraction is motivated by the assumption that N^k is large, and it enables the use of continuous and convex analysis techniques throughout the rest of the book.

As discussed in Chap. 1, in many applications, the society's strategic distribution $\mathbf{x}$ may be subject to global constraints that go beyond the individual feasibility of each population. While the set Δ defines all possible aggregate strategic distributions across the entire society, it is often necessary to restrict attention to a subset that satisfies additional coupling constraints. Let $\Omega \subseteq \mathbb{R}^n$ denote the *constraint set*, which captures such system-wide requirements. The set of *feasible society states* is then defined as

$$\mathcal{X} = \Delta \cap \Omega.$$

In this way, a strategic distribution $\tilde{\mathbf{x}} \in \Delta$ is considered feasible if and only if $\tilde{\mathbf{x}} \in \mathcal{X}$. Throughout this book, it is assumed that Ω satisfies Assumption 2.1.

Assumption 2.1 The set Ω is of the form

$$\Omega = \left\{ \mathbf{x} \in \mathbb{R}^n : \begin{array}{r} \mathbf{A}\mathbf{x} - \mathbf{b} = \mathbf{0}_{c_=} \\ \mathbf{g}(\mathbf{x}) \preceq \mathbf{0}_{c_\leq} \end{array} \right\},$$

where $\mathbf{A} \in \mathbb{R}^{c_= \times n}$, $\mathbf{b} \in \mathbb{R}^{c_=}$, $\mathbf{g} : \mathbb{R}^n_{\geq 0} \to \mathbb{R}^{c_\leq}$, for some $c_=, c_\leq \in \mathbb{Z}_{\geq 1}$. Moreover, it holds that:

(i) The set $\mathrm{int}(\mathcal{X}) = \left\{ \mathbf{x} \in \Delta \cap \Omega \cap \mathbb{R}^n_{>0} : \mathbf{g}(\mathbf{x}) \prec \mathbf{0}_{c_\leq} \right\}$ is nonempty.

(ii) The matrix $\tilde{\mathbf{A}} = \left[\mathbf{A}^\top, \mathbf{A}_\Delta^\top \right]^\top \in \mathbb{R}^{c_= + P \times n}$ is full-row rank. Here,

$$\mathbf{A} = \left[\mathbf{A}^1, \cdots, \mathbf{A}^P \right] \in \mathbb{R}^{c_= \times n},$$

and $\mathbf{A}_\Delta \in \mathbb{R}^{P \times n}$ is such that Δ can be written as

$$\Delta = \left\{ \mathbf{x} \in \mathbb{R}^n_{\geq 0} : \mathbf{A}_\Delta \mathbf{x} = \mathrm{col}\left(m^k \right)_{k \in \mathcal{P}} \right\}.$$

(iii) $\mathbf{g}(\cdot) = \mathrm{col}\left(g_z(\cdot) \right)_{z \in \{1,2,\ldots,c_\leq\}}$, where $g_z : \mathbb{R}^n_{\geq 0} \to \mathbb{R}$ is continuously differentiable and convex, for all $z \in \{1, 2, \ldots, c_\leq\}$. ◊

By Assumption 2.1, it follows that $\mathcal{X}$ is a compact convex set with nonempty relative interior. In addition, since society states describe the aggregate behavior of populations, the considered constraint set Ω does not depend on the identities of individual agents, but rather on the overall distribution of strategic choices.

Continuing with the description of the modeling framework, each strategy $i \in \mathcal{S}^k$ is associated with a fitness function $f_i^k : \Delta \to \mathbb{R}$, which quantifies the intrinsic desirability of the strategy as a function of the society state $\mathbf{x}$. These fitness functions capture the underlying objective of the population game and are defined independently of any external constraints. In particular, they do not account for feasibility with respect to Ω, but rather reflect how favorable each strategy is under the current strategic distribution of the society. Throughout this book, fitness functions are assumed to be continuous.

Assumption 2.2 The function $f_i^k(\cdot)$ is continuous, for all $i \in \mathcal{S}^k$ and all $k \in \mathcal{P}$. ◊

Besides, the overall fitness vector is denoted by

$$\mathbf{f}(\cdot) = \mathrm{col}\left(\mathbf{f}^k(\cdot) \right) k \in \mathcal{P},$$

where each population-specific fitness vector is defined as

$$\mathbf{f}^k(\cdot) = \operatorname{col}\left(f_i^k(\cdot)\right) i \in \mathcal{S}^k.$$

Remark 2.2 (*A Note on the Fitness Functions*) The fitness functions are problem-dependent and define the strategic landscape of the population game. In some scenarios (e.g., Example 1.1), they may represent the partial derivatives of utility functions that each population seeks to maximize. In such cases, the overall fitness vector corresponds to the so-called *pseudo-gradient mapping* of the game. Additionally, when fitness functions are said to be continuously differentiable, they are assumed to be defined on the extended domain $\mathbb{R}^n_{\geq 0} \supset \Delta$, ensuring the existence of partial derivatives at all relevant points. ◊

With this notation in place, a population game can be compactly represented in normal form as the tuple

$$G = (\mathcal{P}, \mathcal{X}, \mathbf{f}(\cdot)),$$

which specifies the key components of the considered multi-agent decision-making problem: the set of populations $\mathcal{P}$, the feasible strategic distributions $\mathcal{X} \subseteq \Delta$, and the overall fitness vector $\mathbf{f}(\cdot)$. This formalization defines the strategic environment by identifying the decision-making agents, their available choices and collective constraints, and the underlying objectives driving their behavior.

Depending on the structure and properties of $\mathbf{f}(\cdot)$, different classes of population games can be distinguished. This book focuses on two such classes: contractive games and full-potential games. These classes possess desirable analytical properties and are particularly relevant for the design and stability analysis of large-scale multi-agent systems. Their formal definitions are presented next.

Definition 2.1 (*Contractive Games*) A population game $G = (\mathcal{P}, \mathcal{X}, \mathbf{f}(\cdot))$ is a (merely) contractive game if $\mathbf{f} : \Delta \to \mathbb{R}^n$ satisfies that

$$(\mathbf{x} - \tilde{\mathbf{x}})^\top (\mathbf{f}(\mathbf{x}) - \mathbf{f}(\tilde{\mathbf{x}})) \leq 0, \qquad \forall \mathbf{x}, \tilde{\mathbf{x}} \in \Delta.$$

The game G is said to be strictly contractive if the inequality above is strict for all $\mathbf{x} \neq \tilde{\mathbf{x}}$. Moreover, G is said to be μ-strongly contractive if there exists a $\mu \in \mathbb{R}_{>0}$ such that

$$(\mathbf{x} - \tilde{\mathbf{x}})^\top (\mathbf{f}(\mathbf{x}) - \mathbf{f}(\tilde{\mathbf{x}})) \leq -\mu \left\|\mathbf{x} - \tilde{\mathbf{x}}\right\|_2^2,$$

for all $\mathbf{x}, \tilde{\mathbf{x}} \in \Delta$. Here, $\|\cdot\|_2$ denotes the Euclidean norm. ◊

Remark 2.3 Notice that the (mere/strict/μ-strong) contractivity of $\mathbf{f}(\cdot)$ is equivalent to the (mere/strict/μ-strong) monotonicity of $-\mathbf{f}(\cdot)$. ◊

Definition 2.2 (*Full-Potential Games*) A population game $G = (\mathcal{P}, \mathcal{X}, \mathbf{f}(\cdot))$ is a full-potential game if there exists a continuously differentiable function $\varphi : \mathbb{R}^n_{\geq 0} \to \mathbb{R}$ such that

$$\nabla_{\mathbf{x}}\varphi(\cdot) = \mathbf{f}(\cdot).$$

The function $\varphi(\cdot)$ is referred to as the *potential function* of the game G. ◊

Based on Definitions 2.1 and 2.2, the next lemmas are straightforward to check.

Lemma 2.1 *Let* $\mathbf{f} : \mathbb{R}^n_{\geq 0} \to \mathbb{R}^n$ *be continuously differentiable. Then, the game* $G = (\mathcal{P}, \mathcal{X}, \mathbf{f}(\cdot))$ *is (merely) contractive if and only if*[3]

$$\mathbf{z}^\top \mathrm{D}_{\mathbf{x}}\,\mathbf{f}(\mathbf{x})\mathbf{z} \leq 0, \quad \forall\,(\mathbf{x}, \mathbf{z})\;\Delta \times \mathrm{T}\,\Delta.$$

In contrast, G is strictly contractive if and only if

$$\mathbf{z}^\top \mathrm{D}_{\mathbf{x}}\,\mathbf{f}(\mathbf{x})\mathbf{z} < 0, \quad \forall\,(\mathbf{x}, \mathbf{z}) \in \Delta \times (\mathrm{T}\,\Delta \setminus \{\mathbf{0}_n\}).$$

Similarly, G is μ*-strongly contractive if and only if*

$$\mathbf{z}^\top \mathrm{D}_{\mathbf{x}}\,\mathbf{f}(\mathbf{x})\mathbf{z} \leq -\mu\mathbf{z}^\top\mathbf{z}, \quad \forall\,(\mathbf{x}, \mathbf{z}) \in \Delta \times \mathrm{T}\,\Delta,$$

for some $\mu \in \mathbb{R}_{>0}$. ◊

Lemma 2.2 *Let G be a full-potential game. If* $\varphi(\cdot)$ *is (merely/strictly/*μ*-strongly) concave, then G is (merely/strictly/*μ*-strongly) contractive.* ◊

Finally, based on the notion of contractive games, one can characterize a general population game G by its lack of contractivity. This fact is formalized in Definition 2.3.

Definition 2.3 (*Contractivity Deficit*) An arbitrary population game $G = (\mathcal{P}, \mathcal{X}, \mathbf{f}(\cdot))$ is said to have a contractivity deficit of ν^*, if there exists a ν^* such that

$$\nu^* = \inf\left\{\nu \in \mathbb{R}_{>0} : (\mathbf{x} - \tilde{\mathbf{x}})^\top (\mathbf{f}(\mathbf{x}) - \mathbf{f}(\tilde{\mathbf{x}})) \leq \nu\,\|\mathbf{x} - \tilde{\mathbf{x}}\|_2^2,\ \forall \mathbf{x}, \tilde{\mathbf{x}} \in \Delta\right\}.$$

In other words, G has a contractivity deficit of ν^* if and only if the auxiliary game $\tilde{G} = \left(\mathcal{P}, \mathcal{X}, \tilde{\mathbf{f}}(\cdot)\right)$, with $\tilde{\mathbf{f}}(\mathbf{x}) = \mathbf{f}(\mathbf{x}) - \nu\mathbf{x}$, is merely contractive for every $\nu \geq \nu^*$. ◊

Definitions 2.1–2.3 are further illustrated through the next examples.

Example 2.1 Consider a population game $G = (\mathcal{P}, \mathcal{X}, \mathbf{f}(\cdot))$ characterized by an affine fitness vector of the form

$$\mathbf{f}(\mathbf{x}) = \mathbf{Q}\mathbf{x} + \mathbf{r}, \tag{2.1}$$

where $\mathbf{Q} \in \mathbb{R}^{n\times n}$ and $\mathbf{r} \in \mathbb{R}^n$. Notice that

[3] $\mathrm{T}\,\Delta$ denotes the tangent space of Δ.

$$
\begin{aligned}
\mathbf{z}^{\top} \mathrm{D}_{\mathbf{x}} \mathbf{f}(\mathbf{x}) \mathbf{z} &= \mathbf{z}^{\top} \mathbf{Q} \mathbf{z} \\
&= \frac{1}{2} \mathbf{z}^{\top} \left(\mathbf{Q} + \mathbf{Q}^{\top}\right) \mathbf{z} \\
&\preceq \frac{1}{2} \lambda_{\max} \left(\mathbf{Q} + \mathbf{Q}^{\top}\right) \mathbf{z}^{\top} \mathbf{z},
\end{aligned}
$$

for every $\mathbf{z} \in \mathbb{R}^n$. Here, $\lambda_{\max}\left(\mathbf{Q} + \mathbf{Q}^{\top}\right) \in \mathbb{R}$ is the maximum eigenvalue of $\left(\mathbf{Q} + \mathbf{Q}^{\top}\right)$. Consequently, it follows that the contractivity properties of G are directly linked to the maximum eigenvalue of $\left(\mathbf{Q} + \mathbf{Q}^{\top}\right)$. More formally,

$$
\begin{aligned}
\lambda_{\max}\left(\mathbf{Q} + \mathbf{Q}^{\top}\right) = 0 \quad &\Leftrightarrow \quad G \text{ is contractive} \\
\lambda_{\max}\left(\mathbf{Q} + \mathbf{Q}^{\top}\right) < 0 \quad &\Leftrightarrow \quad G \text{ is strictly contractive} \\
\exists \mu \in \mathbb{R}_{>0} : \lambda_{\max}\left(\mathbf{Q} + \mathbf{Q}^{\top}\right) = -2\mu \quad &\Leftrightarrow \quad G \text{ is } \mu\text{-strongly contractive} \\
\exists \nu^* \in \mathbb{R}_{>0} : \lambda_{\max}\left(\mathbf{Q} + \mathbf{Q}^{\top}\right) = 2\nu^* \quad &\Leftrightarrow \quad G \text{ has a contractivity deficit of } \nu^*.
\end{aligned}
$$

On the other hand, note that if $\mathbf{Q} = \mathbf{Q}^{\top}$, then it follows that

$$
\varphi(\mathbf{x}) = \frac{1}{2} \mathbf{x}^{\top} \mathbf{Q} \mathbf{x} + \mathbf{r} \mathbf{x}
$$

satisfies $\nabla_{\mathbf{x}} \varphi(\cdot) = \mathbf{f}(\cdot)$, implying that G is a full-potential game with potential function $\varphi(\cdot)$. In fact, $\mathbf{Q} = \mathbf{Q}^{\top}$ is both a sufficient and necessary condition for G to be a full-potential game.

Population games with affine fitness vectors of the form (2.1) are relevant for several practical applications. Some examples include the *load-balancing* problem described in Sect. 1.2.2; the *multi-robot garbage collection* task of Sect. 1.2.3; the *competition and coordination in ride-hailing apps* scenario introduced in Sect. 1.2.4; and the *microgrid energy management* problem of Sect. 1.2.6. It is straightforward to check that, with the exception of the ride-hailing apps context, the underlying game G is at least merely contractive for all of these examples. Consider for instance the ride-hailing apps context with only three apps for simplicity of notation, i.e., $\hat{n} = 3$. In such a scenario, the matrix $\mathbf{Q}$ is of the form

$$
\mathbf{Q} = \begin{bmatrix}
-\left(1 - \epsilon_1^1\right) & 0 & 0 & \epsilon_1^1 & 0 & 0 \\
0 & -\left(1 - \epsilon_2^1\right) & 0 & 0 & \epsilon_2^1 & 0 \\
0 & 0 & -\left(1 - \epsilon_3^1\right) & 0 & 0 & \epsilon_3^1 \\
\epsilon_1^2 & 0 & 0 & -\left(1 - \epsilon_1^2\right) & 0 & 0 \\
0 & \epsilon_2^2 & 0 & 0 & -\left(1 - \epsilon_2^2\right) & 0 \\
0 & 0 & \epsilon_3^2 & 0 & 0 & -\left(1 - \epsilon_3^2\right)
\end{bmatrix},
$$

where the parameters $\epsilon_i^k \in (0, 1)$ define a convex combination weighting the importance of competition and coordination. Note that if $\epsilon_i^k < 0.5$, for all i and all k, then $-\mathbf{Q}$ is symmetric and strictly diagonally dominant with positive diagonal entries, which implies that $\mathbf{Q}$ is symmetric negative definite. Therefore, if $\epsilon_i^k < 0.5$, for all i and all k, then the corresponding game G is strictly contractive. Similar analyses yield that setting $\epsilon_i^k \leq 0.5$, for all i and all k, with at least one $\epsilon_i^k = 0.5$ results in a merely contractive game, while every configuration with at least one $\epsilon_i^k > 0.5$ is not contractive. Finally, G is a full-potential game if and only if $\epsilon_i^1 = \epsilon_i^2$, for all i. ◊

Example 2.2 Consider the *traffic routing in urban networks* application discussed in Sect. 1.2.5. Each driver population $k \in \mathcal{P}$ must choose among a set of available routes to commute from their origin $\mathcal{O}^k$ to their destination $\mathcal{D}^k$. While origins and destinations differ across populations, their available route sets may overlap, leading to route congestion effects that couple their decisions.

The fitness function for each population k is given by

$$\mathbf{f}^k(\mathbf{x}) = -\boldsymbol{\alpha}^k - \boldsymbol{\beta}^k(\mathbf{x}), \quad \forall k \in \mathcal{P},$$

where $\boldsymbol{\alpha}^k \in \mathbb{R}^{n^k}_{\geq 0}$ denotes nominal travel times, and $\boldsymbol{\beta}^k : \mathbb{R}^n_{\geq 0} \to \mathbb{R}^{n^k}_{\geq 0}$ represents congestion-induced delays that depend on the aggregate traffic over all routes.

Suppose there are $r \in \mathbb{Z}_{\geq 2}$ total routes. For each $k \in \mathcal{P}$, let $\mathbf{C}^k \in \mathbb{R}^{r \times n^k}$ be a binary matrix formed by selecting columns from the $r \times r$ identity matrix corresponding to the routes available to population k. Let $\mathbf{C} = \left[\mathbf{C}^1, \mathbf{C}^2, \ldots, \mathbf{C}^P\right] \in \mathbb{R}^{r \times n}$, and define the aggregate route flow as $\mathbf{Cx}$.

Let congestion costs for each route $z \in \{1, \ldots, r\}$ be modeled by differentiable, Lipschitz continuous functions $\overline{\beta}_z : \mathbb{R} \to \mathbb{R}$, with the overall congestion vector given by $\overline{\boldsymbol{\beta}}(\cdot) = \mathrm{col}\left(\overline{\beta}_1(\cdot), \ldots, \overline{\beta}_r(\cdot)\right)$. Then, the congestion term for population k can be written as

$$\boldsymbol{\beta}^k(\mathbf{x}) = \mathbf{D}^k \mathbf{x}^k + \mathbf{C}^{k\top} \overline{\boldsymbol{\beta}}(\mathbf{Cx}),$$

where $\mathbf{D}^k \in \mathbb{R}^{n^k \times n^k}_{\geq 0}$ is diagonal and models route-dependent preferences or travel effort.

As an illustrative case, consider affine congestion functions of the form $\overline{\beta}_z(\omega) = \gamma_z \omega + \epsilon_z$ for all z. For simplicity, let $\gamma_z = \gamma \geq 0$ and $\epsilon_z = \epsilon$ for all z, so that $\overline{\boldsymbol{\beta}}(\mathbf{Cx}) = \gamma \mathbf{Cx} + \boldsymbol{\epsilon}$, with $\boldsymbol{\epsilon} = \epsilon \mathbf{1}_r$. The resulting fitness function becomes

$$\mathbf{f}^k(\mathbf{x}) = -\boldsymbol{\alpha}^k - \mathbf{D}^k \mathbf{x}^k - \gamma \mathbf{C}^{k\top} \mathbf{Cx} - \mathbf{C}^{k\top} \boldsymbol{\epsilon}.$$

It follows that the population game $G = (\mathcal{P}, \mathcal{X}, \mathbf{f}(\cdot))$ has an affine fitness structure of the form (2.1). As in Example 2.1, it is straightforward to verify that such a game is at least merely contractive. Naturally, alternative contractivity properties may arise under different congestion functions $\overline{\beta}_z(\cdot)$. ◊

Example 2.3 Recall the *research topic selection* problem introduced in Sect. 1.2.1. To ease the illustration, consider the simplified scenario with only one population of researchers, i.e., $P = 1$, and let fitness functions be of the form

$$f_i(x_i) = a_i \exp\left(-\left(\frac{x_i - b_i}{c_i}\right)^2\right),$$

with parameters $a_i, b_i, c_i \in \mathbb{R}_{>0}$, for all $i \in \mathcal{S}$. Since the fitness functions are decoupled between different strategies, the Jacobian matrix $\mathrm{D}_{\mathbf{x}}\,\mathbf{f}(\cdot)$ is a diagonal matrix with diagonal entries given by

$$\frac{df_i(x_i)}{dx_i} = -2a_i\left(\frac{x_i - b_i}{(c_i)^2}\right)\exp\left(-\left(\frac{x_i - b_i}{c_i}\right)^2\right), \quad \forall i \in \mathcal{S}.$$

Clearly, for every i it holds that

$$\frac{df_i(x_i)}{dx_i} > 0, \quad \forall x_i \in [0, b_i),$$
$$\frac{df_i(x_i)}{dx_i} \leq 0, \quad \forall x_i \in [b_i, m].$$

Thus, the corresponding population game G is not contractive for every $b_i > 0$, and its contractivity deficit is given by

$$\nu^* = \max_{i\in\mathcal{S}} \max_{x_i\in[0,b_i)} -2a_i\left(\frac{x_i - b_i}{(c_i)^2}\right)\exp\left(-\left(\frac{x_i - b_i}{c_i}\right)^2\right).$$

As will be discussed later in the book, this lack of contractivity has serious implications in the convergence of the society toward an equilibrium state. ◊

2.2 Payoff Mechanisms

As discussed in Chap. 1, the population game G unfolds dynamically over time, making it necessary to formalize how agents evaluate their strategic options as the system evolves.

While fitness functions encode the intrinsic desirability of each strategy based on the current state of the society, they do not necessarily reflect the actual feedback agents receive during gameplay. This role is played by the payoff mechanism, which determines the strategy-specific signals perceived by the agents. In essence, payoffs translate the evolving strategic landscape into the information that guides agents' decisions.

Formally, the payoff mechanism is described by a continuous-time dynamical system called the *payoff dynamics model* (PDM), which establishes a causal relationship between the evolving society state and the payoffs perceived by agents.

Let $t \in \mathbb{R}_{\geq 0}$ denote the continuous-time index, and let $\mathbf{x}(t)$ represent the society state at time t. For every strategy $i \in \mathcal{S}^k$ in population $k \in \mathcal{P}$, let $p_i^k(t) \in \mathbb{R}$ denote the payoff perceived by agents choosing strategy i at time t. Collecting these for all strategies in population k yields the payoff vector

$$\mathbf{p}^k(t) = \mathrm{col}\left(p_i^k(t)\right)_{i\in\mathcal{S}^k},$$

and stacking across all populations gives the overall payoff vector

$$\mathbf{p}(t) = \mathrm{col}\left(\mathbf{p}^k(t)\right)_{k\in\mathcal{P}}.$$

The PDM governs how these payoff signals evolve over time in response to changes in the society state $\mathbf{x}(t)$. It is formally stated in Definition 2.4.

Definition 2.4 (*Payoff Dynamics Model (PDM)*) A payoff dynamics model (PDM) is a continuous-time dynamical system of the form

$$\begin{aligned}\dot{\mathbf{q}}(t) &= \mathcal{W}(\mathbf{q}(t), \mathbf{x}(t)), \quad \mathbf{q}(0) \in \mathbb{R}^d,\\ \mathbf{p}(t) &= \mathcal{H}(\mathbf{q}(t), \mathbf{x}(t)),\end{aligned}$$

where

- $\mathbf{q}(t) \in \mathbb{R}^d$ is the internal state of the PDM at time t.
- $\mathcal{W} : \mathbb{R}^d \times \Delta \to \mathbb{R}^d$ is a Lipschitz continuous function describing the PDM's internal state evolution.
- $\mathcal{H} : \mathbb{R}^d \times \mathbb{R}^n_{\geq 0} \to \mathbb{R}^n$ is a continuously differentiable and Lipschitz continuous function that maps the PDM's state and input to the resulting payoffs.

The input space of a PDM consists of all differentiable and bounded society state trajectories $\mathbf{x}(t)$ such that $\mathbf{x}(t) \in \Delta$ and $\|\dot{\mathbf{x}}(t)\|_\infty < \infty$ for all $t \geq 0$. Besides, a PDM is said to be *bounded* if $\|\mathbf{q}(t)\|_\infty < \infty$ for all $t \geq 0$. ◊

Remark 2.4 Note that when the PDM is bounded, the Lipschitz continuity of $\mathcal{W}(\cdot,\cdot)$ and $\mathcal{H}(\cdot,\cdot)$, and the compactness of Δ, imply that $\mathbf{p}(t)$ remains within a bounded set $\mathcal{B} = \mathcal{B}^1 \times \mathcal{B}^2 \times \cdots \times \mathcal{B}^P$, for all times $t \geq 0$.

Following Definition 2.4, it is useful to distinguish once again between the concepts of fitness and payoffs. While both relate to how agents evaluate strategies, they play fundamentally different roles in the modeling framework.

Fitness functions determine the strategic environment faced by the agents and define the underlying game-theoretic problem: they specify the desirability of each strategy independent of the constraints in Ω. *Payoff signals*, on the other hand, represent the actual information perceived by the agents and guide their strategy updates over time. In the case where payoffs are directly derived from the fitness functions through a memoryless relationship, such as $\mathbf{p}(t) = \mathbf{f}(\mathbf{x}(t))$, the notions of fitness and payoff coincide. However, in the more general setting where payoff signals are generated by a PDM, fitness and payoffs are not necessarily the same.

Allowing payoffs to depend on dynamic rather than static maps of the society state enables the model to capture important effects such as anticipation, delay, estimation, and inertia; features that often arise in learning and adaptation processes. Additionally, as illustrated in Example 1.2, dynamic payoff mechanisms allow agents to respond to constraint violations with respect to the set Ω, making them a powerful tool for modeling constrained decision-making environments. Notably, the considered PDM framework is flexible enough to encompass the simpler memoryless approach $\mathbf{p}(t) = \mathbf{f}(\mathbf{x}(t))$ as a special case. This can be achieved by setting $d = 0$, neglecting the internal dynamics, and defining the output map as $\mathcal{H}(\cdot) = \mathbf{f}(\cdot)$.

To ensure that the payoff signals remain aligned with the underlying game structure, a formal connection between the PDM and the population game G is required. This relationship is formally characterized through Assumption 2.3.

Assumption 2.3 For every $(\mathbf{x}^*, \mathbf{q}^*) \in \Delta \times \mathbb{R}^d$, the PDM satisfies that

$$\mathcal{W}\left(\mathbf{q}^*, \mathbf{x}^*\right) = \mathbf{0}_d \Rightarrow \begin{cases} \mathcal{H}\left(\mathbf{q}^*, \mathbf{x}^*\right) = \mathbf{f}\left(\mathbf{x}^*\right) - \mathbf{A}^\top \mathbf{M}_{c_=} \mathbf{q}^* - \mathrm{D}_{\mathbf{x}}\, \mathbf{g}\left(\mathbf{x}^*\right)^\top \mathbf{M}_{c_\leq} \mathbf{q}^* \\ \left(\mathbf{M}_{c_\leq} \mathbf{q}^*\right) \odot \mathbf{g}\left(\mathbf{x}^*\right) = \mathbf{0}_{c_\leq} \\ \mathbf{M}_{c_\leq} \mathbf{q}^* \succeq \mathbf{0}_{c_\leq} \\ \mathbf{x}^* \in \Omega, \end{cases}$$

for some $\mathbf{M}_{c_=} \in \mathbb{R}^{c_= \times d}$ and some $\mathbf{M}_{c_\leq} \in \mathbb{R}^{c_\leq \times d}$. Here, $\odot$ and $\preceq$ denote the elementwise product and inequality, respectively. ◊

Assumption 2.3 is a technical condition on the stationary regime of the PDM, and it effectively relates the payoffs to the fitness functions and constraints of the underlying population game G. Specifically, in the unconstrained case where $\Omega = \mathbb{R}^n$ (so that $\mathbf{A}$, $\mathbf{b}$, and $\mathbf{g}(\cdot)$ can be trivially set to zero), Assumption 2.3 simply requires that at every equilibrium state $(\mathbf{q}^*, \mathbf{x}^*)$ of the PDM, the following equality holds:

$$\mathcal{H}\left(\mathbf{q}^*, \mathbf{x}^*\right) = \mathbf{f}\left(\mathbf{x}^*\right).$$

That is, the payoff signals must coincide with the fitness values at equilibrium, thereby aligning the dynamic feedback received by the agents with the underlying objectives of the population game. On the other hand, in the more general case where $\Omega \subset \mathbb{R}^n$ and $\mathcal{X} = \Omega \cap \Delta \subset \Delta$, Assumption 2.3 requires that the payoff signals align with the fitness values, but are augmented with additional terms associated with the constraint set Ω. These additional terms effectively act as dual, Lagrange-like multipliers, embedded within the payoff structure to incentivize agents toward constraint satisfaction.

To better clarify the presented ideas, two examples of PDMs are now discussed.

Example 2.4 (*Smoothing-Anticipatory PDM*) Consider a population game $G = (\mathcal{P}, \mathcal{X}, \mathbf{f}(\cdot))$ under the unconstrained setting with $\Omega = \mathbb{R}^n$. A flexible class of PDMs that captures a variety of learning behaviors is the *smoothing-anticipatory* PDM, defined as

$$\begin{aligned} \dot{\mathbf{q}}(t) &= \tau\left(\mathbf{f}\left(\mathbf{x}(t)\right) - \mathbf{q}(t)\right), \\ \mathbf{p}(t) &= \alpha_0 \mathbf{f}\left(\mathbf{x}(t)\right) + \alpha_1 \mathbf{q}(t) + \alpha_2 \dot{\mathbf{q}}(t), \end{aligned}$$

where $\tau \in \mathbb{R}_{>0}$ and $\alpha_0, \alpha_1, \alpha_2 \in \mathbb{R}_{\geq 0}$ are tuning parameters subject to $\alpha_0 + \alpha_1 = 1$. This PDM structure enables the modeling of different types of payoff dynamics depending on the choice of parameters:

- $\alpha_0 = 1, \alpha_1 = \alpha_2 = 0$ yield a *memoryless* PDM.
- $\alpha_0 = 0, \alpha_1 = 1, \alpha_2 = 0$ yield a *smoothing* PDM.
- $\alpha_0 = 1, \alpha_1 = 0, \alpha_2 > 0$ yield an *anticipatory* PDM.

This class of payoff mechanisms is particularly useful for capturing dynamic effects such as delay, inertia, and anticipation in how agents perceive and respond to payoffs throughout the evolution of the game. Furthermore, if $\mathbf{f}(\cdot)$ is continuously differentiable, then these PDMs are known to be bounded (see the *Notes* at the end of this chapter) and they trivially satisfy Assumption 2.3, since $\dot{\mathbf{q}}(t) = \mathbf{0}_d$ implies $\mathbf{p}(t) = \mathbf{f}(\mathbf{x}(t))$, for every $t \geq 0$. $\diamondsuit$

Example 2.5 (*Dual-Lagrangian-Based PDM*) Consider a population game $G = (\mathcal{P}, \mathcal{X}, \mathbf{f}(\cdot))$ under the constrained setup, where Ω is any set of constraints satisfying Assumption 2.1. In the context of enforcing constraint satisfaction through the payoff mechanism, a particularly relevant class of PDMs is the so-called *dual-Lagrangian-based* PDM, defined as

$$\begin{aligned}
\dot{\mathbf{q}}_1(t) &= \hat{\vartheta}\left(\mathbf{A}\mathbf{x}(t) - \mathbf{b}\right) - \check{\vartheta}\left(\mathbf{b} - \mathbf{A}\mathbf{x}(t)\right), \\
\dot{\mathbf{q}}_2(t) &= \hat{\theta}\left(\mathbf{g}\left(\mathbf{x}(t)\right)\right) - \mathbf{q}_2(t) \odot \check{\theta}\left(-\mathbf{g}\left(\mathbf{x}(t)\right)\right), \\
\mathbf{p}(t) &= \mathbf{f}\left(\mathbf{x}(t)\right) - \mathbf{A}^\top \mathbf{q}_1(t) - \mathrm{D}_{\mathbf{x}}\, \mathbf{g}\left(\mathbf{x}(t)\right)^\top \mathbf{q}_2(t),
\end{aligned}$$

where $\mathbf{q} := \left[\mathbf{q}_1^\top,\ \mathbf{q}_2^\top\right]^\top$, $\odot$ denotes the Hadamard product, and each function

$$z(\cdot) \in \left\{\hat{\vartheta}(\cdot),\ \check{\vartheta}(\cdot),\ \hat{\theta}(\cdot),\ \check{\theta}(\cdot)\right\}$$

is a Lipschitz continuous scalar function $z : \mathbb{R} \to \mathbb{R}_{\geq 0}$, applied element-wise as

$$z(\boldsymbol{\zeta}) := \begin{bmatrix} z(\zeta_1) \\ z(\zeta_2) \\ \vdots \\ z(\zeta_{d_z}) \end{bmatrix}, \qquad \forall \boldsymbol{\zeta} = \begin{bmatrix} \zeta_1 \\ \zeta_2 \\ \vdots \\ \zeta_{d_z} \end{bmatrix} \in \mathbb{R}^{d_z},$$

while satisfying $z(\alpha) > 0$ if $\alpha > 0$, and $z(\alpha) = 0$ if $\alpha \leq 0$, for all $\alpha \in \mathbb{R}$. A canonical example satisfying these properties would be $z(\cdot) = \max\{\cdot,\ 0\}$, applied element-wise as described.

In some settings, the output of these PDMs can be interpreted in terms of the dual-Lagrangian function associated with an underlying optimization problem, where $\mathbf{q}_1$ and $\mathbf{q}_2$ correspond to the dual variables (Lagrange multipliers) for the equality and inequality constraints in Ω, respectively. This statement makes the dual-Lagrangian-based PDM a useful mechanism for encouraging constraint satisfaction via payoff signals. While this class of PDMs is generally not bounded, it does satisfy Assumption 2.3 (see the *Notes* at the end of this chapter for further discussion). ◊

2.3 Revision Processes

The third fundamental element of the modeling framework is the revision process through which agents update their decisions. As briefly introduced in Chap. 1, the revision process consists of two components:

- An *alarm clock*, which provides revision opportunities for agents to evaluate and potentially update their strategies.
- A *revision protocol*, which defines the decision rule that agents use to choose their strategy when given a revision opportunity.

Together, these two components define how individual decisions evolve over time in response to the perceived payoffs. The alarm clock dictates the timing of possible strategy revisions, while the revision protocol governs the manner in which agents react to payoff information. The combined operation of these elements constitutes what is referred to as the *revision process*.

Specifically, the alarm clock determines the probability distribution of the time intervals between successive revision events for every agent. Being t_1 and t_2, with $t_2 > t_1$, two consecutive times at which an arbitrary agent revises their strategy, the alarm clock is the mathematical abstraction that governs the probability distribution for the inter-event time T, defined as

$$T = t_2 - t_1.$$

That is, T is a random variable whose probability distribution is determined by the alarm clock.

The revision protocol, on the other hand, is defined as a matrix-valued map $\boldsymbol{\rho} := \operatorname{diag}\left(\boldsymbol{\rho}^k\right)_{k\in\mathcal{P}}$, where

$$\boldsymbol{\rho}^k : \mathbb{R}^{n^k} \times \Delta^k \to \mathbb{R}_{\geq 0}^{n^k \times n^k}, \quad \forall k \in \mathcal{P}.$$

For each $i, j \in \mathcal{S}^k$, with $i \neq j$, the (i, j)th entry of $\boldsymbol{\rho}^k\left(\tilde{\mathbf{p}}^k, \tilde{\mathbf{x}}^k\right)$, denoted $\rho_{ij}^k\left(\tilde{\mathbf{p}}^k, \tilde{\mathbf{x}}^k\right)$, gives the *conditional switching rate* at which agents in population k move from strategy i to j, given the local payoff vector $\tilde{\mathbf{p}}^k$ and strategic distribution $\tilde{\mathbf{x}}^k$. Thus, the revision protocol encodes how agents respond to the current state of the system.

Remark 2.5 As will be explained in brief, in general, the diagonal entries $\rho_{ii}^k(\cdot, \cdot)$ play no formal role in the model and are simply used as placeholders to simplify the notation.

While many types of alarm clocks and revision protocols may be motivated by practical applications, for the sake of analytical tractability, this book focuses on a specific class of alarm clocks and assumes certain regularity conditions on the revision protocols. These properties are formalized in Assumption 2.4.

Assumption 2.4 For every population $k \in \mathcal{P}$, each agent $a \in \mathcal{A}^k$ is equipped with a *Poisson alarm clock* with rate $R^k \in \mathbb{R}_{>0}$. That is, all inter-event times T between revision opportunities, for every agent in population k, are independently and identically distributed according to an exponential distribution with rate R^k, i.e., $T \sim \text{Exp}\left(R^k\right)$. Furthermore, the clocks of different agents are assumed to be mutually independent.

For the revision protocols, all agents within a population $k \in \mathcal{P}$ share the same rule $\boldsymbol{\rho}^k(\cdot,\cdot)$, which is Lipschitz continuous and satisfies the condition

$$\sum_{j \in \mathcal{S}^k \setminus \{i\}} \rho_{ij}^k\left(\mathbf{p}^k, \mathbf{x}^k\right) \leq 1, \quad \forall \left(i, \mathbf{p}^k, \mathbf{x}^k\right) \in \mathcal{S}^k \times \mathbb{R}^{n^k} \times \Delta^k.$$

As a result, the matrix

$$\boldsymbol{\Gamma}^k\left(\mathbf{p}^k, \mathbf{x}^k\right) = \boldsymbol{\rho}^k\left(\mathbf{p}^k, \mathbf{x}^k\right) + \operatorname{diag}\left(1 - \sum_{j \in \mathcal{S}^k} \rho_{ij}^k\left(\mathbf{p}^k, \mathbf{x}^k\right)\right)_{i \in \mathcal{S}^k}$$

is row-stochastic for every $\left(\mathbf{p}^k, \mathbf{x}^k\right) \in \mathbb{R}^{n^k} \times \Delta^k$. ◊

Assumption 2.4 is motivated by both practical considerations and analytical tractability. First, the use of Poisson clocks models asynchronous decision-making environments, where each agent revises their strategy at a constant average rate, but with randomness in the actual realized timing. This randomness captures uncertainty and heterogeneity in agents' response times.

Second, the conditions on the revision protocols ensure that they define valid probabilistic switching behaviors for agents. These probabilistic models reflect the bounded rationality of agents, acknowledging that agents may make suboptimal or noisy decisions, and thus better capture real-world behavior.

Finally, and crucially for analysis, Assumption 2.4 enables the stochastic dynamics of large populations to be approximated by deterministic differential equations, greatly simplifying the mathematical treatment. This connection is further discussed in Sect. 2.4.

The remainder of this section formalizes the revision process at a microscopic level and introduces three representative classes of revision protocols.

2.3.1 Microscopic Decision-Making Dynamics

Based on Assumption 2.4, the *microscopic* decision-making process of agents is established as follows.

Let $a \in \mathcal{A}^k$ be an arbitrary agent from population $k \in \mathcal{P}$, and suppose that such an agent a receives a strategy revision opportunity at time t. That is, agent a's Poisson alarm clock "rings" at time t, granting them a strategy revision opportunity.

Now, let $\tilde{t} < t$ denote an arbitrary time between the previous strategy revision time of any agent in the society and time t. Since $\tilde{t}$ and t can be arbitrarily close, there is no loss of generality in assuming that a is the only agent of the entire society revising their strategy at time t, and that the strategy selection of all agents, including a, remains fixed during the interval $[\tilde{t}, t)$.

Suppose that agent a's selected strategy at time $\tilde{t}$ is $i \in \mathcal{S}^k$. That is, agent a is currently playing strategy i when their revision opportunity arrives at time t. Then, agent a's chooses their new strategy at time t according to the following probabilities:

$$\begin{aligned}
\mathbb{P}(j|i) &= \rho_{ij}^k\left(\mathbf{p}^k(\tilde{t}), \mathbf{x}^k(\tilde{t})\right), \quad \forall j \in \mathcal{S}^k \setminus \{i\} \\
\mathbb{P}(i|i) &= 1 - \sum_{\ell \in \mathcal{S}^k \setminus \{i\}} \rho_{i\ell}^k\left(\mathbf{p}^k(\tilde{t}), \mathbf{x}^k(\tilde{t})\right).
\end{aligned}$$

Specifically, agent a revising their strategy at time t switches from strategy i to j with probability $\mathbb{P}(j|i)$, for all $j \in \mathcal{S}^k \setminus \{i\}$, or remains selecting strategy i with probability $\mathbb{P}(i|i)$. By Assumption 2.4, all of these probabilities are well defined.

As mentioned in Remark 2.5, note that the elements $\rho_{ii}^k\left(\mathbf{p}^k(\tilde{t}), \mathbf{x}^k(\tilde{t})\right)$ play no role within the decision-making process. Such diagonal entries of $\boldsymbol{\rho}^k\left(\mathbf{p}^k(\tilde{t}), \mathbf{x}^k(\tilde{t})\right)$ become meaningful only when

$$\sum_{j \in \mathcal{S}^k} \rho_{ij}^k\left(\mathbf{p}^k(\tilde{t}), \mathbf{x}^k(\tilde{t})\right) = 1,$$

in which case it readily follows that

$$\mathbb{P}(i|i) = \rho_{ii}^k\left(\mathbf{p}^k(\tilde{t}), \mathbf{x}^k(\tilde{t})\right).$$

Revision protocols satisfying such a property are often referred to as *exact*.

When the number of agents N^k in each population k is sufficiently large and Assumption 2.4 holds, the aggregate behavior of the society (resulting from agents individually following the microscopic decision-making process described above) can be accurately approximated by a set of deterministic ordinary differential equations. The justification behind such a result is discussed in Sect. 2.4.

2.3.2 *Representative Classes of Revision Protocols*

This section introduces three classes of revision protocols, each capturing distinct types of decision-making behavior.

Perturbed Best-Response Protocols

To model the scenario where agents always choose the strategy yielding the highest perceived payoff, one may consider a revision protocol of the form

$$\rho_{ij}^k\left(\mathbf{p}^k, \mathbf{x}^k\right) = \begin{cases} \frac{1}{c^k}, & \text{if } j \in \arg\max_{\ell \in \mathcal{S}^k} p_\ell^k \\ 0, & \text{otherwise,} \end{cases} \tag{2.2}$$

for all $i, j \in \mathcal{S}^k$, where c^k is the cardinality of the set $\left\{j : j \in \arg\max_{\ell \in \mathcal{S}^k} p_\ell^k\right\}$. While intuitive, the protocol defined in (2.2) suffers from several limitations. First, it is discontinuous, violating the regularity conditions required in Assumption 2.4, which complicates the analysis. More importantly, it assumes that agents are fully rational, perfectly informed, and make no errors when selecting strategies. These assumptions are often unrealistic in practice, where bounded rationality, imperfect information, and uncertain stochastic behavior are common.

To address these shortcomings, more realistic revision protocol models should account for randomness in the decision-making process, leading to the class of *perturbed best-response* (PBR) protocols, which smooth out the decision rule and better capture real-world behavior. The canonical example of PBR protocols is the *softmax* or *logit-choice* protocol defined as

$$\rho_{ij}^k\left(\mathbf{p}^k, \mathbf{x}^k\right) = \frac{\exp\left(\frac{p_j^k}{\eta^k}\right)}{\sum_{\ell \in \mathcal{S}^k} \exp\left(\frac{p_\ell^k}{\eta^k}\right)}, \tag{2.3}$$

where $\eta^k \in \mathbb{R}_{>0}$ is a parameter modeling the decision-making noise of the agents in population k. In fact, taking the limit $\eta^k \to \infty$ results in a uniform distribution over $\mathcal{S}^k$, where

$$\rho_{ij}^k\left(\mathbf{p}^k, \mathbf{x}^k\right) \to \frac{1}{n^k}, \quad \forall i, j \in \mathcal{S}^k,$$

while $\eta^k \to 0$ recovers (2.2) as

$$\rho_{ij}^k\left(\mathbf{p}^k, \mathbf{x}^k\right) \to \frac{1}{c^k}, \quad \forall j \in \arg\max_{\ell \in \mathcal{S}^k} p_\ell^k$$
$$\rho_{ij}^k\left(\mathbf{p}^k, \mathbf{x}^k\right) \to 0, \quad \forall j \notin \arg\max_{\ell \in \mathcal{S}^k} p_\ell^k,$$

for all $i \in \mathcal{S}^k$. Namely, the parameter η^k models the degree at which agents may make mistakes and do not always choose the strategy with the highest payoff. This relationship between the revision protocols (2.2) and (2.3) can be further understood through an optimization-based perspective as follows.

Consider an arbitrary payoff vector $\mathbf{p}^k$ and the auxiliary optimization problem

$$\max_{\mathbf{z}^k} \mathbf{z}^{k\top}\mathbf{p}^k, \quad \text{s.t.} \quad \mathbf{z}^k \in \mathbb{S}^k, \tag{2.4}$$

where $\mathbb{S}^k = \left\{ \mathbf{z}^k \in \mathbb{R}^{n^k}_{\geq 0} : \mathbf{1}^\top_{n^k} \mathbf{z}^k = 1 \right\}$ is the probability simplex of dimension n^k. The set of solutions of (2.4) is given by

$$\left\{ \mathbf{z}^k \in \mathbb{S}^k : z_i^k > 0 \Rightarrow p_i^k = \max_{j \in \mathcal{S}^k} p_j^k, \quad \forall i \in \mathcal{S}^k \right\}.$$

Consequently, the revision protocol (2.2) corresponds to a particular solution of (2.4), where probability is assigned uniformly over the strategies with the maximum payoff. As such, (2.2) is a *best-response* decision-making rule.

On the other hand, consider the *entropy-regularized* auxiliary optimization problem

$$\max_{\mathbf{z}^k} \mathbf{z}^{k\top} \mathbf{p}^k - \eta^k \sum_{i \in \mathcal{S}^k} z_i^k \ln\left(z_i^k\right), \quad \text{s.t.} \quad \mathbf{z}^k \in \mathbb{S}^k, \tag{2.5}$$

for some $\eta^k > 0$. Note that the domain of $\ln(\cdot)$ acts as an implicit constraint that excludes the boundary of $\mathbb{S}^k$. Therefore, without loss of generality, (2.5) can be equivalently written as

$$\max_{\mathbf{z}^k} \mathbf{z}^{k\top} \mathbf{p}^k - \eta^k \sum_{i \in \mathcal{S}^k} z_i^k \ln\left(z_i^k\right), \quad \text{s.t.} \quad \mathbf{1}^\top_{n^k} \mathbf{z}^k = 1. \tag{2.6}$$

The dual-Lagrangian function of (2.6) is given by

$$L\left(\mathbf{z}^k, \lambda\right) = \mathbf{z}^{k\top} \mathbf{p}^k - \eta^k \sum_{i \in \mathcal{S}^k} z_i^k \ln\left(z_i^k\right) - \lambda\left(\mathbf{1}^\top_{n^k} \mathbf{z}^k - 1\right),$$

and the (necessary and sufficient) Karush–Kuhn–Tucker (KKT) optimality conditions for (2.6) are then given by

$$\begin{aligned} p_i^k - \eta^k\left(\ln\left(z_i^{k*}\right) + 1\right) - \lambda^* = 0, \quad \forall i \in \mathcal{S}^k \\ \sum_{i \in \mathcal{S}^k} z_i^{k*} = 1. \end{aligned}$$

In consequence,

$$\begin{aligned} z_i^{k*} &= \exp\left(\frac{p_i^k - \lambda^*}{\eta^k} - 1\right) \\ &= \frac{1}{\exp\left(\frac{\lambda}{\eta^k} + 1\right)} \exp\left(\frac{p_i^k}{\eta^k}\right), \end{aligned}$$

for all $i \in \mathcal{S}^k$, and

$$\frac{1}{\exp\left(\frac{\lambda}{\eta^k} + 1\right)} \sum_{i \in \mathcal{S}^k} \exp\left(\frac{p_i^k}{\eta^k}\right) = 1.$$

Therefore,

$$z_i^{k*} = \frac{\exp\left(\frac{p_i^k}{\eta^k}\right)}{\sum_{i\in\mathcal{S}^k}\exp\left(\frac{p_i^k}{\eta^k}\right)}, \quad \forall i \in \mathcal{S}^k.$$

Marshaling all these facts yields that the *softmax* revision protocol (2.3) corresponds to the closed form solution of the optimization problem (2.5), which is an entropy-regularized version of (2.4). Thus, while (2.2) is a *best-response* protocol, (2.3) can be interpreted as a PBR, where the perturbation is given by the entropy regularization with parameter η^k.

Based on this regularized optimization-based perspective, one can extend the definition of PBR protocols to more general regularizations. Namely, PBR protocols are revision protocols of the form

$$\rho_{ij}^k\left(\mathbf{p}^k, \mathbf{x}^k\right) = \left[\arg\max_{\mathbf{z}^k\in\mathrm{int}\left(\mathbb{S}^k\right)}\left(\mathbf{z}^{k\top}\mathbf{p}^k - \tilde{\omega}^k\left(\mathbf{z}^k\right)\right)\right]_j, \quad \forall i, j \in \mathcal{S}^k, \tag{2.7}$$

where $[\cdot]_j$ extracts the jth element of its input vector, i.e., $[\boldsymbol{\alpha}]_j = \alpha_j$, and the regularization (perturbation) function $\tilde{\omega}^k : \mathbb{R}^{n^k}_{>0} \to \mathbb{R}$ is assumed twice continuously differentiable and satisfies:

(i) $\mathrm{D}_{\mathbf{z}^k}\,\tilde{\omega}^k\left(\mathbf{z}^k\right) \succ 0$, for all $\mathbf{z}^k \in \mathrm{int}\left(\mathbb{S}^k\right)$;
(ii) $\left\|\nabla_{\mathbf{z}^k}\tilde{\omega}^k\left(\mathbf{z}^k\right)\right\|_\infty \to \infty$ whenever $\mathbf{z}^k \to \mathrm{bd}\left(\mathbb{S}^k\right)$.

That is, $\tilde{\omega}^k(\cdot)$ is smooth, strictly convex, and steep near $\mathrm{bd}\left(\mathbb{S}^k\right)$. For the particular case of the softmax revision protocol (2.3), $\tilde{\omega}^k(\cdot)$ is defined as

$$\tilde{\omega}^k\left(\mathbf{z}^k\right) = \eta^k \sum_{i\in\mathcal{S}^k} z_i^k \ln\left(z_i^k\right),$$

with $\eta^k \in \mathbb{R}_{>0}$ being the regularization parameter.

Remark 2.6 Note that in (2.7), the function $\tilde{\omega}^k(\cdot)$ takes as input the probability vector $\mathbf{z}^k \in \mathrm{int}\left(\mathbb{S}^k\right)$. To ease the notation in forthcoming formulations, it is often convenient to consider the function

$$\omega^k\left(\mathbf{x}^k\right) = \tilde{\omega}^k\left(\frac{1}{m^k}\mathbf{x}^k\right),$$

and regard the PBR protocol (2.7) in its equivalent form

$$\rho_{ij}^k\left(\mathbf{p}^k, \mathbf{x}^k\right) = \frac{1}{m^k}\left[\arg\max_{\tilde{\mathbf{x}}^k\in\mathrm{int}\left(\Delta^k\right)}\left(\tilde{\mathbf{x}}^{k\top}\mathbf{p}^k - \omega^k\left(\tilde{\mathbf{x}}^k\right)\right)\right]_j, \quad \forall i, j \in \mathcal{S}^k. \tag{2.8}$$

In both cases, $\rho_{ij}^k(\mathbf{p}^k, \mathbf{x}^k)$ yields a valid switching probability, which depends only on $\mathbf{p}^k$. The input argument $\mathbf{x}^k$ acts simply as a placeholder for uniformity of notation. Therefore, one can then introduce the auxiliary notation

$$C_i^{\omega^k}\left(\mathbf{p}^k\right) := \left[\arg\max_{\tilde{\mathbf{x}}^k \in \operatorname{int}(\Delta^k)} \left(\tilde{\mathbf{x}}^{k\top}\mathbf{p}^k - \omega^k\left(\tilde{\mathbf{x}}^k\right)\right)\right]_i, \tag{2.9}$$

such that $\rho_{ij}^k\left(\mathbf{p}^k, \mathbf{x}^k\right) = \left(1/m^k\right) C_j^{\omega^k}\left(\mathbf{p}^k\right)$, for all $i, j \in \mathcal{S}^k$. ◊

Separable Excess Payoff Target Protocols

Separable excess payoff target (SEPT) protocols are revision protocols of the form

$$\rho_{ij}^k\left(\mathbf{p}^k, \mathbf{x}^k\right) = \phi_j^k\left(p_j^k - \frac{1}{m^k}\sum_{\ell \in \mathcal{S}^k} x_\ell^k p_\ell^k\right), \quad \forall i, j \in \mathcal{S}^k, \tag{2.10}$$

where $\phi_j^k : \mathbb{R} \to \mathbb{R}_{\geq 0}$ is a Lipschitz continuous function that satisfies

$$\begin{cases} \phi_j^k(\alpha) > 0, & \text{if } \alpha > 0 \\ \phi_j^k(\alpha) = 0, & \text{if } \alpha \leq 0. \end{cases}$$

In addition, it is required that

$$\sum_{j \in \mathcal{S}^k} \phi_j^k\left(p_j^k - \frac{1}{m^k}\sum_{\ell \in \mathcal{S}^k} x_\ell^k p_\ell^k\right) \leq 1, \qquad \forall\left(\mathbf{x}^k, \mathbf{p}^k\right) \in \Delta^k \times \mathbb{R}^{n^k}. \tag{2.11}$$

When the payoff mechanism under consideration is characterized by a bounded PDM, condition (2.11) is only required to hold for all $\left(\mathbf{x}^k, \mathbf{p}^k\right) \in \Delta^k \times \mathcal{B}^k$ (recall Remark 2.4).

SEPT protocols model scenarios where agents tend to favor strategies that offer payoffs above the population-wide average, with this average weighted by the popularity of each strategy. This formulation captures situations in which agents evaluate not only how well a strategy performs in isolation but also how it compares to what is typical across the population. The underlying idea is that agents are more likely to adopt strategies that are outperforming the societal norm, even if they are not currently selecting them. This behavior is especially relevant in social and economic contexts where individuals are influenced by collective trends or benchmarks. For instance, the drivers operating in ride-hailing apps (see Sect. 1.2.4) may gravitate toward strategies (apps) that exhibit a competitive edge relative to the average performance.

Some canonical examples of SEPT protocols are characterized by $\phi_j^k(\cdot)$ functions given by

$$\phi_j^k(\alpha) = \frac{1}{n^k\gamma^k}\min\left\{\max\{\alpha, 0\}, \gamma^k\right\},$$

for some arbitrary $\gamma^k \in \mathbb{R}_{>0}$; or

$$\phi_j^k(\alpha) = \frac{1}{\varsigma^k}\max\{\alpha, 0\}, \tag{2.12}$$

for some sufficiently large $\zeta^k \in \mathbb{R}_{>0}$, and subject to the requirement that the underlying PDM is bounded.

Note that, in contrast with the protocols (2.2) and (2.3), SEPT protocols are in general not exact, meaning that their diagonal entries $\rho^k_{ii}(\cdot,\cdot)$ play no role in the decision-making process of the agents (recall Remark 2.5 and the discussion in Sect. 2.3.1). Therefore, for the case of SEPT protocols, $\rho^k_{ii}(\cdot,\cdot)$ are just taken as placeholders.

Impartial Pairwise Comparison Protocols

The final representative class of revision protocols considered in this book is the so-called impartial pairwise comparison (IPC) protocols. These revision protocols are of the form

$$\rho^k_{ij}\left(\mathbf{p}^k, \mathbf{x}^k\right) = \phi^k_j\left(p^k_j - p^k_i\right), \quad \forall i, j \in \mathcal{S}^k, \tag{2.13}$$

where $\phi^k_j : \mathbb{R} \to \mathbb{R}_{\geq 0}$ is again a Lipschitz continuous function that satisfies

$$\begin{cases} \phi^k_j(\alpha) > 0, & \text{if } \alpha > 0 \\ \phi^k_j(\alpha) = 0, & \text{if } \alpha \leq 0. \end{cases}$$

Similar to SEPT protocols, in the case of IPC protocols it is required that

$$\sum_{j \in \mathcal{S}^k} \phi^k_j\left(p^k_j - p^k_i\right) \leq 1, \qquad \forall \left(\mathbf{x}^k, \mathbf{p}^k\right) \in \Delta^k \times \mathbb{R}^{n^k}, \tag{2.14}$$

so that the revision protocol yields valid switching probabilities. Again, when the payoff mechanism under consideration is characterized by a bounded PDM, condition (2.14) is only required to hold for all $\left(\mathbf{x}^k, \mathbf{p}^k\right) \in \Delta^k \times \mathcal{B}^k$ (recall Remark 2.4).

IPC protocols model applications where agents make decisions by performing pairwise comparisons between their currently selected strategy and each of the other available options. This type of decision-making rule is relevant in scenarios where agents exhibit a preference for maintaining their current choice but remain open to switching if an alternative provides a sufficiently better payoff. In particular, agents compare their current payoff to the payoff they would receive from each alternative strategy and switch with a probability that increases with the relative gain. Such behavior naturally arises in environments where individual decision-making is driven by incremental improvements rather than global optimization. For instance, researchers choosing a topic for their upcoming scientific endeavors (see Sect. 1.2.1), or drivers deciding on which route to take for their daily commute from home to work (see Sect. 1.2.5), may each compare their current preferred option against others and only switch if the perceived improvement justifies the change.

Canonical examples of IPC protocols resemble those of SEPTs and are characterized by $\phi^k_j(\cdot)$ functions of the form

$$\phi^k_j(\alpha) = \frac{1}{n^k \gamma^k} \min\left\{\max\{\alpha, 0\}, \gamma^k\right\},$$

for some arbitrary $\gamma^k \in \mathbb{R}_{>0}$; or

$$\phi_j^k(\alpha) = \frac{1}{\zeta^k}\max\{\alpha, 0\},$$

for some sufficiently large $\zeta^k \in \mathbb{R}_{>0}$, and subject to the requirement that the underlying PDM is bounded.

Similar to SEPT protocols, IPC protocols are in general not exact, and so their diagonal entries $\rho_{ii}^k(\cdot, \cdot)$ are mostly taken as placeholders.

2.4 Deterministic Approximations

To analyze the interplay between population games, payoff mechanisms, and revision processes, it is convenient to consider a deterministic approximation of the stochastic revision process in the limit as the number of agents tends to infinity, i.e., $N^k \to \infty$, for all $k \in \mathcal{P}$.

This deterministic approximation involves two components. First, the revision process is approximated by a deterministic model in the limit of large populations. Second, the overall closed-loop system is formulated by interconnecting the deterministic revision dynamics with the payoff mechanism. Both components are now described in detail.

2.4.1 Evolutionary Dynamics Models

In the limit, where the number of agents within each population becomes arbitrarily large (i.e., $N^k \to \infty$, for all $k \in \mathcal{P}$), the expected behavior of the stochastic revision process followed by the agents can *often*[4] be accurately approximated by a set of deterministic ordinary differential equations. This approximation is referred to as the *evolutionary dynamics model (EDM)*, formally defined below.

Definition 2.5 (*Evolutionary Dynamics Model (EDM)*) An evolutionary dynamics model (EDM) is a continuous-time dynamical system of the form

$$\dot{\mathbf{x}}(t) = \mathcal{V}(\mathbf{x}(t), \mathbf{p}(t)), \quad \mathbf{x}(0) \in \Delta,$$

where $\mathcal{V} : \Delta \times \mathbb{R}^n \to \mathbb{R}^n$ is Lipschitz continuous and satisfies

[4] Tractable deterministic approximations are known to exist for various classes of revision processes, subject to different types of alarm clocks. The microscopic decision-making process described in Sect. 2.3.1 is one such example. Additional examples and references can be found in the *Notes* at the end of this chapter.

$$\|\boldsymbol{\mathcal{V}}(\mathbf{x}(t), \mathbf{p}(t))\|_\infty < \infty$$
$$\boldsymbol{\mathcal{V}}(\mathbf{x}(t), \mathbf{p}(t)) \in \mathrm{T}\,\Delta\,(\mathbf{x}(t)),$$

for all $t \geq 0$. Here, $\mathrm{T}\,\Delta\,(\mathbf{x}(t))$ denotes the tangent cone of Δ at $\mathbf{x}(t)$.

The input space of an EDM includes all differentiable and bounded $\mathbf{p}(t) \in \mathbb{R}^n$ such that $\|\mathbf{p}(t)\|_\infty < \infty$ and $\|\dot{\mathbf{p}}(t)\|_\infty < \infty$, for all $t \geq 0$. In addition, if an EDM satisfies

$$\boldsymbol{\mathcal{V}}(\mathbf{x}(t), \mathbf{p}(t)) = \mathbf{0}_n \Leftrightarrow \mathbf{x}(t) \in \arg\max_{\tilde{\mathbf{x}} \in \Delta} \tilde{\mathbf{x}}^\top \mathbf{p}(t),$$

for every $t \geq 0$, then such an EDM is said to be *Nash stationary*. ◊

The EDM abstraction of Definition 2.5 is general, as it makes no specific assumptions about how agents revise their strategies. When focusing on the microscopic decision-making process introduced in Sect. 2.3.1, however, one can obtain a particular EDM instance that clearly reflects the roles of the Poisson alarm clocks and the revision protocol $\boldsymbol{\rho}(\cdot, \cdot)$. This structured version is referred to as the *Poisson EDM*, and it is introduced next.

Definition 2.6 (*Poisson EDM*) The Poisson EDM has the form

$$\dot{x}_i^k(t) = R^k \left(\sum_{j \in \mathcal{S}^k} x_j^k(t) \rho_{ji}^k \left(\mathbf{p}^k(t), \mathbf{x}^k(t)\right) - \sum_{j \in \mathcal{S}^k} x_i^k(t) \rho_{ij}^k \left(\mathbf{p}^k(t), \mathbf{x}^k(t)\right) \right), \tag{2.15}$$

for all $i \in \mathcal{S}^k$ and all $k \in \mathcal{P}$. ◊

How is the Poisson EDM Deduced?

An intuitive explanation is provided below to clarify how the Poisson EDM is indeed an approximation of the microscopic decision-making dynamics of Sect. 2.3.1. References to formal justifications and technical results supporting this derivation can be found in the *Notes* at the end of this chapter.

According to the model introduced in Sect. 2.3.1, the time between consecutive "rings" of each agent's alarm clock follows an exponential distribution with rate R^k, independently across agents, for every population $k \in \mathcal{P}$. By standard results from probability theory, the expected number of alarm rings within an infinitesimal interval $[t, t + dt]$ is $R^k dt$ per agent, where $dt > 0$ denotes an arbitrarily small time increment. If the state of population k at time t is given by $\mathbf{x}^k(t)$, then the expected number of agents currently choosing strategy $i \in \mathcal{S}^k$ who receive a revision opportunity during the interval $[t, t + dt]$ is approximately

$$N^k \chi_i^k(t) R^k dt = N^k \frac{x_i^k(t)}{m^k} R^k dt.$$

This expression is approximate, since the value of $\chi_i^k(t)$ (equivalently, $x_i^k(t)$) may change over the interval, but such variations are negligible when dt is small.

Given that each agent using strategy $i \in \mathcal{S}^k$ who receives a revision opportunity at time $t + dt$ switches to strategy $j \in \mathcal{S}^k$ with probability $\rho^k_{ij}\left(\mathbf{p}^k(t), \mathbf{x}^k(t)\right)$, the expected number of switches from i to j during the interval $[t, t + dt]$ is approximately

$$N^k \frac{x_i^k(t)}{m^k} R^k \rho^k_{ij}\left(\mathbf{p}^k(t), \mathbf{x}^k(t)\right) dt.$$

Thus, the net expected change in the use of strategy i over the interval $[t, t + dt]$ is given by

$$R^k \frac{N^k}{m^k} \left(\sum_{j \in \mathcal{S}^k} x_j^k(t) \rho^k_{ji}\left(\mathbf{p}^k(t), \mathbf{x}^k(t)\right) - \sum_{j \in \mathcal{S}^k} x_i^k(t) \rho^k_{ij}\left(\mathbf{p}^k(t), \mathbf{x}^k(t)\right) \right) dt,$$

where the first term accounts for switches to strategy i, and the second term accounts for switches away from strategy i. Dividing this expression by N^k, multiplying by the population mass m^k, and taking the limit as $dt \to 0$, yields that the expected instantaneous rate of change in the mass of agents choosing strategy i is

$$\dot{x}_i^k(t) = R^k \left(\sum_{j \in \mathcal{S}^k} x_j^k(t) \rho^k_{ji}\left(\mathbf{p}^k(t), \mathbf{x}^k(t)\right) - \sum_{j \in \mathcal{S}^k} x_i^k(t) \rho^k_{ij}\left(\mathbf{p}^k(t), \mathbf{x}^k(t)\right) \right),$$

for all $i \in \mathcal{S}^k$ and $k \in \mathcal{P}$. Notably, the Poisson clock rate R^k acts as a time-scaling factor, influencing the speed at which agents in population k update their strategies.

This derivation provides intuition behind the Poisson EDM as a deterministic approximation of the microscopic revision process outlined in Sect. 2.3.1. In fact, in the limit where the number of agents in each population tends to infinity (i.e., $N^k \to \infty$, for all $k \in \mathcal{P}$), the stochastic behavior described by the microscopic revision process can be arbitrarily well approximated, over any finite-time horizon, by the deterministic Poisson EDM presented in Definition 2.5.

Representative Poisson Evolutionary Dynamics Models

It is interesting to characterize the Poisson EDMs that arise when considering the representative classes of revision protocols described in Sect. 2.3.2. These three instances of dynamics are defined and characterized below.

Definition 2.7 (*PBR EDM*) Replacing the PBR revision protocol (2.8) into (2.15) yields the so-called PBR EDM given by

$$\dot{x}_i^k(t) = R^k \left(C_i^{\omega^k}\left(\mathbf{p}^k(t)\right) - x_i^k(t) \right),$$

for all $i \in \mathcal{S}^k$ and all $k \in \mathcal{P}$. Here, $C_i^{\omega^k}(\cdot)$ is defined as in (2.9). As a canonical example, taking $\omega^k\left(\mathbf{x}^k\right) = \eta^k \sum_{i \in \mathcal{S}^k} \left(x_i^k/m^k\right) \ln\left(x_i^k/m^k\right)$, with $\eta^k \in \mathbb{R}_{>0}$, yields the so-called *logit-choice* EDM given by

$$\dot{x}_i^k(t) = R^k \left(m^k \frac{\exp\left(\frac{p_i^k(t)}{\eta^k}\right)}{\sum_{\ell \in \mathcal{S}^k} \exp\left(\frac{p_\ell^k(t)}{\eta^k}\right)} - x_i^k(t) \right),$$

for all $i \in \mathcal{S}^k$ and all $k \in \mathcal{P}$. ◊

Definition 2.8 (*SEPT EDM*) Replacing the SEPT revision protocol (2.10) into (2.15) results in the SEPT EDM defined as

$$\dot{x}_i^k(t) = R^k \left(m^k \phi_i^k\left(\hat{p}_i^k(t)\right) - x_i^k(t) \sum_{j \in \mathcal{S}^k} \phi_j^k\left(\hat{p}_j^k(t)\right) \right),$$

for all $i \in \mathcal{S}^k$ and all $k \in \mathcal{P}$. Here,

$$\hat{p}_i^k(t) := p_i^k(t) - \frac{1}{m^k} \sum_{\ell \in \mathcal{S}^k} x_\ell^k(t) p_\ell^k(t), \quad \forall i \in \mathcal{S}^k.$$

As a canonical example, taking $\phi_j^k(\cdot)$ as in (2.12) with $\zeta^k = R^k$, and assuming that R^k is sufficiently large so that (2.11) holds,[5] yields the so-called *Brown–von Neumann–Nash* SEPT EDM given by

$$\dot{x}_i^k(t) = m^k \max\left\{\hat{p}_i^k(t), 0\right\} - x_i^k(t) \sum_{j \in \mathcal{S}^k} \max\left\{\hat{p}_j^k(t), 0\right\},$$

for all $i \in \mathcal{S}^k$ and all $k \in \mathcal{P}$. ◊

Definition 2.9 (*IPC EDM*) Replacing the IPC revision protocol (2.13) into (2.15) yields the IPC EDM given by

$$\dot{x}_i^k(t) = R^k \sum_{j \in \mathcal{S}^k} \left(x_j^k(t) \phi_i^k\left(p_i^k(t) - p_j^k(t)\right) - x_i^k(t) \phi_j^k\left(p_j^k(t) - p_i^k(t)\right) \right),$$

for all $i \in \mathcal{S}^k$ and all $k \in \mathcal{P}$. The canonical example of such an EDM is the so-called *Smith* EDM, which results from setting $\phi_j^k(\cdot)$ as in (2.12) with $\zeta^k = R^k$, and assuming that R^k is sufficiently large so that (2.14) holds. The resulting dynamics are given by

$$\dot{x}_i^k(t) = \sum_{j \in \mathcal{S}^k} \left(x_j^k(t) \max\left\{p_i^k(t) - p_j^k(t), 0\right\} - x_i^k(t) \max\left\{p_j^k(t) - p_i^k(t), 0\right\} \right),$$

for all $i \in \mathcal{S}^k$ and all $k \in \mathcal{P}$. ◊

[5] Such an assumption implicitly requires for the underlying PDM to be bounded.

It turns out that while every SEPT and IPC EDM is Nash stationary (see Definition 2.11), the PBR EDM is not Nash stationary. These properties are formalized in the next lemmas.

Lemma 2.3 *Every SEPT EDM and every IPC EDM satisfy the Nash stationarity property:*

$$\boldsymbol{\mathcal{V}}\left(\mathbf{x}(t), \mathbf{p}(t)\right) = \mathbf{0}_n \Leftrightarrow \mathbf{x}(t) \in \arg\max_{\tilde{\mathbf{x}} \in \Delta} \tilde{\mathbf{x}}^\top \mathbf{p}(t),$$

for every $t \geq 0$. ◊

Proof Refer to [21, Theorem 5.5.2] and [21, Theorem 5.6.2], respectively. □

Lemma 2.4 *The PBR EDM is not Nash stationary. Instead, such an EDM satisfies the so-called perturbed Nash stationarity property defined as*

$$\boldsymbol{\mathcal{V}}\left(\mathbf{x}(t), \mathbf{p}(t)\right) = \mathbf{0}_n \Leftrightarrow \mathbf{x}(t) \in \arg\max_{\tilde{\mathbf{x}} \in \mathrm{int}(\Delta)} \left(\tilde{\mathbf{x}}^\top \mathbf{p}(t) - \omega\left(\tilde{\mathbf{x}}\right)\right),$$

for every $t \geq 0$. *Here,* $\omega(\mathbf{x}) = \sum_{k \in \mathcal{P}} \omega^k\left(\mathbf{x}^k\right)$. ◊

Proof Denote

$$\mathbf{R} = \mathrm{col}\left(R^k \mathbf{1}_{n^k}\right)_{k \in \mathcal{P}}$$

$$\boldsymbol{C}^\omega(\mathbf{p}) = \mathrm{col}\left(\mathrm{col}\left(C_i^{\omega^k}\left(\mathbf{p}^k\right)\right)_{i \in \mathcal{S}^k}\right)_{k \in \mathcal{P}}.$$

Then, the PBR EDM for the entire society can be written in compact form as

$$\dot{\mathbf{x}}(t) = \mathbf{R} \odot \left(\boldsymbol{C}^\omega\left(\mathbf{p}(t)\right) - \mathbf{x}(t)\right),$$

where $\odot$ is the Hadamard (element-wise) product. By (2.9) it follows that

$$\begin{aligned}
\dot{\mathbf{x}}(t) &= \mathbf{R} \odot \left(\mathrm{col}\left(\arg\max_{\tilde{\mathbf{x}}^k \in \mathrm{int}(\Delta^k)} \left(\tilde{\mathbf{x}}^{k\top} \mathbf{p}^k(t) - \omega^k\left(\tilde{\mathbf{x}}^k\right)\right)\right)_{k \in \mathcal{P}} - \mathbf{x}(t)\right) \\
&= \mathbf{R} \odot \left(\arg\max_{\tilde{\mathbf{x}} \in \mathrm{int}(\Delta)} \left(\tilde{\mathbf{x}}^\top \mathbf{p}(t) - \omega\left(\tilde{\mathbf{x}}\right)\right) - \mathbf{x}(t)\right),
\end{aligned}$$

which yields the desired result. □

Notice that Definitions 2.7–2.9 implicitly assume that all agents of the society operate following the same class of revision protocol. That is, the PBR EDM in Definition 2.7 implies that all agents of the society operate following PBR protocols (2.8); the SEPT EDM in Definition 2.8 considers that all agents of the society use SEPT protocols (2.10); and the IPC EDM in Definition 2.9 implicitly assumes that all agents of the society execute IPC protocols (2.13). In many practical applications, however, it is more appropriate to model different populations under different revision protocols, enabling the consideration of mixed EDMs.

To ease the formulation and analysis of mixed societies where each population implements a different revision protocol, it is convenient to encapsulate the Nash and perturbed Nash stationarity properties of Lemmas 2.3 and 2.4 into a single unified property, as formalized next.

Definition 2.10 (*ϖ-Perturbed Nash Stationarity*) An arbitrary EDM is said to be ϖ-perturbed Nash stationary if it satisfies that

$$\boldsymbol{\mathcal{V}}(\mathbf{x}(t), \mathbf{p}(t)) = \mathbf{0}_n \Leftrightarrow \mathbf{x}(t) \in \arg\max_{\tilde{\mathbf{x}} \in \mathrm{int}_{\varpi}(\Delta)} \left(\tilde{\mathbf{x}}^\top \mathbf{p}(t) - \varpi(\tilde{\mathbf{x}})\right),$$

for all $t \geq 0$; and for some

$$\varpi(\mathbf{x}) = \sum_{k \in \mathcal{P}} \varpi^k\left(\mathbf{x}^k\right),$$

where each function $\varpi^k(\cdot)$ satisfies exactly one of the following two conditions:

(i) $\varpi^k : \mathbb{R}^{n^k} \to \mathbb{R}$ is the zero function, i.e., $\varpi^k\left(\mathbf{x}^k\right) = 0$, for all $\mathbf{x}^k \in \mathbb{R}^{n^k}$.
(ii) $\varpi^k : \mathbb{R}^{n^k}_{>0} \to \mathbb{R}$ is twice continuously differentiable, strictly convex, and steep near $\mathrm{bd}\left(\Delta^k\right)$, i.e., $\left\| \nabla_{\mathbf{x}^k} \varpi^k\left(\mathbf{x}^k\right) \right\|_\infty \to \infty$ whenever $\mathbf{x}^k \to \mathrm{bd}\left(\Delta^k\right)$.

Moreover, the set $\mathrm{int}_{\varpi}(\Delta)$ is defined as

$$\mathrm{int}_{\varpi}(\Delta) = \prod_{k \in \mathcal{P}} \mathrm{int}_{\varpi^k}\left(\Delta^k\right),$$

where

$$\mathrm{int}_{\varpi^k}\left(\Delta^k\right) = \begin{cases} \Delta^k, & \text{if } \varpi^k(\cdot) \text{ satisfies Condition (i)}, \\ \mathrm{int}\left(\Delta^k\right), & \text{if } \varpi^k(\cdot) \text{ satisfies Condition (ii)}; \end{cases}$$

for all $k \in \mathcal{P}$. ◊

Remark 2.7 According to Definition 2.10, the Nash stationarity property in Definition 2.5 and Lemma 2.3 is equivalent to 0-perturbed Nash stationarity; while the perturbed Nash stationarity property of Lemma 2.4 is equivalent to ω-perturbed Nash stationarity with

$$\omega(\mathbf{x}) = \sum_{k \in \mathcal{P}} \omega^k\left(\mathbf{x}^k\right),$$

where each $\omega^k(\cdot)$ is the function characterizing the PBR revision protocol of the agents of population k. Consequently, the notion of ϖ-perturbed Nash stationarity in Definition 2.10 indeed captures both of the stationarity notions of Lemmas 2.3 and 2.4 in a unified fashion. ◊

Based on Definition 2.10, Corollary 2.1 extends Lemmas 2.3 and 2.4 to the case of *mixed* societies where each population implements a different class of revision protocol out of the three representative classes of protocols of Sect. 2.3.2.

Corollary 2.1 *Consider a society where each population behaves according to either a SEPT, IPC, or PBR revision protocol. The resulting EDM satisfies the ϖ-perturbed Nash stationarity property of Definition 2.10 with*

$$\varpi^k\left(\mathbf{x}^k\right) = \begin{cases} 0, & \textit{if population k uses either a SEPT or IPC protocol,} \\ \omega^k\left(\mathbf{x}^k\right), & \textit{if population k uses a PBR protocol,} \end{cases}$$

for all $k \in \mathcal{P}$. ◊

Proof The proof follows immediately from Lemmas 2.3 and 2.4. □

2.4.2 Closed-Loop Interconnection

The EDM abstraction introduced in Definition 2.5 is a dynamical system with input $\mathbf{p}(t)$ and output $\mathbf{x}(t)$. In contrast, the PDM in Definition 2.4, which characterizes the underlying payoff mechanism, is a dynamical system with input $\mathbf{x}(t)$ and output $\mathbf{p}(t)$. Consequently, under the considered framework, the EDM and PDM form a closed-loop interconnection as depicted in Fig. 2.1. This *closed-loop interconnection* is formally defined as follows.

Definition 2.11 (*EDM–PDM*) An EDM–PDM is the positive feedback interconnection between an EDM and a PDM, where the EDM takes $\mathbf{p}(t)$ as input and provides $\mathbf{x}(t)$ as output, while the PDM takes $\mathbf{x}(t)$ as input and provides $\mathbf{p}(t)$ as output (see Fig. 2.1). More precisely, an EDM–PDM is characterized by the dynamics

$$\begin{aligned} \dot{\mathbf{x}}(t) &= \mathcal{V}\left(\mathbf{x}(t), \mathcal{H}\left(\mathbf{q}(t), \mathbf{x}(t)\right)\right), & \mathbf{x}(0) \in \Delta \\ \dot{\mathbf{q}}(t) &= \mathcal{W}\left(\mathbf{q}(t), \mathbf{x}(t)\right), & \mathbf{q}(0) \in \mathbb{R}^d. \end{aligned}$$

Consequently, the set of equilibria of an EDM–PDM is

$$\mathcal{E} = \left\{ \left(\mathbf{x}^*, \mathbf{q}^*\right) \in \Delta \times \mathbb{R}^d : \begin{array}{l} \mathcal{V}\left(\mathbf{x}^*, \mathcal{H}(\mathbf{q}^*, \mathbf{x}^*)\right) = \mathbf{0}_n \\ \mathcal{W}(\mathbf{q}^*, \mathbf{x}^*) = \mathbf{0}_d \end{array} \right\}.$$

Moreover, an EDM–PDM is said to be *admissible* if, for every initial condition $(\mathbf{x}(0), \mathbf{q}(0)) \in \Delta \times \mathbb{R}^d$, it satisfies that $\|\mathbf{q}(t)\|_\infty < \infty$, for all $t \geq 0$. ◊

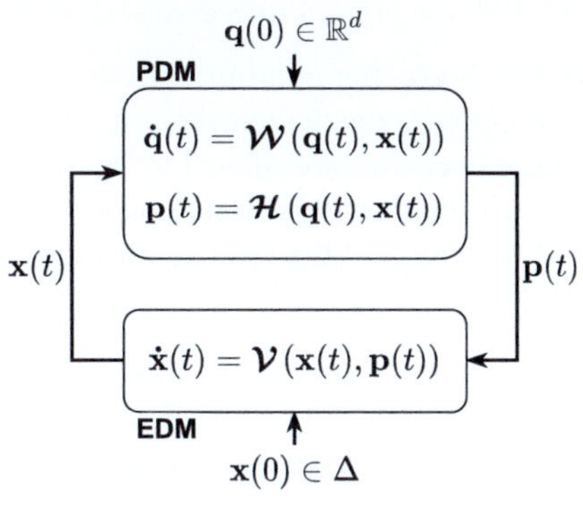

Fig. 2.1 Closed-loop interconnection between an EDM and a PDM

Remark 2.8 (*Boundedness and Admissibility*) Based on Definitions 2.4, 2.5, and 2.11, if the PDM is bounded, then the resulting EDM–PDM interconnection is admissible. Moreover, if the EDM–PDM is admissible, then for every initial condition $(\mathbf{x}(0), \mathbf{q}(0)) \in \Delta \times \mathbb{R}^d$, there exists a unique solution trajectory $\{(\mathbf{x}(t), \mathbf{q}(t))\}_{t \geq 0}$. This trajectory satisfies that $\mathbf{x}(t)$ remains within the input space of the PDM and $\mathbf{p}(t)$ remains within the input space of the EDM, for all $t \geq 0$. $\diamondsuit$

The EDM–PDM pair defines the deterministic mathematical model that underpins the analysis of large-scale multi-agent systems throughout this book. By studying the set of equilibria $\mathcal{E}$ of the EDM–PDM and its stability properties, one can derive rigorous conclusions about the expected behavior of the society. In particular, this approach enables the characterization of conditions under which a generalized Nash equilibrium exists, and provides insight into whether the collective behavior of the agents is expected to converge toward such an outcome. Building on this foundation, the next section establishes the meaning of control within the considered framework, and formally introduces the generalized Nash equilibrium problem along with an optimization-based perspective.

2.5 Formalizing the Meaning of Control

What is the meaning of control under the considered framework?
What does it mean to control a large-scale multi-agent system as the ones considered?

When dealing with large-scale multi-agent systems, where each agent acts as a decentralized and autonomous decision-maker, it becomes essential to formalize what it means to *control* such systems. In the considered context, control is not about imposing specific decisions on individual agents, as that would contradict the decentralized nature of the framework. Instead, control is about steering the system toward a self-enforced equilibrium state, one where no agent has incentives to unilaterally deviate from, even when having the freedom to do so.

As discussed in Chap. 1, two fundamental questions naturally arise:

Q1 *Does such an equilibrium state exist?*
Q2 *Will the populations converge to such an equilibrium, provided one exists?*

In this book, the equilibrium concept of interest is the *generalized Nash equilibrium* (GNE), which represents an agreement outcome among non-cooperative agents whose decisions are subject to shared constraints. Accordingly, the control objective is to ensure that the society converges to a GNE, provided one exists.

Ensuring that the system converges to an existent GNE can be accomplished in two ways:

- If the agents are engineering systems whose decentralized decision-making policies can be designed, then one approach is to devise their revision processes to ensure convergence to a desired outcome.

- In contrast, if agents have fixed, immutable decision-making rules which cannot be designed or modified, then control is exerted by designing the payoff mechanism to steer the society toward the desired equilibrium.

Both of these perspectives are explored in detail in Chap. 5. Before that, however, the remainder of this section addresses Question Q1 by formally introducing the GNE notion, which serves as the foundation for understanding the equilibrium behavior of the multi-agent systems considered in this book.

2.5.1 The Generalized Nash Equilibrium Problem

Throughout this book, the set of generalized Nash equilibria of a population game G is formally defined as follows.

Definition 2.12 (*Generalized Nash Equilibria*) The set of generalized Nash equilibria of a population game $G = (\mathcal{P}, \mathcal{X}, \mathbf{f}(\cdot))$ is defined as

$$\mathrm{GNE}(\mathbf{f}) = \mathrm{fix}\left(\arg\max_{\mathbf{x}\in\mathcal{X}} \mathbf{x}^\top \mathbf{f}(\cdot)\right).$$

Here, $\mathrm{fix}(F)$ denotes the fixed point set of the operator $F(\cdot) = \arg\max_{\mathbf{x}\in\mathcal{X}} \mathbf{x}^\top \mathbf{f}(\cdot)$. That is, $\mathbf{x}^* \in \mathrm{GNE}(\mathbf{f})$ if and only if $\mathbf{x}^* \in \arg\max_{\mathbf{x}\in\mathcal{X}} \mathbf{x}^\top \mathbf{f}(\mathbf{x}^*)$. ♢

Remark 2.9 The notion of GNE adopted in Definition 2.12 is commonly referred to in the literature as a *variational* GNE (vGNE). For simplicity of notation, it is denoted simply as GNE throughout this book. This notion corresponds to the solution of the variational inequality $\mathrm{VI}(\mathcal{X}, -\mathbf{f})$, defined as finding $\mathbf{x}^* \in \mathcal{X}$ such that $(\mathbf{x} - \mathbf{x}^*)^\top(-\mathbf{f}(\mathbf{x}^*)) \geq 0$ for all $\mathbf{x} \in \mathcal{X}$. This concept is particularly suited to population games, where the focus is placed on the aggregate strategic distribution $\mathbf{x}$ rather than on the individual strategies of each agent. ♢

Remark 2.10 In Definition 2.12, the dependence on the fitness vector $\mathbf{f}(\cdot)$ in $\mathrm{GNE}(\mathbf{f})$ is made explicit. This is done both to accommodate with the common notation of the related literature, as well as to ease the notation when referring to the sets of generalized Nash equilibria of two population games, G_1 and G_2, that only differ on their fitness vectors, i.e., $G_1 = (\mathcal{P}, \mathcal{X}, \mathbf{f}_1(\cdot))$ and $G_2 = (\mathcal{P}, \mathcal{X}, \mathbf{f}_2(\cdot))$. As will be discussed in brief, such a scenario is particularly relevant when studying the set of equilibria of PBR EDMs. ♢

Based on Definition 2.12, it follows that a GNE is a strategic distribution

$$\mathbf{x}^* \in \mathcal{X} = \Delta \cap \Omega,$$

where no agent can increase its fitness value by unilaterally changing its strategy while still satisfying the constraints imposed by Ω. In the absence of explicit coupling constraints, i.e., when $\Omega = \mathbb{R}^n$ (or even when $\Omega \supseteq \Delta$), the set of generalized Nash equilibria reduces to the set of Nash equilibria, as defined next.

Definition 2.13 (*Nash Equilibria*) The set of Nash equilibria of a population game $G = (\mathcal{P}, \mathcal{X}, \mathbf{f}(\cdot))$ is defined as

$$\mathrm{NE}(\mathbf{f}) = \mathrm{fix}\left(\arg\max_{\mathbf{x}\in\Delta} \mathbf{x}^\top \mathbf{f}(\cdot)\right).$$

Or equivalently,

$$\mathrm{NE}(\mathbf{f}) = \left\{\mathbf{x} \in \Delta : x_i^k > 0 \Rightarrow f_i^k(\mathbf{x}) = \max_{j\in\mathcal{S}^k} f_j^k(\mathbf{x}), \forall i \in \mathcal{S}^k,\ \forall k \in \mathcal{P}\right\}.$$

That is, $\mathbf{x}^* \in \mathrm{NE}(\mathbf{f})$ if and only if $\mathbf{x}^* \in \arg\max_{\mathbf{x}\in\Delta} \mathbf{x}^\top \mathbf{f}\,(\mathbf{x}^*)$. ◊

Clearly, if $\Omega \supseteq \Delta$, then $\mathcal{X} = \Delta$, $\mathrm{NE}(\mathbf{f}) = \mathrm{GNE}(\mathbf{f})$, and the notions of NE and GNE coincide. Consequently, an NE is a particular instance of a GNE.

Based on the properties of the fitness vector $\mathbf{f}(\cdot)$ and the set $\mathcal{X}$, one can characterize the set $\mathrm{GNE}(\mathbf{f})$ as follows.

Lemma 2.5 *Let* $\mathbf{f}(\cdot)$ *be continuous and let* $\mathcal{X}$ *be nonempty, convex, and compact.*[6] *Then, the set* $\mathrm{GNE}\,(\mathbf{f})$ *is nonempty and compact.* ◊

Proof Notice that $\mathrm{GNE}(\mathbf{f})$ in Definition 2.12 is equivalent to

$$\mathrm{GNE}(\mathbf{f}) = \left\{\mathbf{x} \in \mathcal{X} : \mathbf{x} \in \arg\max_{\tilde{\mathbf{x}}\in\mathcal{X}} \tilde{\mathbf{x}}^\top \mathbf{f}(\mathbf{x})\right\}, \tag{2.16}$$

and thus $\mathbf{x}^* \in \mathrm{GNE}\,(\mathbf{f}) \Leftrightarrow (\mathbf{x}^*)^\top \mathbf{f}\,(\mathbf{x}^*) \geq \mathbf{x}^\top \mathbf{f}\,(\mathbf{x}^*)$, for all $\mathbf{x} \in \mathcal{X}$. Consequently, the considered set $\mathrm{GNE}(\mathbf{f})$ coincides with the set of solutions of the variational inequality $\mathrm{VI}\,(\mathcal{X}, -\mathbf{f})$ given by: find $\mathbf{x}^* \in \mathcal{X}$ such that $(\mathbf{x} - \mathbf{x}^*)^\top\,(-\mathbf{f}(\mathbf{x}^*)) \geq 0$, for all $\mathbf{x} \in \mathcal{X}$. Since $\mathbf{f}(\cdot)$ is continuous and $\mathcal{X}$ is nonempty, convex, and compact, it follows from [4, Corollary 2.2.5] that $\mathrm{GNE}(\mathbf{f})$ is nonempty and compact. □

Lemma 2.6 *Let* $\mathbf{f}(\cdot)$ *be continuous and let* $\mathcal{X}$ *be nonempty, convex, and compact. If the corresponding population game G is strictly contractive, then there exists a unique GNE for the game G, i.e., the set* $\mathrm{GNE}(\mathbf{f})$ *is a singleton.* ◊

Proof The existence of a GNE follows from Lemma 2.5, and its uniqueness holds from [4, Theorem 2.3.3]. □

Lemma 2.7 *Let G be a full-potential game with concave potential function* $\varphi(\cdot)$. *Then,* $\mathbf{x}^* \in \mathrm{GNE}(\mathbf{f})$ *if and only if* $\mathbf{x}^* \in \arg\max_{\mathbf{x}\in\mathcal{X}} \varphi(\mathbf{x})$. ◊

---

[6] Note that Assumption 2.1 implies that $\mathcal{X}$ is indeed nonempty, convex, and compact. Similarly, Assumption 2.2 implies the continuity of $\mathbf{f}(\cdot)$.

Proof Recall that GNE($\mathbf{f}$) in Definition 2.12 can be equivalently defined as in (2.16). Since $\varphi(\cdot)$ is concave and differentiable and $\varphi(\cdot) = \nabla \mathbf{f}(\cdot)$, it follows that

$$\mathbf{x}^* \in \arg\max_{\mathbf{x}\in\mathcal{X}} \varphi(\mathbf{x})$$

if and only if $(\mathbf{x} - \mathbf{x}^*)^\top \mathbf{f}(\mathbf{x}^*) \leq 0$, for all $\mathbf{x} \in \mathcal{X}$, which is equivalent to

$$(\mathbf{x}^*)^\top \mathbf{f}(\mathbf{x}^*) \geq \mathbf{x}^\top \mathbf{f}(\mathbf{x}^*), \quad \forall \mathbf{x} \in \mathcal{X}.$$

Clearly, $(\mathbf{x}^*)^\top \mathbf{f}(\mathbf{x}^*) \geq \mathbf{x}^\top \mathbf{f}(\mathbf{x}^*)$, for all $\mathbf{x} \in \mathcal{X}$, if and only if $\mathbf{x}^* \in$ GNE($\mathbf{f}$). □

Granted that at least one GNE exists, the so-called GNE-seeking problem is formally stated next.

Definition 2.14 (*The Generalized Nash Equilibrium Problem*) Consider a population game $G = (\mathcal{P}, \mathcal{X}, \mathbf{f}(\cdot))$. The GNE (seeking) problem consists in designing either the revision processes of the agents or the payoff mechanism (or both) so that the society converges asymptotically to a strategic distribution $\mathbf{x}^* \in$ GNE($\mathbf{f}$). ◊

The GNE problem introduced in Definition 2.14 is the central focus of this book. To evaluate whether a particular design of revision process or payoff mechanism effectively addresses the GNE problem, the analysis focuses on the stability properties of the equilibrium set $\mathcal{E}$ of the associated EDM–PDM system, which models the expected behavior of the society. The remainder of this section establishes a formal connection between the set GNE($\mathbf{f}$), associated with a given population game G, and the equilibrium set $\mathcal{E}$ of the corresponding EDM–PDM pair.

Lemma 2.8 *Consider a population game G characterized by a fitness vector* $\mathbf{f}(\cdot)$, *and consider an EDM–PDM comprised of a Nash stationary EDM and a PDM satisfying Assumption 2.3. Then, it holds that*

$$\left(\mathbf{x}^*, \mathbf{q}^*\right) \in \mathcal{E} \quad \Rightarrow \quad \mathbf{x}^* \in \mathrm{GNE}(\mathbf{f}).$$

That is, if $(\mathbf{x}^*, \mathbf{q}^*)$ *is an equilibrium point of the EDM–PDM feedback system, then* $\mathbf{x}^*$ *is a GNE of the underlying population game G.* ◊

Proof Recall that GNE($\mathbf{f}$) $=$ SOL $(\mathcal{X}, -\mathbf{f})$, where SOL$(\mathcal{X}, -\mathbf{f})$ is the set of solutions of the *variational inequality* VI $(\mathcal{X}, -\mathbf{f})$ given by: find $\mathbf{x}^* \in \mathcal{X}$ such that $(\mathbf{x} - \mathbf{x}^*)^\top (-\mathbf{f}(\mathbf{x}^*)) \geq 0$, for all $\mathbf{x} \in \mathcal{X}$. Therefore, from [4, Proposition 1.3.4] it follows that $\mathbf{x}^* \in$ GNE($\mathbf{f}$) if and only if there exist some $\boldsymbol{\mu}^*_\Omega \in \mathbb{R}^{c_=}$, $\boldsymbol{\lambda}^*_\Omega \in \mathbb{R}^{c_\leq}$, $\boldsymbol{\mu}^*_\Delta \in \mathbb{R}^P$, and $\boldsymbol{\lambda}^*_\Delta \in \mathbb{R}^n$, such that

$$\mathbf{f}\left(\mathbf{x}^*\right) - \mathbf{A}^\top \boldsymbol{\mu}_\Omega^* - \mathrm{D}_\mathbf{x}\, \mathbf{g}\left(\mathbf{x}^*\right)^\top \boldsymbol{\lambda}_\Omega^* - \mathbf{A}_\Delta^\top \boldsymbol{\mu}_\Delta^* + \boldsymbol{\lambda}_\Delta^* = \mathbf{0}_n \tag{2.17a}$$

$$\boldsymbol{\lambda}_\Omega^* \odot \mathbf{g}\left(\mathbf{x}^*\right) = \mathbf{0}_{c_\leq} \tag{2.17b}$$

$$\boldsymbol{\lambda}_\Delta^* \odot \mathbf{x}^* = \mathbf{0}_n \tag{2.17c}$$

$$\boldsymbol{\lambda}_\Omega^* \succeq \mathbf{0}_{c_\leq} \tag{2.17d}$$

$$\boldsymbol{\lambda}_\Delta^* \succeq \mathbf{0}_n \tag{2.17e}$$

$$\mathbf{x}^* \in \mathcal{X}. \tag{2.17f}$$

Namely, (2.17) are the KKT conditions associated to SOL($\mathcal{X}$, $-\mathbf{f}$). Therefore, to prove the desired claim one must show that $(\mathbf{x}^*, \mathbf{q}^*) \in \mathcal{E}$ implies that $\mathbf{x}^*$ satisfies (2.17) for some $\boldsymbol{\mu}_\Omega^*$, $\boldsymbol{\lambda}_\Omega^*$, $\boldsymbol{\mu}_\Delta^*$, and $\boldsymbol{\lambda}_\Delta^*$.

Let $(\mathbf{x}^*, \mathbf{q}^*) \in \mathcal{E}$. Given that the EDM is Nash stationary, and using Assumption 2.3, it holds that

$$\mathbf{x}^* \in \arg\max_{\tilde{\mathbf{x}} \in \Delta} \left(\tilde{\mathbf{x}}^\top \mathbf{f}\left(\mathbf{x}^*\right) - \tilde{\mathbf{x}}^\top \mathbf{A}^\top \mathbf{M}_{c_=} \mathbf{q}^* - \tilde{\mathbf{x}}^\top \mathrm{D}_\mathbf{x}\, \mathbf{g}\left(\mathbf{x}^*\right)^\top \mathbf{M}_{c_\leq} \mathbf{q}^*\right), \tag{2.18}$$

with $\mathbf{M}_{c_\leq} \mathbf{q}^* \succeq \mathbf{0}_{c_\leq}$ and $\left(\mathbf{M}_{c_\leq} \mathbf{q}^*\right) \odot \mathbf{g}\left(\mathbf{x}^*\right) = \mathbf{0}_{c_=}$. Thus, from the necessity of the KKT conditions associated to the optimization problem in (2.18), it holds that $(\mathbf{x}^*, \mathbf{q}^*)$ satisfies

$$\mathbf{f}\left(\mathbf{x}^*\right) - \mathbf{A}^\top \mathbf{M}_{c_=} \mathbf{q}^* - \mathrm{D}_\mathbf{x}\, \mathbf{g}\left(\mathbf{x}^*\right)^\top \mathbf{M}_{c_\leq} \mathbf{q}^* - \mathbf{A}_\Delta^\top \tilde{\boldsymbol{\mu}}_\Delta^* + \tilde{\boldsymbol{\lambda}}_\Delta^* = \mathbf{0}_n \tag{2.19a}$$

$$\tilde{\boldsymbol{\lambda}}_\Delta^* \odot \mathbf{x}^* = \mathbf{0}_n \tag{2.19b}$$

$$\tilde{\boldsymbol{\lambda}}_\Delta^* \succeq \mathbf{0}_n, \tag{2.19c}$$

for some $\tilde{\boldsymbol{\mu}}_\Delta^* \in \mathbb{R}^P$ and $\tilde{\boldsymbol{\lambda}}_\Delta^* \in \mathbb{R}^n$. Consequently, by setting $\boldsymbol{\mu}_\Omega^* = \mathbf{M}_{c_=} \mathbf{q}^*$, $\boldsymbol{\lambda}_\Omega^* = \mathbf{M}_{c_\leq} \mathbf{q}^*$, $\boldsymbol{\mu}_\Delta^* = \tilde{\boldsymbol{\mu}}_\Delta^*$, and $\boldsymbol{\lambda}_\Delta^* = \tilde{\boldsymbol{\lambda}}_\Delta^*$, it follows that (2.19) implies that (2.17a)–(2.17e) hold. Furthermore, since Δ is invariant under any EDM (see Definition 2.5) and $\mathcal{X} = \Delta \cap \Omega$ (see Definition 2.12), it follows that Assumption 2.3 implies that (2.17f) holds. In summary, the Nash stationarity of the EDM and Assumption 2.3 imply that (2.17) holds for some $\boldsymbol{\mu}_\Omega^*$, $\boldsymbol{\lambda}_\Omega^*$, $\boldsymbol{\mu}_\Delta^*$, and $\boldsymbol{\lambda}_\Delta^*$, yielding the desired result. □

Lemma 2.8 establishes a correspondence between the set of equilibria of the EDM–PDM system and the set of generalized Nash equilibria of the underlying population game G. Specifically, it shows that solving the GNE problem of Definition 2.14 reduces to designing an EDM–PDM pair that satisfies three key properties: (i) the EDM is Nash stationary, (ii) the PDM satisfies Assumption 2.3, and (iii) the resulting EDM–PDM's equilibrium set $\mathcal{E}$ is asymptotically stable. Achieving these properties requires either designing the agents' revision process (i.e., the EDM in prescriptive control settings) or the payoff mechanism (i.e., the PDM in non-prescriptive approaches). While sufficient conditions for the asymptotic stability of $\mathcal{E}$ are developed in Chap. 4, the design of prescriptive and non-prescriptive control strategies is addressed in Chap. 5.

It is worth to highlight that Lemma 2.8 does not apply to the PBR EDM of Definition 2.7, as such an EDM is not Nash stationary (see Lemma 2.4). Instead, Lemma 2.9 provides a result analogous to Lemma 2.8 but tailored for the PBR EDM.

Lemma 2.9 *Consider a population game $G = (\mathcal{P}, \mathcal{X}, \mathbf{f}(\cdot))$, and an EDM–PDM comprised of a PBR EDM and a PDM satisfying Assumption 2.3. Let $\tilde{G}$ be a perturbed version of G defined as $\tilde{G} = \left(\mathcal{P}, \mathcal{X}, \tilde{\mathbf{f}}(\cdot)\right)$, where*

$$\tilde{\mathbf{f}}(\mathbf{x}) = \mathbf{f}(\mathbf{x}) - \nabla_{\mathbf{x}}\omega(\mathbf{x}),$$

with $\omega(\cdot)$ defined as in Lemma 2.4. Then, it holds that

$$\left(\mathbf{x}^*, \mathbf{q}^*\right) \in \mathcal{E} \quad \Rightarrow \quad \mathbf{x}^* \in \mathrm{GNE}\left(\tilde{\mathbf{f}}\right).$$

That is, if $(\mathbf{x}^*, \mathbf{q}^*)$ is an equilibrium point of the EDM–PDM pair, then $\mathbf{x}^*$ is a GNE of the perturbed population game $\tilde{G} = \left(\mathcal{P}, \mathcal{X}, \tilde{\mathbf{f}}(\cdot)\right)$. ◊

Proof The proof is immediate from Lemmas 2.4 and 2.8. □

Finally, Corollary 2.2 encapsulates Lemmas 2.8 and 2.9 in a unified fashion, allowing also consideration of EDMs describing mixed societies with populations implementing different classes of revision protocols.

Corollary 2.2 *Consider a population game $G = (\mathcal{P}, \mathcal{X}, \mathbf{f}(\cdot))$, and an EDM–PDM comprised of a ϖ-perturbed Nash stationary EDM and a PDM satisfying Assumption 2.3. Let $\tilde{G}$ be a perturbed version of G defined as $\tilde{G} = \left(\mathcal{P}, \mathcal{X}, \tilde{\mathbf{f}}(\cdot)\right)$, where*

$$\tilde{\mathbf{f}}(\mathbf{x}) = \mathbf{f}(\mathbf{x}) - \nabla_{\mathbf{x}}\varpi(\mathbf{x}),$$

with $\varpi(\cdot)$ taken as in Definition 2.10. Then, it holds that

$$\left(\mathbf{x}^*, \mathbf{q}^*\right) \in \mathcal{E} \quad \Rightarrow \quad \mathbf{x}^* \in \mathrm{GNE}\left(\tilde{\mathbf{f}}\right).$$

That is, if $(\mathbf{x}^*, \mathbf{q}^*)$ is an equilibrium point of the EDM–PDM pair, then $\mathbf{x}^*$ is a GNE of the ϖ-perturbed population game $\tilde{G} = \left(\mathcal{P}, \mathcal{X}, \tilde{\mathbf{f}}(\cdot)\right)$. ◊

Proof The result follows from Corollary 2.1 and Lemmas 2.8–2.9. □

2.5.2 An Optimization-Based Perspective

This section provides an optimization-based interpretation of GNE problems, highlighting their connection to a set of interdependent optimization problems, one for each population.

For the following discussion, let $\mathbf{x} = \text{col}\left(\mathbf{x}^1, \ldots, \mathbf{x}^P\right)$ be equivalently written as $\left(\mathbf{x}^k, \mathbf{x}^{-k}\right)$, for all $k \in \mathcal{P}$. Here, $\mathbf{x}^{-k} = \text{col}\left(\mathbf{x}^1, \ldots, \mathbf{x}^{k-1}, \mathbf{x}^{k+1}, \ldots, \mathbf{x}^P\right) \in \mathbb{R}^{n-n^k}_{\geq 0}$ denotes the strategic distribution of all populations other than k. That is, for each k, it always holds that $\left(\mathbf{x}^k, \mathbf{x}^{-k}\right) = \text{col}\left(\mathbf{x}^1, \ldots, \mathbf{x}^P\right) = \mathbf{x}$, meaning that the ordering is preserved regardless of the population index.

Consider a scenario where the agents of each population $k \in \mathcal{P}$ aim to collectively solve an optimization problem of the form

$$\max_{\mathbf{x}^k \in \mathbb{R}^{n^k}} \psi^k\left(\mathbf{x}^k, \mathbf{x}^{-k}\right), \quad \text{s.t.} \quad \mathbf{x}^k \in \Omega^k\left(\mathbf{x}^{-k}\right), \tag{2.20}$$

where $\psi^k : \mathbb{R}^n_{\geq 0} \to \mathbb{R}$ is the concave and differentiable objective function of population k, and

$$\Omega^k\left(\mathbf{x}^{-k}\right) = \left\{\mathbf{x}^k \in \mathbb{R}^{n^k} : \left(\mathbf{x}^k, \mathbf{x}^{-k}\right) \in \mathcal{X}\right\} \tag{2.21}$$

is the feasible set for population k as a function of the strategic distributions of the remaining populations. Note that since

$$\mathbf{x}^k = \frac{m^k}{N^k} \sum_{a \in \mathcal{A}^k} \mathbf{s}^k_a,$$

and given that every agent $a \in \mathcal{A}^k$ is responsible for choosing their own strategy $\mathbf{s}^k_a$, the optimization problem (2.20) corresponds to a decentralized large-scale integer programming problem, coupled among populations through the constraints set Ω.

Define the pseudo-gradient mapping $\mathbf{f}(\mathbf{x})$ associated with the functions ψ^k as

$$\mathbf{f}(\mathbf{x}) = \text{col}\left(\nabla_{\mathbf{x}^1}\psi^1(\mathbf{x}), \nabla_{\mathbf{x}^2}\psi^2(\mathbf{x}), \ldots, \nabla_{\mathbf{x}^P}\psi^P(\mathbf{x})\right). \tag{2.22}$$

If $\mathbf{x}^* \in \text{GNE}(\mathbf{f})$, with $\mathbf{x}^* = \text{col}\left(\mathbf{x}^{1*}, \mathbf{x}^{2*}, \ldots, \mathbf{x}^{P*}\right)$, then each $\mathbf{x}^{k*}$ solves the optimization problem in (2.20) for the corresponding population $k \in \mathcal{P}$. To see this fact, observe that $\mathbf{x}^* \in \text{GNE}(\mathbf{f})$ implies $\mathbf{x}^* \in \mathcal{X}$ and

$$\left(\tilde{\mathbf{x}} - \mathbf{x}^*\right)^\top \mathbf{f}(\mathbf{x}^*) \leq 0, \quad \forall \tilde{\mathbf{x}} \in \mathcal{X}.$$

Therefore, for any $k \in \mathcal{P}$, it follows that $\mathbf{x}^{k*} \in \Omega^k(\mathbf{x}^{-k*})$, where

$$\mathbf{x}^{-k*} = \text{col}\left(\mathbf{x}^{1*}, \ldots, \mathbf{x}^{(k-1)*}, \mathbf{x}^{(k+1)*}, \ldots, \mathbf{x}^{P*}\right),$$

and that

$$\left(\tilde{\mathbf{x}}^k - \mathbf{x}^{k*}\right)^\top \nabla_{\mathbf{x}^k}\psi^k(\mathbf{x}^*) \leq 0, \quad \forall \tilde{\mathbf{x}}^k \in \Omega^k(\mathbf{x}^{-k*}),$$

where $\tilde{\mathbf{x}} = \text{col}(\tilde{\mathbf{x}}^1, \tilde{\mathbf{x}}^2, \ldots, \tilde{\mathbf{x}}^P)$. Hence, any $\mathbf{x}^* \in \text{GNE}(\mathbf{f})$ satisfies the necessary and sufficient optimality conditions to solve the convex optimization problem in (2.20) for each population. Solving the GNE problem associated with the fitness vector $\mathbf{f}(\cdot)$

therefore simultaneously solves the optimization problems in (2.20) for all populations.

2.6 Simulation Examples

To illustrate the concepts discussed throughout the chapter, this section presents some simulation examples, implemented using the `popgames` package in *Python*.

Competition and Coordination in Ride-Hailing Apps

Consider the *ride-hailing apps* scenario introduced in Sect. 1.2.4. In this context, a population of 2000 users and a population of 1000 drivers interact to decide which app to use out of a set of three available apps, i.e., $\mathcal{S}^1 = \mathcal{S}^2 = \{1, 2, 3\}$ (and so $n = 6$). To model the fact that there are twice as many users as drivers, the population masses are set as $m^1 = 2$ and $m^2 = 1$, respectively. Let $\Omega = \mathbb{R}^n$ and define the fitness function as $\mathbf{f}(\mathbf{x}) = \mathbf{Q}\mathbf{x} + \mathbf{r}$, where

$$\mathbf{Q} = \begin{bmatrix} -0.7 & 0 & 0 & 0.3 & 0 & 0 \\ 0 & -0.8 & 0 & 0 & 0.2 & 0 \\ 0 & 0 & -0.7 & 0 & 0 & 0.3 \\ 0.1 & 0 & 0 & -0.9 & 0 & 0 \\ 0 & 0.3 & 0 & 0 & -0.7 & 0 \\ 0 & 0 & 0.1 & 0 & 0 & -0.9 \end{bmatrix}, \qquad \mathbf{r} = \begin{bmatrix} -0.6 \\ -0.7 \\ -0.8 \\ 0.1 \\ 0.2 \\ 0.3 \end{bmatrix}.$$

Here, the matrix $\mathbf{Q}$ captures the interaction effects of coordination or competition between users and drivers, while the vector $\mathbf{r} := \left[\gamma_1^1, \gamma_1^2, \gamma_1^3, \gamma_2^1, \gamma_2^2, \gamma_3^2\right]^\top$ models the nominal fares and incentives offered by each app to users and drivers, respectively.

Both users and drivers are assumed to revise their strategies at constant, yet distinct, average rates. Specifically, strategy revision times are governed by Poisson alarm clocks with rates $R^1 = 0.1$ and $R^2 = 1$ for users and drivers, respectively. This implies that, on average, drivers revise their strategies ten times more frequently than users.

Regarding the revision protocols, users are assumed to follow IPC-type rules of the form

$$\rho_{ij}^1\left(\mathbf{p}^1, \mathbf{x}^1\right) = \frac{1}{\zeta^1} \max\left\{p_j^1 - p_i^1, 0\right\}, \quad \forall i, j \in \mathcal{S}^1,$$

while drivers adopt SEPT-type protocols given by

$$\rho_{ij}^2\left(\mathbf{p}^2, \mathbf{x}^2\right) = \frac{1}{\zeta^2} \max\left\{p_j^2 - \sum_{\ell \in \mathcal{S}^2} x_\ell^2 p_\ell^2, 0\right\}, \quad \forall i, j \in \mathcal{S}^2.$$

Both parameters $\zeta^1, \zeta^2 \in \mathbb{R}_{>0}$ are assumed to be sufficiently large to ensure that the IPC and SEPT conditions in (2.14) and (2.11) hold, respectively.

In the simplest scenario, where the apps do not implement dynamic pricing or incentive schemes, the payoffs perceived by users and drivers coincide with the fitness values. That is,

$$p_i^k(t) = f_i^k(\mathbf{x}(t)) . \tag{2.23}$$

Figure 2.2 shows a simulation of this setup, using the initial conditions $\mathbf{x}^1(0) = [1, 1, 0]^\top$ and $\mathbf{x}^2(0) = [0, 1, 0]^\top$. The black solid lines depict the trajectory of $\mathbf{x}(t)$ for finite populations of 2000 users (Fig. 2.2a) and 1000 drivers (Fig. 2.2b). The magenta dotted lines represent the deterministic approximation of $\mathbf{x}(t)$ provided by the Poisson EDM. The label "GNE" refers to the projection of the six-dimensional GNE onto the three-dimensional subspace associated with each population. It is observed that the society state converges to the unique GNE of the underlying population game G, which for the chosen parameters is

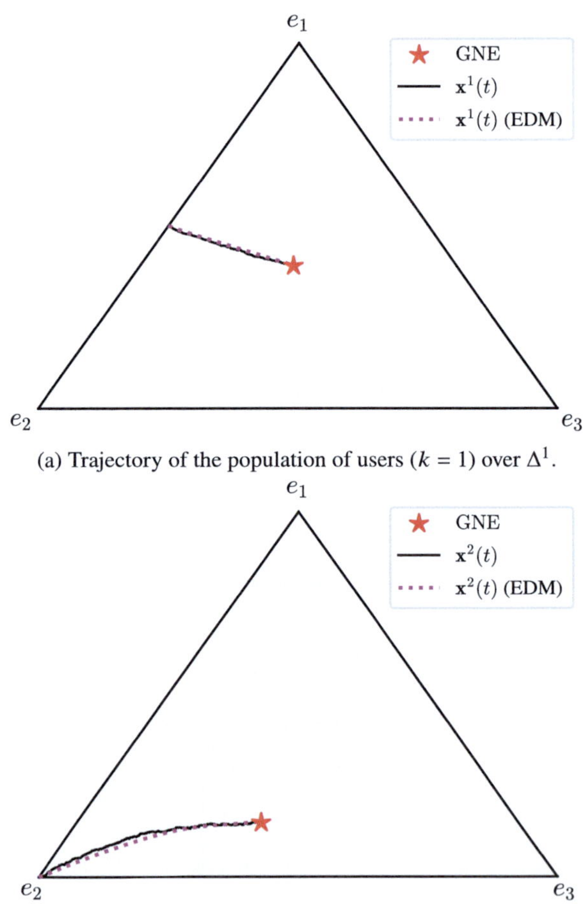

(a) Trajectory of the population of users ($k = 1$) over Δ^1.

(b) Trajectory of the population of drivers ($k = 2$) over Δ^2.

Fig. 2.2 Trajectories for the populations of users and drivers in the considered ride-hailing app scenario with all apps implementing the memoryless payoff mechanism (2.23). Here, "GNE" denotes the projection of the *six*-dimensional GNE to the *three*-dimensional space of each population

$$\mathbf{x}^* \approx \begin{bmatrix} 0.78, & 0.63, & 0.59, & 0.15, & 0.50, & 0.35 \end{bmatrix}. \tag{2.24}$$

Moreover, the deterministic approximation provided by the EDM closely matches the behavior of the corresponding finite-population simulation.

Now, suppose that the company managing app $i = 1$ seeks to dynamically increase both its fares and incentives in an attempt to attract more drivers, at the expense of higher costs for users. To this end, the company implements the payoff mechanism

$$\dot{q}_1(t) = \left[\overline{q}_1 - q_1(t)\right]_+ \left[\underline{x}_1 + x_1^1(t) - x_1^2(t)\right]_+ - q_1(t)\left[-\underline{x}_1 - x_1^1(t) + x_1^2(t)\right]_+, \tag{2.25a}$$

$$p_1^1(t) = f_1^1\left(\mathbf{x}(t)\right) - q_1(t), \tag{2.25b}$$

$$p_1^2(t) = f_1^2\left(\mathbf{x}(t)\right) + q_1(t), \tag{2.25c}$$

where $[\cdot]_+ := \max\{\cdot, 0\}$; the parameter $\overline{q}_1 \in \mathbb{R}_{\geq 0}$ denotes an upper bound on the dynamic adjustment of fares and incentives (e.g., a value set by a regulatory authority) and is set to $\bar{q}_1 = 0.2$; and the parameter $\underline{x}_1 \in \mathbb{R}$ specifies the constraint

$$x_1^2 - x_1^1 \geq \underline{x}_1, \tag{2.26}$$

which lower bounds the maximum allowed deficit of drivers with respect to users. Note that such a payoff mechanism resembles the dual-Lagrangian-based PDM of Example 2.5, with an additional upper bound saturation on q_i provided by $\overline{q}_i$. In fact, in this setup, the company managing app $i = 1$ dynamically adjusts the perceived payoffs of users and drivers through an additive term $q_1(t) \in [0, \overline{q}_1]$, aiming to enforce the constraint (2.26). Whether the system is able to satisfy this constraint in practice depends on the specific values of $\overline{q}_1$ and $\underline{x}_1$. In any case, with Ω updated to include the constraint (2.26) with $\underline{x}_1 = -0.4$, the new unique GNE of the underlying game G is

$$\mathbf{x}^* \approx \begin{bmatrix} 0.67, & 0.69, & 0.64, & 0.27, & 0.44, & 0.29 \end{bmatrix}. \tag{2.27}$$

Under the new GNE, the deficit of drivers relative to the users of app $i = 1$ is reduced to -0.4, compared to the previous value of -0.63 when the GNE was given by (2.24), enabling app $i = 1$ to better capitalize on its available user base.

Figure 2.3 illustrates the simulation results for the second setup, where app $i = 1$ adopts the dynamic payoff mechanism (2.25), while apps $i = 2$ and $i = 3$ continue to use the memoryless payoff mechanism (2.23). The system once again converges to the unique GNE of the underlying game G, with the deficit of drivers relative to the users of app $i = 1$ now reduced as intended. In addition, Fig. 2.4 shows the temporal evolution of the PDM state q_1 for both the finite-agent simulation and its deterministic EDM–PDM approximation. As observed, the constraint imposed by the regulatory authority is satisfied, with $q_1(t) \leq 0.2$ for all $t \geq 0$.

By comparing the GNEs in (2.24) and (2.27), one can observe that, as a byproduct, the dynamic payoff mechanism implemented by app $i = 1$ has both increased the

Fig. 2.3 Trajectories for the populations of users and drivers in the considered ride-hailing app scenario with app $i = 1$ implementing the dynamic payoff mechanism (2.25), and apps $i = 2$ and $i = 3$ using the memoryless payoff mechanism (2.23). Here, "GNE" denotes the projection of the six-dimensional GNE to the $three$-dimensional space of each population. The initial condition is taken according to (2.24)

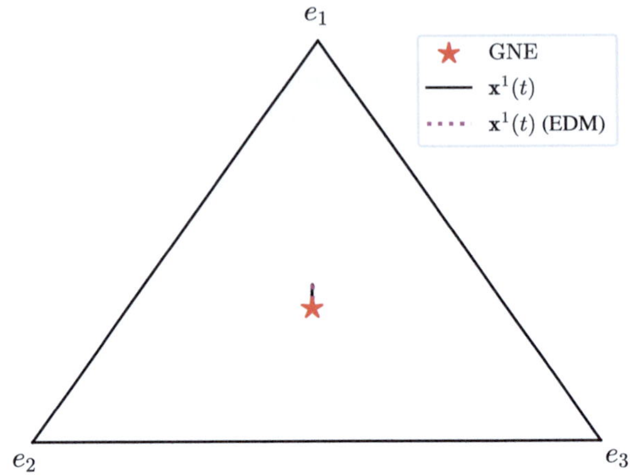

(a) Trajectory of the population of users ($k = 1$) over Δ^1.

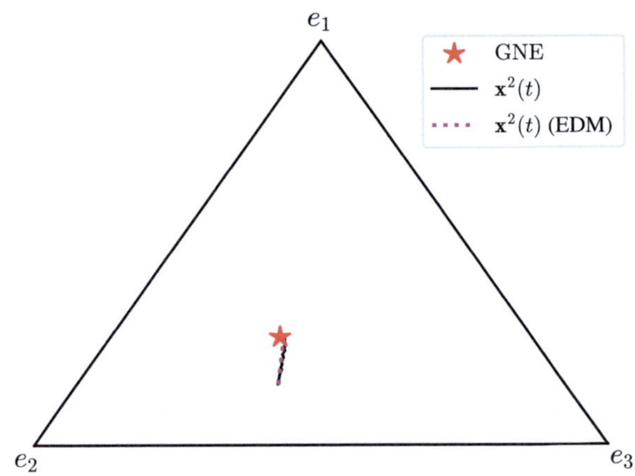

(b) Trajectory of the population of drivers ($k = 2$) over Δ^2.

number of users and decreased the number of drivers choosing apps $i = 2$ and $i = 3$. This new equilibrium state offers an opportunity for apps $i = 2$ and $i = 3$ to further capitalize on their expanded user base by attracting more drivers to meet the increased demand. As a response, the company managing app $i = 2$ decides to implement a dynamic payoff mechanism designed to attract both more users and drivers.

In contrast to the PDM (2.25) employed by app $i = 1$, the PDM introduced by app $i = 2$ is more user-centric. In particular, it avoids increasing user fares to finance the higher driver incentives. The specific form of the PDM for app $i = 2$ is given by

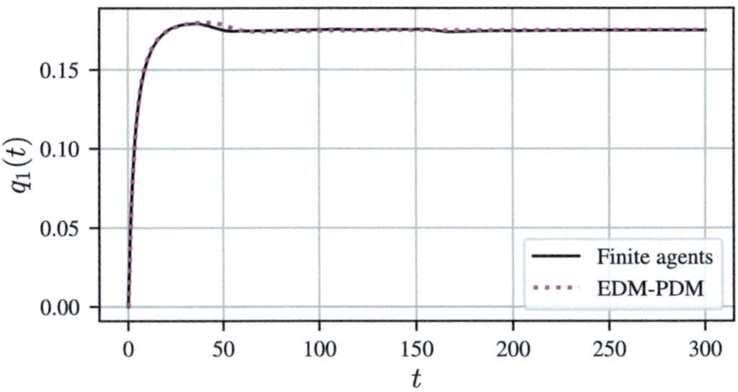

Fig. 2.4 Temporal evolution of the PDM state q_1 for the finite-agent simulation and the EDM–PDM deterministic approximation

$$\dot{q}_2(t) = \left[\overline{q}_2 - q_2(t)\right]_+ \left[\underline{x}_2 - x_2^1(t) - x_2^2(t)\right]_+ - q_2(t)\left[-\underline{x}_2 + x_2^1(t) + x_2^2(t)\right]_+, \tag{2.28a}$$

$$p_2^1(t) = f_2^1(\mathbf{x}(t)) + q_2(t), \tag{2.28b}$$

$$p_2^2(t) = f_2^2(\mathbf{x}(t)) + q_2(t), \tag{2.28c}$$

where

$$\overline{q}_2 = \frac{1}{2}\left(\gamma_2^1 - \gamma_2^2 - \bar{\gamma}_2^*\right),$$

upper-bounds the dynamic adjustment $q_2(t)$ in terms of the nominal user fare $\gamma_2^1 \in \mathbb{R}_{>0}$, the nominal driver incentive $\gamma_2^2 \in \mathbb{R}_{>0}$, and the minimum required operational margin $\bar{\gamma}_2^* \in \mathbb{R}_{>0}$. Clearly, for app $i = 2$ to sustain a positive operational margin, it must hold that $\gamma_2^1 - \gamma_2^2 - \bar{\gamma}_2^* > 0$. For the upcoming simulation, $\bar{\gamma}_2^* = 0.1$, which results in $\overline{q}_2 = 0.2$. On the other hand, the parameter $\underline{x}_2 \in \mathbb{R}_{\geq 0}$ characterizes the constraint

$$x_2^1 + x_2^2 \geq \underline{x}_2, \tag{2.29}$$

which specifies the minimum desired mass of users plus drivers selecting app $i = 2$. As before, whether the system is capable of satisfying this constraint depends on the specific values of $\overline{q}_2$ and $\underline{x}_2$.

With Ω updated to include both constraints (2.26) (with $\underline{x}_1 = -0.4$) and (2.29) (with $\underline{x}_2 = 1.6$), the resulting unique GNE of the underlying game G becomes

$$\mathbf{x}^* \approx \left[0.55, \quad 0.92, \quad 0.53, \quad 0.15, \quad 0.67, \quad 0.18\right]. \tag{2.30}$$

Under this new GNE, the mass of users plus drivers choosing app $i = 2$ surpasses the value of 1.13 observed in both (2.24) and (2.27), reaching the desired threshold of 1.6 and enabling app $i = 2$ to successfully attract more drivers and better capitalize on its growing user base. Furthermore, the mass of drivers choosing app $i = 1$ decreases back to 0.15, as in (2.24), and the mass of users selecting app $i = 1$ falls below the initial value of 0.78 seen in (2.24). These observations indicate that the user-centric PDM implemented by app $i = 2$ outperforms the one adopted by app $i = 1$. However, this dominance comes at the cost of reduced effective operational margins for the company managing app $i = 2$.

Figure 2.5 presents the simulation results for this third scenario, where app $i = 2$ adopts the dynamic payoff mechanism (2.28), while apps $i = 1$ and $i = 3$ continue to use the mechanisms in (2.25) and (2.23), respectively. The system once again converges to the unique GNE of the underlying game G.

In addition, Fig. 2.6 shows the temporal evolution of the PDM states q_1 and q_2 for both the finite-agent simulation and its deterministic EDM–PDM approximation. As seen in Fig. 2.6a, the introduction of dynamic fares and incentives in app $i = 2$ induces a slight perturbation in $q_1(t)$ for app $i = 1$. Nonetheless, the system eventually stabilizes at the corresponding equilibrium value $q_1^* = 0.175$. Moreover, the constraint imposed by the regulatory authority is satisfied, with $q_i(t) \leq 0.2$ for all $t \geq 0$ and all $i \in \{1, 2\}$.

All of the presented simulations can be replicated in *Python* using the `popgames` package and running the scripts of codes in Listings 2.1–2.3 at the end of this chapter. The reader is highly encouraged to replicate the simulations and experiment with the system's parameters.

Traffic Routing in Urban Networks

As a second example, consider the *urban traffic routing* problem introduced in Sect. 1.2.5 and revisited in Example 2.2.

Specifically, consider a single population consisting of 500 drivers who commute daily from an origin A to a destination B. As depicted in Fig. 2.7, the traffic network connecting A to B comprises 7 roads, which form a total of 5 possible strategies. These are defined as

$$\begin{aligned} \mathcal{S} &= \{1, 2, 3, 4, 5\} \\ &\triangleq \{\{r_1, r_2\}, \{r_1, r_3\}, \{r_4\}, \{r_5, r_7\}, \{r_6, r_7\}\} . \end{aligned}$$

While each driver has a preferred strategy to travel from A to B, this preference may change over time based on their revision process.

Each strategy, representing a specific route from A to B, is associated with a congestion cost that depends on the proportion of drivers selecting it. These congestion costs are captured by fitness functions of the form

$$\begin{aligned} f_1(\mathbf{x}) &= -2x_1 - x_2 \\ f_2(\mathbf{x}) &= -x_1 - 2x_2 \\ f_3(\mathbf{x}) &= -2x_3 \\ f_4(\mathbf{x}) &= -2x_4 - x_5 \\ f_5(\mathbf{x}) &= -x_4 - 2x_5 . \end{aligned}$$

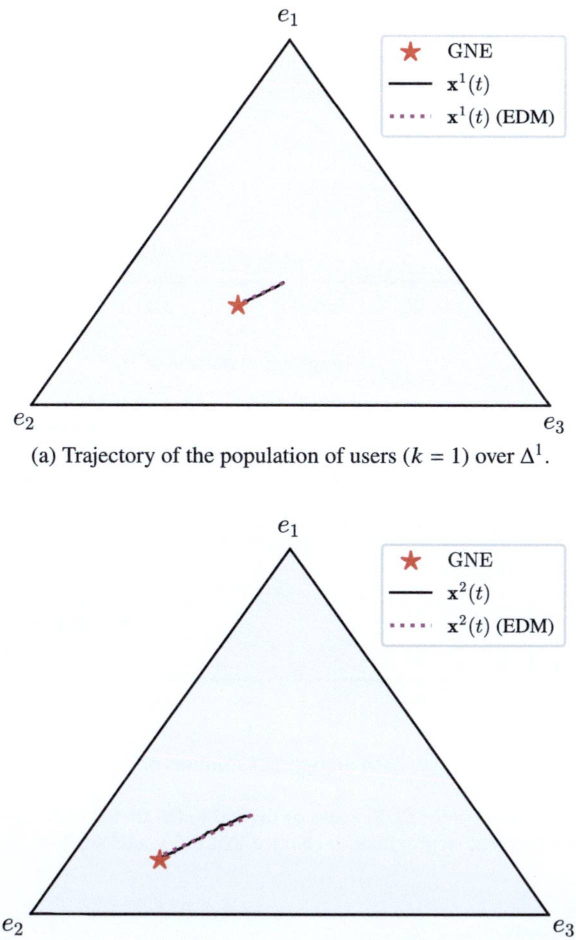

(a) Trajectory of the population of users ($k = 1$) over Δ^1.

(b) Trajectory of the population of drivers ($k = 2$) over Δ^2.

Fig. 2.5 Trajectories for the populations of users and drivers in the considered ride-hailing app scenario with app $i = 1$ implementing the dynamic payoff mechanism (2.25), app $i = 2$ using the dynamic payoff mechanism (2.28), and app $i = 3$ applying the memoryless payoff mechanism (2.23). Here, "GNE" denotes the projection of the six-dimensional GNE to the $three$-dimensional space of each population. The initial condition is taken according to (2.27)

Namely, note that less congested routes yield a higher (less negative) fitness value.

Due to environmental considerations, the city planner aims to limit the number of drivers simultaneously using roads r_1 and r_7. Expressed in terms of unitary population mass, the desired constraints are

$$x_1 + x_2 \leq 0.3$$
$$x_4 + x_5 \leq 0.2.$$

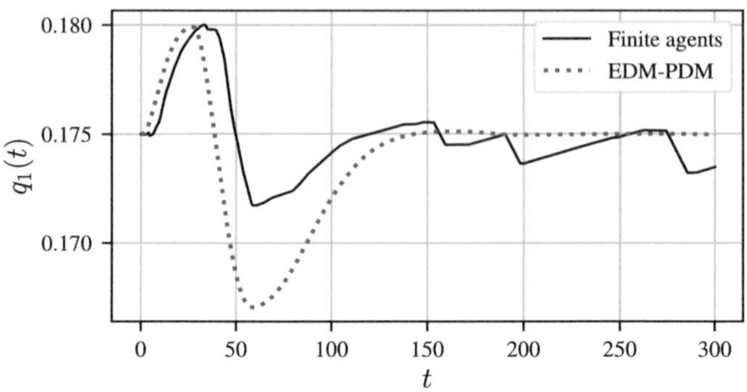

(a) Temporal evolution of q_1.

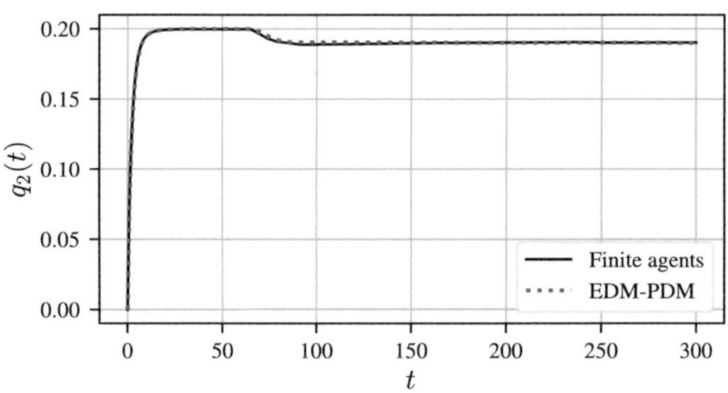

(b) Temporal evolution of q_2.

Fig. 2.6 Temporal evolution of the PDM states q_1 (**a**) and q_2 (**b**) for the finite-agent simulation and the EDM–PDM deterministic approximation. Notice that the y-axis scale of (**a**) is much smaller than that of (**b**)

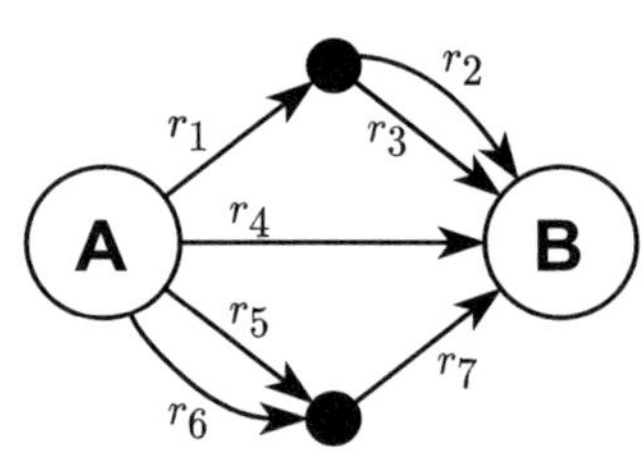

Fig. 2.7 Considered urban traffic network comprised of 7 roads spanning 5 possible routes (strategies) to commute from A to B

To enforce these constraints, two tolls governed by dynamic pricing schemes are introduced. These tolls are indicated by the solid black dots in Fig. 2.7, and their pricing dynamics are given by

$$\dot{q}_1(t) = [x_1(t) + x_2(t) - 0.3]_+ - q_1(t)\,[0.3 - x_1(t) - x_2(t)]_+$$
$$\dot{q}_2(t) = [x_4(t) + x_5(t) - 0.5]_+ - q_2(t)\,[0.2 - x_4(t) - x_5(t)]_+\,,$$

respectively. Here, q_1 denotes the cost incurred by drivers using road r_1, while q_2 corresponds to the cost associated with road r_7. It is worth to highlight that the notion of *price* considered here is not necessarily an immediate fee paid daily by the drivers. Instead, it is an abstract representation of a broader cost, which could take the form of a monthly charge, tax, or another cumulative penalty incurred by users of the respective roads.

Based on this setup, the payoffs perceived by the drivers incorporate both the fitness values (nominal congestion costs) and the prices of the associated tolls (dynamic constraint-related costs). Accordingly, the payoffs perceived by the drivers can be modeled as

$$\begin{aligned} p_1(t) &= f_1(\mathbf{x}(t)) - q_1(t) \\ p_2(t) &= f_2(\mathbf{x}(t)) - q_1(t) \\ p_3(t) &= f_3(\mathbf{x}(t)) \\ p_4(t) &= f_4(\mathbf{x}(t)) - q_2(t) \\ p_5(t) &= f_5(\mathbf{x}(t)) - q_2(t). \end{aligned}$$

With regard to the strategy revision process, it is assumed that drivers revise their preferred routes at a constant, yet potentially distinct, average rate. Without loss of generality, this behavior can be modeled using Poisson alarm clocks with rate $R = 1$ (in arbitrary time units, e.g., in the order of weeks or months).

On the other hand, since the strategies represent preferred routes, it is reasonable to consider IPC-type revision protocols of the form

$$\rho_{ij}(\mathbf{p}, \mathbf{x}) = \frac{1}{\zeta} \max\left\{p_j - p_i, 0\right\}, \quad \forall i, j \in \mathcal{S},$$

where the parameter $\zeta \in \mathbb{R}_{>0}$ is assumed to be sufficiently large to ensure that the IPC condition in (2.14) is satisfied. These protocols reflect the fact that drivers may occasionally try alternative routes seeking to improve their commuting experience, and therefore are prone to update their preferences based on pairwise comparisons.

Figures 2.8 and 2.9 present the numerical simulation results for the considered scenario, where the initial strategic distribution is given by

$$\mathbf{x}(0) = \left[0.5, 0.0, 0.0, 0.0, 0.5\right]^\top,$$

which does not satisfy the desired constraints. Specifically, Fig. 2.8 demonstrates that the drivers' decision-making process converges over time to a GNE of the underlying population game. For this particular game, it can be verified that the unique GNE is

$$\mathbf{x}^* = \left[0.15, 0.15, 0.50, 0.10, 0.10\right]^\top.$$

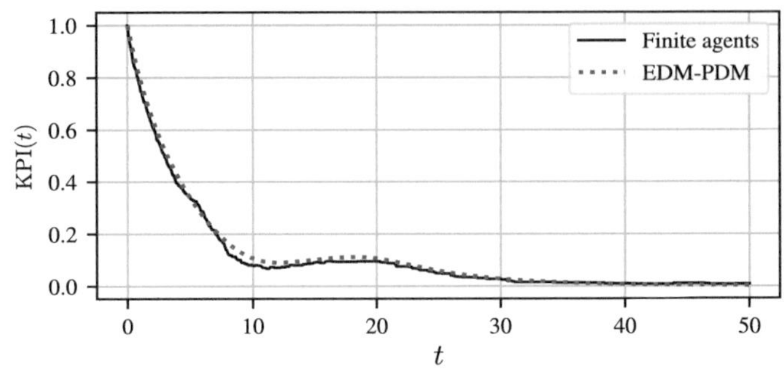

Fig. 2.8 Temporal evolution of the metric $\mathrm{KPI}(t) = \|\mathbf{x}(t) - \mathbf{x}^*\|_2 \,/\, \|\mathbf{x}(0) - \mathbf{x}^*\|_2$, with $\mathbf{x}^* \in \mathrm{GNE}(\mathbf{f})$, for both the finite-agent simulation and the EDM–PDM approximation

Convergence to this GNE is illustrated in Fig. 2.9a. On the other hand, Fig. 2.9b depicts the temporal evolution of the payoff signals. Observe that all payoffs converge to the same value (-1), indicating that no driver has an incentive to unilaterally deviate from $\mathbf{x}^*$, as the perceived payoff is equal across all strategies. Lastly, Fig. 2.9c shows the evolution of the dynamic toll prices. Initially, prices increase due to the fact that $\mathbf{x}(0) \notin \Omega$, and eventually decrease, settling at equilibrium values that ensure constraint satisfaction.

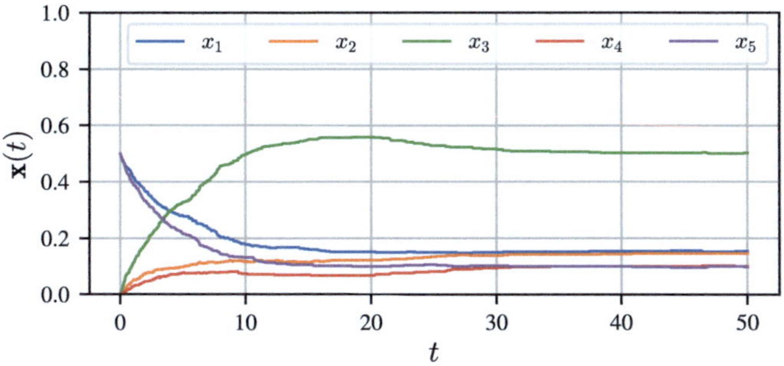

(a) Temporal evolution of **x**.

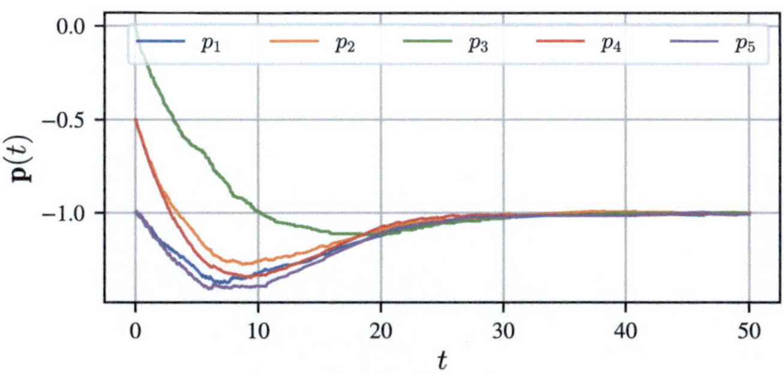

(b) Temporal evolution of **p**.

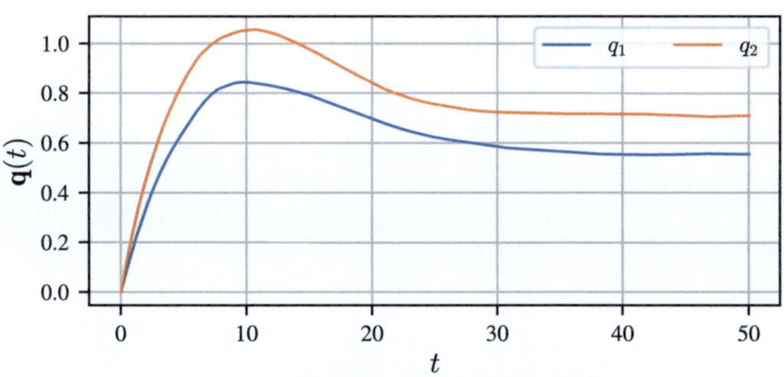

(c) Temporal evolution of **q**.

Fig. 2.9 Temporal evolution of **a** the strategic distribution **x**; **b** the payoff vector **p**; and **c** the dynamic toll prices **q**, for the finite-agent simulation

Python Scripts

```python
import numpy as np
import popgames as pg

Q = -np.diag([0.7, 0.8, 0.7, 0.9, 0.7, 0.9]) \
    + np.diag([0.3, 0.2, 0.3], 3) \
    + np.diag([0.1, 0.3, 0.1], -3)
r = np.array([-0.6, -0.7, -0.8, 0.1, 0.2, 0.3]).reshape(6, 1)

def fitness_function(x):
    return np.dot(Q, x) + r

population_game = pg.PopulationGame(
    num_populations=2, num_strategies=[3, 3],
    fitness_function=fitness_function,
    masses=[2, 1]
)

payoff_mechanism = pg.PayoffMechanism(
    h_map=fitness_function, n=6
)

revision_processes = [
    pg.PoissonRevisionProcess(
        Poisson_clock_rate=0.1,
        revision_protocol=pg.protocol.Smith(scale=0.5)
    ),
    pg.PoissonRevisionProcess(
        Poisson_clock_rate=1.0,
        revision_protocol=pg.protocol.BNN(scale=0.5)
    )
]

sim = pg.Simulator(
    population_game = population_game,
    payoff_mechanism = payoff_mechanism,
    revision_processes = revision_processes,
    num_agents = [2000, 1000]
)

x0 = np.array([1, 1, 0, 0, 1, 0]).reshape(6, 1)
sim.reset(x0, q0=None)
sim.run(T_sim=30)
sim.plot(
    plot_type='ternary',
    plot_deterministic_approximation=True
)
```

Listing 2.1 Python code to replicate the simulation of Fig. 2.2

```
import numpy as np
import popgames as pg
Q = -np.diag([0.7, 0.8, 0.7, 0.9, 0.7, 0.9]) \
    + np.diag([0.3, 0.2, 0.3], 3) \
    + np.diag([0.1, 0.3, 0.1], -3)
r = np.array([-0.6, -0.7, -0.8, 0.1, 0.2, 0.3]).reshape
    (6, 1)
A = np.array([[1, 0, 0, -1, 0, 0]])
b = np.array([[0.4]])
def fitness_function(x):
    return np.dot(Q, x) + r
def w_map(q, x):
    c = np.dot(A, x) - b
    ub = np.maximum(0.2 - q, 0)
    return ub*np.maximum(c, 0) - q*np.maximum(-c, 0)
def h_map(q, x):
    return fitness_function(x) - np.dot(A.T, q)
population_game = pg.PopulationGame(
    num_populations=2, num_strategies=[3, 3],
    fitness_function=fitness_function,
    masses=[2, 1], A_ineq=A, b_ineq=b)
payoff_mechanism = pg.PayoffMechanism(
    h_map=h_map, n=6, w_map=w_map, d=1
)
revision_processes = [
    pg.PoissonRevisionProcess(
        Poisson_clock_rate=0.1,
        revision_protocol=pg.protocol.Smith(scale=0.5)
    ),
    pg.PoissonRevisionProcess(
        Poisson_clock_rate=1.0,
        revision_protocol=pg.protocol.BNN(scale=0.5)
    )
]
sim = pg.Simulator(
    population_game=population_game,
    payoff_mechanism=payoff_mechanism,
    revision_processes=revision_processes,
    num_agents=[2000, 1000]
)
x0 = np.array([0.78, 0.63, 0.59, 0.15, 0.50, 0.35]).
    reshape(6,1)
q0 = np.array([0]).reshape(1, 1)
sim.reset(x0=x0, q0=q0)
sim.run(T_sim=300)
sim.plot(
    plot_type='ternary',
    plot_deterministic_approximation=True
)
sim.plot(
    plot_type='univariate_split',
    plot_deterministic_approximation=True
)
```

Listing 2.2 Python code to replicate the simulation of Figs. 2.3 and 2.4

```
import numpy as np
import popgames as pg
Q = -np.diag([0.7, 0.8, 0.7, 0.9, 0.7, 0.9]) \
    + np.diag([0.3, 0.2, 0.3], 3) \
    + np.diag([0.1, 0.3, 0.1], -3)
r = np.array([-0.6, -0.7, -0.8, 0.1, 0.2, 0.3]).reshape
    (6, 1)
A = np.array([[1, 0, 0, -1, 0, 0], [0, -1, 0, 0, -1,
    0]])
b = np.array([[0.4], [-1.6]])
def fitness_function(x):
    return np.dot(Q, x) + r
def w_map(q, x):
    c = np.dot(A, x) - b
    ub = np.maximum(0.2 - q, 0)
    return ub*np.maximum(c, 0) - q*np.maximum(-c, 0)
def h_map(q, x):
    return fitness_function(x) - np.dot(A.T, q)
population_game = pg.PopulationGame(
    num_populations=2, num_strategies=[3, 3],
    fitness_function=fitness_function,
    masses=[2, 1], A_ineq=A, b_ineq=b
)
payoff_mechanism = pg.PayoffMechanism(
    h_map=h_map, n=6, w_map=w_map, d=2
)
revision_processes = [
    pg.PoissonRevisionProcess(
        Poisson_clock_rate=0.1,
        revision_protocol=pg.protocol.Smith(scale=0.5)
    ),
    pg.PoissonRevisionProcess(
        Poisson_clock_rate=1.0,
        revision_protocol=pg.protocol.BNN(scale=0.5)
    )]
sim = pg.Simulator(
    population_game=population_game,
    payoff_mechanism = payoff_mechanism,
    revision_processes = revision_processes,
    num_agents=[2000, 1000]
)
x0 = np.array([0.67, 0.69, 0.64, 0.27, 0.44, 0.29]).
    reshape(6,1)
q0 = np.array([0.175, 0]).reshape(2, 1)
sim.reset(x0=x0, q0=q0)
sim.run(T_sim=300)
sim.plot(
    plot_type='ternary',
    plot_deterministic_approximation=True
)
sim.plot(
    plot_type='univariate_split',
    plot_deterministic_approximation=True
)
```

Listing 2.3 Python code to replicate the simulation of Figs. 2.5 and 2.6

```python
import numpy as np
import popgames as pg

def fitness_function(x):
    return np.dot(
        -np.array([
            [2, 1, 0, 0, 0],
            [1, 2, 0, 0, 0],
            [0, 0, 2, 0, 0],
            [0, 0, 0, 2, 1],
            [0, 0, 0, 1, 2],
        ]), x
    )

# Constraints: Ax - b <= 0
A = np.array([[1, 1, 0, 0, 0], [0, 0, 0, 1, 1]])
b = np.array([[0.3], [0.2]])

def h_map(q, x):
    return fitness_function(x) - np.dot(A.T, q)

def w_map(q, x):
    c = np.dot(A, x) - b
    return np.maximum(c, 0) - q*np.maximum(-c, 0)

population_game = pg.SinglePopulationGame(
    num_strategies=5, fitness_function=fitness_function,
    A_ineq=A, b_ineq=b
)

payoff_mechanism = pg.PayoffMechanism(
    h_map=h_map, n=5, w_map=w_map, d=2
)

revision_process = pg.PoissonRevisionProcess(
    Poisson_clock_rate=1,
    revision_protocol=pg.protocol.Smith(scale=0.1)
)

sim = pg.Simulator(
    population_game=population_game,
    payoff_mechanism=payoff_mechanism,
    revision_processes=revision_process,
    num_agents=500
)

x0 = np.array([0.5, 0, 0, 0, 0.5]).reshape(5, 1)
sim.reset(x0=x0)
sim.run(T_sim=50)
sim.plot(plot_type='univariate', ylim={'x' : [0, 1]})
sim.plot(plot_type='kpi',
    plot_deterministic_approximation=True)
```

Listing 2.4 Python code to replicate the simulation of Figs. 2.8 and 2.9

Notes

Detailed treatments of population games, revision processes, and their associated deterministic approximations are available in [6, 21], with additional perspectives from evolutionary biology discussed in [22–24]. These foundational models assume memoryless payoff mechanisms determined entirely by the fitness functions. The extension to dynamic payoff mechanisms was first introduced in [18, 19], and subsequently developed in works such as [10, 15, 17], among others. In all of these studies, explicit coupling constraints among agents are not considered, implying that $\Omega = \mathbb{R}^n$, and the equilibrium concept employed is the classical Nash equilibrium (NE), as defined in Definition 2.13. The NE notion was originally introduced by Nash in [14] from the perspective of classical N-player games.

Extensions of population game models incorporating explicit coupling constraints among agents have been explored in [1, 12] for memoryless payoff mechanisms, and in [11, 13] for dynamic ones. In these settings, the appropriate equilibrium notion is the generalized Nash equilibrium (GNE), as established in Definition 2.14. A comprehensive treatment of GNE problems is provided in [3, 5] from the classical N-player game perspective.

Further discussions on revision protocols that lead to best-response and perturbed best-response dynamics can be found in [21, Chap. 6] and [16]. Additional motivation and analyses for protocols such as SEPT and IPC are presented in [20, 22], respectively. Other classes of revision protocols not covered in this book, including for instance imitative revision protocols that give rise to the well-known *replicator dynamics*, are detailed in [21, Chap. 4].

Formal characterizations of the deterministic approximations described in Sect. 2.4 are provided in [2, 7, 19], [21, Chap. 10], all based on the Poisson alarm clock model introduced in Sect. 2.3.1. Generalizations of this model and their corresponding deterministic approximations are discussed in [8, 9].

References

1. Barreiro-Gomez J, Tembine H (2018) Constrained evolutionary games by using a mixture of imitation dynamics. Automatica 97:254–262. https://doi.org/10.1016/j.automatica.2018.08.014
2. Brown G, von Neumann J (1950) Solutions of games by differential equations. In: Kuhn H, Tucker A (eds) Contributions to the theory of games I, Annals of mathematics studies. Princeton University Press
3. Facchinei F, Kanzow C (2010) Generalized Nash equilibrium problems. Ann Oper Res 175(1):177–211
4. Facchinei F, Pang JS (2003) Finite-dimensional variational inequalities and complementarity problems, vol 1. Springer
5. Facchinei F, Pang JS (2009) Nash equilibria: the variational approach. Cambridge University Press, pp 443–493. https://doi.org/10.1017/CBO9780511804458.013
6. Hofbauer J, Sigmund K (1998) Evolutionary games and population dynamics. Cambridge University Press

7. Kara S, Martins NC (2023) Differential equation approximations for population games using elementary probability. arXiv:2312.07598
8. Kara S, Martins NC (2023) Pairwise comparison evolutionary dynamics with strategy-dependent revision rates: stability and δ-passivity. IEEE Trans Control Netw Syst 10(4):1656–1668. https://doi.org/10.1109/TCNS.2023.3237485
9. Kara S, Martins NC, Arcak M (2022) Population games with Erlang clocks: convergence to Nash equilibria for pairwise comparison dynamics. In: Proceedings of the 61st IEEE conference on decision and control (CDC), pp 7688–7695. https://doi.org/10.1109/CDC51059.2022.9993228
10. Martinez-Piazuelo J, Ocampo-Martinez C, Quijano N (2023) Distributed Nash equilibrium seeking in strongly contractive aggregative population games. IEEE Trans Autom Control (Early Access) 1–15. https://doi.org/10.1109/TAC.2023.3321208
11. Martinez-Piazuelo J, Ocampo-Martinez C, Quijano N (2024) Generalized Nash equilibrium seeking in a class of contractive population games over networks. IEEE 63rd conference on decision and control (CDC), Milan, Italy, pp 2319–2324. https://doi.org/10.1109/CDC56724.2024.10886517
12. Martinez-Piazuelo J, Quijano N, Ocampo-Martinez C (2022) Nash equilibrium seeking in full-potential population games under capacity and migration constraints. Automatica 141:110285. https://doi.org/10.1016/j.automatica.2022.110285
13. Martinez-Piazuelo J, Quijano N, Ocampo-Martinez C (2022) A payoff dynamics model for generalized Nash equilibrium seeking in population games. Automatica 140:110227. https://doi.org/10.1016/j.automatica.2022.110227
14. Nash J (1951) Non-cooperative games. Ann Math 54(2):286–295
15. Park S, Barreiro-Gomez J (2023) Payoff mechanism design for coordination in multi-agent task allocation games. In: Proceedings of the 62nd IEEE conference on decision and control (CDC), pp 8116–8121. https://doi.org/10.1109/CDC49753.2023.10383344
16. Park S, Certorio J, Martins NC, La RJ (2025) Epidemic population games and perturbed best response dynamics. IEEE Trans Autom Control 1–13. https://doi.org/10.1109/TAC.2025.3554601
17. Park S, Leonard NE (2024) Learning with delayed payoffs in population games using Kullback-Leibler divergence regularization. arXiv:2306.07535
18. Park S, Martins NC, Shamma JS (2019) From population games to payoff dynamics models: a passivity-based approach. In: Proceedings of the 58th IEEE conference on decision and control (CDC), pp 6584–6601. https://doi.org/10.1109/CDC40024.2019.9029756
19. Park S, Martins NC, Shamma JS (2019) Payoff dynamics model and evolutionary dynamics model: Feedback and convergence to equilibria. https://doi.org/10.48550/ARXIV.1903.02018. arXiv:1903.02018
20. Sandholm WH (2005) Excess payoff dynamics and other well-behaved evolutionary dynamics. J Econ Theory 124(2):149–170. https://doi.org/10.1016/j.jet.2005.02.003
21. Sandholm WH (2010) Population games and evolutionary dynamics. MIT Press
22. Smith MJ (1984) The stability of a dynamic model of traffic assignment: an application of a method of Lyapunov. Transp Sci 18(3):245–252
23. Taylor PD, Jonker LB (1978) Evolutionary stable strategies and game dynamics. Math Biosci 40(1):145–156. https://doi.org/10.1016/0025-5564(78)90077-9
24. Weibull JW (1997) Evolutionary game theory. MIT Press

Part II
Analysis

Chapter 3
Delta-Dissipative Dynamical Systems

This chapter introduces the notion of δ-dissipative dynamical systems, a concept that generalizes classical dissipativity by operating over the time derivatives of the input and output signals of a system. Specifically, the δ-dissipativity framework aligns with traditional energy-based system analysis, while regarding a time-derivative-based perspective.

The notion of δ-dissipativity is closely related to other generalized passivity concepts, such as Krasovskii, differential, and incremental passivity. Furthermore, it also offers a flexible approach for the analysis of interconnected systems.

The relevance of δ-dissipativity to the context of population games stems from the fact that many evolutionary and payoff dynamics models exhibit such a property, enabling a unified treatment for several classes of decision-making processes in large-scale multi-agent systems. Consequently, the concepts introduced in this chapter yield the main theoretical foundations for the forthcoming analyses of Chap. 4.

This chapter is organized as follows. Section 3.1 formally defines δ-dissipative dynamical systems. Section 3.2 establishes sufficient conditions to verify the Lyapunov and asymptotic stability of their equilibria. Section 3.3 discusses how δ-dissipativity can be preserved under certain interconnections between subsystems. Finally, Sect. 3.4 provides a simple example integrating several of the discussed concepts under an optimization-based perspective.

3.1 Delta-Dissipativity

Consider the dynamical system given by

$$\Sigma : \begin{cases} \dot{\mathbf{z}}(t) = \mathcal{Z}\left(\mathbf{z}(t), \mathbf{u}(t)\right) \\ \mathbf{y}(t) = \mathcal{Y}\left(\mathbf{z}(t), \mathbf{u}(t)\right), \end{cases} \tag{3.1}$$

J. Martinez-Piazuelo et al., *Generalized Nash Equilibrium Seeking in Population Games*, Communications and Control Engineering,
https://doi.org/10.1007/978-3-032-06081-5_3

where $\mathbf{z} \in \mathbb{R}^{d_z}$, $\mathbf{u} \in \mathbb{R}^{d_u}$, $\mathbf{y} \in \mathbb{R}^{d_y}$ are the state, input, and output vectors, respectively, $\mathcal{Z} : \mathbb{R}^{d_z} \times \mathbb{R}^{d_u} \to \mathbb{R}^{d_z}$ is Lipschitz continuous, $\mathcal{Y} : \mathbb{R}^{d_z} \times \mathbb{R}^{d_u} \to \mathbb{R}^{d_y}$ is continuously differentiable, and $d_z, d_u, d_y \in \mathbb{Z}_{\geq 1}$ are the corresponding dimensions.

Definition 3.1 (*δ-Dissipativity*) A system Σ of the form (3.1) is said to be δ-dissipative with supply rate $s : \mathbb{R}^{d_u} \times \mathbb{R}^{d_y} \to \mathbb{R}$ if there exist a continuously differentiable non-negative function $S : \mathbb{R}^{d_z} \times \mathbb{R}^{d_u} \to \mathbb{R}_{\geq 0}$ and a non-negative function $\sigma : \mathbb{R}^{d_z} \times \mathbb{R}^{d_u} \times \mathbb{R}^{d_u} \to \mathbb{R}_{\geq 0}$ such that

$$S(\mathbf{z}, \mathbf{u}) = 0 \Leftrightarrow \mathcal{Z}(\mathbf{z}, \mathbf{u}) = \mathbf{0}_{d_z} \tag{3.2a}$$

$$\dot{S}(\mathbf{z}, \mathbf{u}, \mathbf{v}) \leq -\sigma(\mathbf{z}, \mathbf{u}, \mathbf{v}) + s\left(\mathbf{v}, \mathbf{w}(\mathbf{z}, \mathbf{u}, \mathbf{v})\right), \tag{3.2b}$$

for all $\mathbf{z} \in \mathbb{R}^{d_z}$ and all $\mathbf{u}, \mathbf{v} \in \mathbb{R}^{d_u}$. Here, $\dot{S}(\cdot, \cdot, \cdot)$ and $\mathbf{w}(\cdot, \cdot, \cdot)$ are defined as

$$\begin{aligned} \dot{S}\,(\mathbf{z}, \mathbf{u}, \mathbf{v}) &= \nabla_{\mathbf{z}} S(\mathbf{z}, \mathbf{u})^\top \mathcal{Z}(\mathbf{z}, \mathbf{u}) + \nabla_{\mathbf{u}} S(\mathbf{z}, \mathbf{u})^\top \mathbf{v} \\ \mathbf{w}(\mathbf{z}, \mathbf{u}, \mathbf{v}) &= \mathrm{D}_{\mathbf{z}}\, \mathcal{Y}(\mathbf{z}, \mathbf{u}) \mathcal{Z}(\mathbf{z}, \mathbf{u}) + \mathrm{D}_{\mathbf{u}}\, \mathcal{Y}(\mathbf{z}, \mathbf{u})\mathbf{v}, \end{aligned}$$

and $S(\cdot, \cdot)$ is referred to as the *δ-storage function* of Σ. Moreover, the *complementary function* $\sigma(\cdot, \cdot, \cdot)$ is said to be *informative* if it satisfies that

$$\sigma(\mathbf{z}, \mathbf{u}, \mathbf{0}_{d_u}) = 0 \Leftrightarrow \mathcal{Z}\,(\mathbf{z}, \mathbf{u}) = \mathbf{0}_{d_z}, \tag{3.3}$$

for all $\mathbf{z} \in \mathbb{R}^{d_z}$ and all $\mathbf{u} \in \mathbb{R}^{d_u}$. Furthermore, if Σ is δ-dissipative with supply rate

$$\begin{aligned} s\,(\mathbf{v}, \mathbf{w}(\mathbf{z}, \mathbf{u}, \mathbf{v})) &= \left\| \begin{bmatrix} \mathbf{v} \\ \mathbf{w}(\mathbf{z}, \mathbf{u}, \mathbf{v}) \end{bmatrix} \right\|_{\mathbf{\Pi}}^2 \\ &= \begin{bmatrix} \mathbf{v} \\ \mathbf{w}(\mathbf{z}, \mathbf{u}, \mathbf{v}) \end{bmatrix}^\top \mathbf{\Pi} \begin{bmatrix} \mathbf{v} \\ \mathbf{w}(\mathbf{z}, \mathbf{u}, \mathbf{v}) \end{bmatrix}, \end{aligned} \tag{3.4}$$

for some matrix $\mathbf{\Pi} \in \mathbb{R}^{(d_u+d_y)\times(d_u+d_y)}$, then Σ is said to be δ-dissipative characterized by $(\sigma, \mathbf{\Pi})$. ◊

Remark 3.1 If the instance of system (3.1) under consideration exhibits an invariance property of the form $\mathbf{z}(0) \in \mathcal{I} \Rightarrow \mathbf{z}(t) \in \mathcal{I}$ for all $t \geq 0$, where $\mathcal{I} \subset \mathbb{R}^{d_z}$ is an invariant set, then the conditions stated in Definition 3.1 are only required to hold for all $\mathbf{z} \in \mathcal{I}$ and all $\mathbf{u}, \mathbf{v} \in \mathrm{T}\mathcal{I}$, with $\mathrm{T}\mathcal{I}$ denoting the tangent space of $\mathcal{I}$. For simplicity, however, Definition 3.1 is stated with respect to the entire space $\mathbb{R}^{d_z}$ and, when any invariance properties are leveraged, this fact will be made explicit by context. ◊

The notion of δ-dissipativity is equivalent to the standard dissipativity property introduced by Willems,[1] but in terms of the time derivatives of the input and output signals of the dynamical system. Specifically, δ-dissipativity of a system Σ of the form (3.1) is equivalent to the standard dissipativity of the extended system

[1] See [11, 12] and the Section *Notes* at the end of this chapter.

$$\hat{\Sigma}:\begin{cases}\dot{\mathbf{u}}(t)=\mathbf{v}(t)\\ \dot{\mathbf{z}}(t)=\mathcal{Z}(\mathbf{z}(t),\mathbf{u}(t))\\ \hat{\mathbf{y}}(t)=\mathbf{w}(\mathbf{z}(t),\mathbf{u}(t),\mathbf{v}(t)),\end{cases}$$

defined with respect to the input $\mathbf{v}$ and output $\hat{\mathbf{y}} := \mathbf{w}$. Namely, $\mathbf{v}$ and $\mathbf{w}$ denote the time derivatives of the input and output signals of the original system Σ, respectively.

Example 3.1 Consider the system (3.1) with $d_z = 0$, i.e., a memoryless system, and with output given by $\mathbf{y}(t) = \mathcal{Y}(\mathbf{u}(t))$, where $\mathcal{Y}\colon \mathbb{R}^{d_u} \to \mathbb{R}^{d_y}$ is continuously differentiable. For such a system it follows that

$$\mathbf{w} := \mathbf{w}(\mathbf{u},\mathbf{v}) = \mathrm{D}_{\mathbf{u}}\,\mathcal{Y}(\mathbf{u})\mathbf{v}.$$

Let $\mathbf{H} \in \mathbb{R}^{d_u \times d_y}$ and $\nu \in \mathbb{R}$ satisfy

$$\frac{1}{2}\left(\mathbf{H}\,\mathrm{D}_{\mathbf{u}}\,\mathcal{Y}(\mathbf{u}) + \mathrm{D}_{\mathbf{u}}\,\mathcal{Y}(\mathbf{u})^{\top}\mathbf{H}^{\top}\right) \preceq -\nu\mathbf{I}_{d_u}, \tag{3.5}$$

for all $\mathbf{u} \in \mathbb{R}^{d_u}$. Then, for every $\alpha \in \mathbb{R}_{\geq 0}$, the considered system is δ-dissipative characterized by $(\sigma_\alpha, \mathbf{\Pi}_\alpha)$ with $\sigma_\alpha(\mathbf{v}) = \alpha\mathbf{v}^{\top}\mathbf{v}$ and

$$\mathbf{\Pi}_\alpha = \begin{bmatrix} (\alpha-\nu)\mathbf{I}_{d_u} & -\frac{1}{2}\mathbf{H} \\ -\frac{1}{2}\mathbf{H}^{\top} & \mathbf{0}_{d_y \times d_y} \end{bmatrix}.$$

To prove such a claim, take $S(\mathbf{u},\mathbf{v}) = 0$ (hence, $\dot{S}(\mathbf{u},\mathbf{v}) = 0$), and note that

$$\begin{aligned}\begin{bmatrix}\mathbf{v}\\ \mathbf{w}\end{bmatrix}^{\top}\mathbf{\Pi}_\alpha\begin{bmatrix}\mathbf{v}\\ \mathbf{w}\end{bmatrix} &= (\alpha-\nu)\mathbf{v}^{\top}\mathbf{v} - \frac{1}{2}\mathbf{v}^{\top}\mathbf{H}\mathbf{w} - \frac{1}{2}\mathbf{w}^{\top}\mathbf{H}^{\top}\mathbf{v}\\ &= (\alpha-\nu)\mathbf{v}^{\top}\mathbf{v} - \frac{1}{2}\mathbf{v}^{\top}\left(\mathbf{H}\,\mathrm{D}_{\mathbf{u}}\,\mathcal{Y}(\mathbf{u}) + \mathrm{D}_{\mathbf{u}}\,\mathcal{Y}(\mathbf{u})^{\top}\mathbf{H}^{\top}\right)\mathbf{v}\\ &\geq (\alpha-\nu)\mathbf{v}^{\top}\mathbf{v} + \mathbf{v}^{\top}\left(\nu\mathbf{I}_{d_u}\right)\mathbf{v}\\ &= \sigma_\alpha(\mathbf{v}).\end{aligned}$$

Thus, $0 \leq -\sigma_\alpha(\mathbf{v}) + \|\mathrm{col}\,(\mathbf{v},\mathbf{w})\|^2_{\mathbf{\Pi}_\alpha}$, and (3.2b) holds. Evidently, for every memoryless δ-dissipative system, the associated σ function is always informative, as (3.3) trivially holds.

Recall Definitions 2.1 and 2.3. Notice that if $d_u = d_y$, $\mathbf{H} = \mathbf{I}_{d_u}$, and $\mathcal{Y}(\cdot)$ represents the fitness vector of the population game $G = (\mathcal{P}, \mathcal{X}, \mathcal{Y}(\cdot))$, then (3.5) is readily satisfied with $\nu = 0$ if G is a merely contractive game, with some $\nu > 0$ if G is a strongly contractive game, or with some $\nu < 0$ if G has deficit of contractivity. ◊

Example 3.2 Consider the system (3.1) with $\mathcal{Z}(\mathbf{z},\mathbf{u}) = \gamma\mathbf{z} + \mathbf{u}$ and $\mathcal{Y}(\mathbf{z},\mathbf{u}) = \mathbf{z}$, where $\gamma \in \mathbb{R}$ (thus, $d_z = d_u = d_y = d$). That is, consider the dynamics given by

$$\begin{aligned}\dot{\mathbf{z}}(t) &= \gamma \mathbf{z}(t) + \mathbf{u}(t)\\ \mathbf{y}(t) &= \mathbf{z}(t).\end{aligned}$$

For such a system, it holds that

$$\mathbf{w} := \mathbf{w}\,(\mathbf{z}, \mathbf{u}) = \gamma \mathbf{z} + \mathbf{u}.$$

Besides, for every $\beta \in \mathbb{R}_{\geq 0}$, the considered system is δ-dissipative characterized by $(\sigma_\beta, \mathbf{\Pi}_\beta)$ with

$$\sigma_\beta\,(\mathbf{z}, \mathbf{u}) = \beta\,(\gamma \mathbf{z} + \mathbf{u})^\top\,(\gamma \mathbf{z} + \mathbf{u})$$

and

$$\mathbf{\Pi}_\beta = \begin{bmatrix} \mathbf{0}_{d\times d} & \frac{1}{2}\mathbf{I}_d \\ \frac{1}{2}\mathbf{I}_d & (\beta+\gamma)\mathbf{I}_d \end{bmatrix}.$$

To prove this claim, note that

$$\begin{aligned}\begin{bmatrix} \mathbf{v} \\ \mathbf{w}(\mathbf{z}, \mathbf{u}) \end{bmatrix}^\top \mathbf{\Pi}_\beta \begin{bmatrix} \mathbf{v} \\ \mathbf{w}(\mathbf{z}, \mathbf{u}) \end{bmatrix} &= \mathbf{v}^\top\,(\gamma \mathbf{z} + \mathbf{u}) + (\beta+\gamma)\,(\gamma \mathbf{z} + \mathbf{u})^\top\,(\gamma \mathbf{z} + \mathbf{u})\\ &= \mathbf{v}^\top\,(\gamma \mathbf{z} + \mathbf{u}) + \sigma_\beta(\mathbf{z}, \mathbf{u}) + \gamma\,(\gamma \mathbf{z} + \mathbf{u})^\top\,(\gamma \mathbf{z} + \mathbf{u})\,.\end{aligned}$$

On the other hand, take

$$S(\mathbf{z}, \mathbf{u}) = \frac{1}{2}\,(\gamma \mathbf{z} + \mathbf{u})^\top\,(\gamma \mathbf{z} + \mathbf{u})\,,$$

and note that

$$\begin{aligned}\dot{S}(\mathbf{z}, \mathbf{u}, \mathbf{v}) &= \left(\gamma^2 \mathbf{z} + \gamma \mathbf{u}\right)^\top\,(\gamma \mathbf{z} + \mathbf{u}) + (\gamma \mathbf{z} + \mathbf{u})^\top\,\mathbf{v}\\ &= \gamma\,(\gamma \mathbf{z} + \mathbf{u})^\top\,(\gamma \mathbf{z} + \mathbf{u}) + \mathbf{v}^\top\,(\gamma \mathbf{z} + \mathbf{u})\,.\end{aligned}$$

Consequently, it holds that

$$\dot{S}\,(\mathbf{z}, \mathbf{u}, \mathbf{v}) = -\sigma_\beta\,(\mathbf{z}, \mathbf{u}) + \begin{bmatrix} \mathbf{v} \\ \mathbf{w}(\mathbf{z}, \mathbf{u}) \end{bmatrix}^\top \mathbf{\Pi}_\beta \begin{bmatrix} \mathbf{v} \\ \mathbf{w}(\mathbf{z}, \mathbf{u}) \end{bmatrix},$$

which implies the δ-dissipativity of the considered system. In addition, if $\beta > 0$, then σ_β is informative.

It is worth to highlight that stability is not a necessary condition for a dynamical system to be δ-dissipative. In fact, notice that the considered system is unstable for every $\gamma > 0$ and under every fixed input $\mathbf{u}(t) = \mathbf{u}^*$. However, it is indeed δ-dissipative as proven above. ◊

A general sufficient condition to certify the δ-dissipativity of a differentiable system (3.1) is provided by Lemma 3.1.

Lemma 3.1 *Consider the system (3.1), with* $\mathcal{Z}(\cdot,\cdot)$ *being differentiable, and a matrix* $\mathbf{\Pi} \in \mathbb{R}^{(d_u+d_y)\times(d_u+d_y)}$. *Suppose that there exist a positive-definite matrix* $\mathbf{P} \in \mathbb{S}^{d_z}_{\succ 0}$ *and a matrix-valued function* $\mathbf{G} : \mathbb{R}^{d_z} \times \mathbb{R}^{d_u} \times \mathbb{R}^{d_u} \to \mathbb{S}^{(d_z+d_u)}_{\succeq 0}$ *such that*

$$\begin{bmatrix} \left(\mathbf{P}\,\mathrm{D}_{\mathbf{z}}\,\mathcal{Z}(\mathbf{z},\mathbf{u}) + \mathrm{D}_{\mathbf{z}}\,\mathcal{Z}(\mathbf{z},\mathbf{u})^\top\mathbf{P}\right) & \mathbf{P}\,\mathrm{D}_{\mathbf{u}}\,\mathcal{Z}(\mathbf{z},\mathbf{u}) \\ \mathrm{D}_{\mathbf{u}}\,\mathcal{Z}(\mathbf{z},\mathbf{u})^\top\mathbf{P} & \mathbf{0}_{d_z\times d_u} \end{bmatrix}$$
$$+\,\mathbf{G}(\mathbf{z},\mathbf{u},\mathbf{v}) - \begin{bmatrix} \mathbf{0}_{d_u\times d_z} & \mathbf{I}_{d_u} \\ \mathrm{D}_{\mathbf{z}}\,\mathcal{Y}(\mathbf{z},\mathbf{u}) & \mathrm{D}_{\mathbf{u}}\,\mathcal{Y}(\mathbf{z},\mathbf{u}) \end{bmatrix}^\top \mathbf{\Pi} \begin{bmatrix} \mathbf{0}_{d_u\times d_z} & \mathbf{I}_{d_u} \\ \mathrm{D}_{\mathbf{z}}\,\mathcal{Y}(\mathbf{z},\mathbf{u}) & \mathrm{D}_{\mathbf{u}}\,\mathcal{Y}(\mathbf{z},\mathbf{u}) \end{bmatrix} \preceq 0, \tag{3.6}$$

for all $\mathbf{z} \in \mathbb{R}^{d_z}$ *and all* $\mathbf{u}, \mathbf{v} \in \mathbb{R}^{d_u}$. *Then, the system (3.1) is* δ*-dissipative characterized by* $(\sigma, \mathbf{\Pi})$*, with*

$$\sigma(\mathbf{z},\mathbf{u},\mathbf{v}) = \left\| \begin{bmatrix} \mathcal{Z}(\mathbf{z},\mathbf{u}) \\ \mathbf{v} \end{bmatrix} \right\|^2_{\mathbf{G}(\mathbf{z},\mathbf{u},\mathbf{v})},$$

and $S(\mathbf{z},\mathbf{u}) = \|\mathcal{Z}(\mathbf{z},\mathbf{u})\|^2_{\mathbf{P}}$ *is a valid* δ*-storage function.* ◊

Proof Since $\mathbf{P}$ is positive definite it follows that $S(\mathbf{z},\mathbf{u}) = 0 \Leftrightarrow \mathcal{Z}(\mathbf{z},\mathbf{u}) = \mathbf{0}_{d_z}$. To show that (3.2b) holds, it suffices to multiply (3.6) from the left by $\mathrm{col}\,(\mathcal{Z}(\mathbf{z},\mathbf{u}),\mathbf{v})^\top$ and from the right by $\mathrm{col}\,(\mathcal{Z}(\mathbf{z},\mathbf{u}),\mathbf{v})$, and then apply the definition $\mathbf{w}(\mathbf{z},\mathbf{u},\mathbf{v}) = \mathrm{D}_{\mathbf{z}}\,\mathcal{Y}(\mathbf{z},\mathbf{u})\mathcal{Z}(\mathbf{z},\mathbf{u}) + \mathrm{D}_{\mathbf{u}}\,\mathcal{Y}(\mathbf{z},\mathbf{u})\mathbf{v}$. The claim then follows from Definition 3.1. □

Lemma 3.1 is particularly useful when dealing with linear systems. Namely, if the system (3.1) is linear and $\mathbf{G}(\cdot,\cdot,\cdot)$ is set as a constant matrix $\mathbf{G}$, then the sufficient condition (3.6) can be verified by solving a linear matrix inequality with $\mathbf{P}$ and $\mathbf{G}$ taken as the variables. This fact is formalized in Corollary 3.1.

Corollary 3.1 *Consider a system of the form*

$$\begin{aligned} \dot{\mathbf{z}}(t) &= \mathbf{A}\mathbf{z}(t) + \mathbf{B}\mathbf{u}(t) \\ \mathbf{y}(t) &= \mathbf{C}\mathbf{z}(t) + \mathbf{D}\mathbf{u}(t), \end{aligned}$$

where $\mathbf{A} \in \mathbb{R}^{d_z\times d_z}$, $\mathbf{B} \in \mathbb{R}^{d_z\times d_u}$, $\mathbf{C} \in \mathbb{R}^{d_y\times d_z}$, *and* $\mathbf{D} \in \mathbb{R}^{d_y\times d_u}$; *and a matrix* $\mathbf{\Pi} \in \mathbb{R}^{(d_u+d_y)\times(d_u+d_y)}$. *Suppose that there exist a positive-definite matrix* $\mathbf{P} \in \mathbb{S}^{d_z}_{\succ 0}$ *and a positive semi-definite matrix* $\mathbf{G} \in \mathbb{S}^{(d_z+d_u)}_{\succeq 0}$ *such that*

$$\begin{bmatrix} (\mathbf{P}\mathbf{A} + \mathbf{A}^\top\mathbf{P}) & \mathbf{P}\mathbf{B} \\ \mathbf{B}^\top\mathbf{P} & \mathbf{0}_{d_z\times d_u} \end{bmatrix} + \mathbf{G} - \begin{bmatrix} \mathbf{0}_{d_u\times d_z} & \mathbf{I}_{d_u} \\ \mathbf{C} & \mathbf{D} \end{bmatrix}^\top \mathbf{\Pi} \begin{bmatrix} \mathbf{0}_{d_u\times d_z} & \mathbf{I}_{d_u} \\ \mathbf{C} & \mathbf{D} \end{bmatrix} \preceq 0,$$

for all $\mathbf{z} \in \mathbb{R}^{d_z}$ *and all* $\mathbf{u}, \mathbf{v} \in \mathbb{R}^{d_u}$. *Then, the system (3.1) is* δ*-dissipative characterized by* $(\sigma, \mathbf{\Pi})$*, with*

$$\sigma(\mathbf{z},\mathbf{u},\mathbf{v}) = \left\| \begin{bmatrix} \mathcal{Z}(\mathbf{z},\mathbf{u}) \\ \mathbf{v} \end{bmatrix} \right\|^2_{\mathbf{G}},$$

and $S(\mathbf{z},\mathbf{u}) = \|\mathbf{A}\mathbf{z} + \mathbf{B}\mathbf{u}\|^2_{\mathbf{P}}$ *is a valid* δ*-storage function.* ◊

Delta-Passivity and Delta-Antipassivity

Two particularly important cases of δ-dissipativity arise under some specific supply rates of the form (3.4).

Definition 3.2 (*δ-Passivity*) The system (3.1) is said to be δ-passive characterized by σ if it is δ-dissipative characterized by $(\sigma, \mathbf{\Pi})$ with

$$\mathbf{\Pi} = \begin{bmatrix} \mathbf{0}_{d_u \times d_u} & \frac{1}{2}\mathbf{I}_{d_u} \\ \frac{1}{2}\mathbf{I}_{d_u} & \mathbf{0}_{d_u \times d_u} \end{bmatrix}.$$

Clearly, for (3.1) to be δ-Apassive, it is required that $d_u = d_y$. ◊

Definition 3.3 (*δ-Antipassivity*) The system (3.1) is said to be δ-antipassive characterized by σ if it is δ-dissipative characterized by $(\sigma, \mathbf{\Pi})$ with

$$\mathbf{\Pi} = \begin{bmatrix} \mathbf{0}_{d_u \times d_u} & -\frac{1}{2}\mathbf{I}_{d_u} \\ -\frac{1}{2}\mathbf{I}_{d_u} & \mathbf{0}_{d_u \times d_u} \end{bmatrix}.$$

Evidently, for (3.1) to be δ-antipassive, it is required that $d_u = d_y$. ◊

Specifically, δ-passivity is equivalent to δ-dissipativity with supply rate $s = \mathbf{w}^\top \mathbf{v}$, while δ-antipassivity is equivalent to δ-dissipativity with supply rate $s = -\mathbf{w}^\top \mathbf{v}$, where $s := s(\mathbf{v}, \mathbf{w}(\mathbf{z}, \mathbf{u}, \mathbf{v}))$ and $\mathbf{w} := \mathbf{w}(\mathbf{z}, \mathbf{u}, \mathbf{v})$. Observe that the memoryless system in Example 3.1 is either δ-passive, if $\mathbf{H} = -\mathbf{I}_{d_u}$ and $\nu \geq 0$; or δ-antipassive, if $\mathbf{H} = \mathbf{I}_{d_u}$ and $\nu \geq 0$. To show such claims, it suffices to take $\alpha = \nu$, which yields the desired supply rates. On the other hand, the dynamical system in Example 3.2 is δ-passive if $\gamma \leq 0$, as it suffices to take $\beta = -\gamma$ to obtain the desired supply rate.

When targeting the particular case of δ-passivity, one can deduce an alternative characterization of this property in the form of sufficient conditions on the maps $\mathcal{Z}(\cdot, \cdot)$ and $\mathcal{Y}(\cdot, \cdot)$. Such a result is established in Lemma 3.2.

Lemma 3.2 *Consider the system Σ in (3.1), and suppose that next conditions hold:*

(i) For every $(\mathbf{z}, \mathbf{u}) \in \mathbb{R}^{d_z} \times \mathbb{R}^{d_u}$, the output map of Σ satisfies:

$$\mathrm{D}_\mathbf{u}\, \mathcal{Y}(\mathbf{z}, \mathbf{u}) + \mathrm{D}_\mathbf{u}\, \mathcal{Y}(\mathbf{z}, \mathbf{u})^\top \succ 0.$$

(ii) There exist some functions $S : \mathbb{R}^{d_z} \times \mathbb{R}^{d_u} \to \mathbb{R}_{\geq 0}$ and[2] $\sigma : \mathbb{R}^{d_z} \times \mathbb{R}^{d_u} \to \mathbb{R}_{\geq 0}$, with $S(\cdot, \cdot)$ continuously differentiable, such that: (3.2a) holds and

$$-\nabla_\mathbf{z} S(\mathbf{z}, \mathbf{u})^\top \mathcal{Z}(\mathbf{z}, \mathbf{u}) - \frac{1}{2}\mathbf{\Psi}^\top \left(\mathrm{D}_\mathbf{u}\, \mathcal{Y}(\mathbf{z}, \mathbf{u}) + \mathrm{D}_\mathbf{u}\, \mathcal{Y}(\mathbf{z}, \mathbf{u})^\top\right)^{-1} \mathbf{\Psi} \geq \sigma(\mathbf{z}, \mathbf{u}),$$

for all $(\mathbf{z}, \mathbf{u}) \in \mathbb{R}^d \times \Delta$. Here, $\mathbf{\Psi} := \mathrm{D}_\mathbf{z}\, \mathcal{Y}(\mathbf{z}, \mathbf{u})\mathcal{Z}(\mathbf{z}, \mathbf{u}) - \nabla_\mathbf{u} S(\mathbf{z}, \mathbf{u})$.

Then, the system Σ is δ-passive characterized by σ. ◊

---

[2] Note that σ here is restricted to be a function only of $(\mathbf{z}, \mathbf{u})$; not of $\mathbf{v}$.

Proof First, recall the following fact: given a quadratic form

$$\psi(\mathbf{v}) = \frac{1}{2}\mathbf{v}^\top \mathbf{Q}\mathbf{v} + \mathbf{b}^\top \mathbf{v} + c,$$

with $\mathbf{v} \in \mathbb{R}^{d_u}$, $\mathbf{Q} \in \mathbb{R}^{d_u \times d_u}$, $\mathbf{Q} = \mathbf{Q}^\top$, $\mathbf{b} \in \mathbb{R}^{d_u}$, and $c \in \mathbb{R}$; it holds that

$$\left.\begin{array}{r}\mathbf{Q} \succ 0 \\ c - \frac{1}{2}\mathbf{b}^\top \mathbf{Q}^{-1}\mathbf{b} \geq \underline{\psi}\end{array}\right\} \quad \Rightarrow \quad \underline{\psi} \leq \min_{\mathbf{v}\in\mathbb{R}^{d_u}} \psi(\mathbf{v}), \tag{3.7}$$

for any $\underline{\psi} \in \mathbb{R}$. To prove this fact, let $\mathbf{Q} \succ 0$ and note that

$$\begin{aligned}\mathbf{v}^* &= \arg\min_{\mathbf{v}\in\mathbb{R}^{d_u}} \psi(\mathbf{v}) \\ &= -\mathbf{Q}^{-1}\mathbf{b}.\end{aligned}$$

Therefore,

$$\begin{aligned}\psi\left(\mathbf{v}^*\right) &= \frac{1}{2}\mathbf{b}^\top \mathbf{Q}^{-1}\mathbf{b} - \mathbf{b}^\top \mathbf{Q}^{-1}\mathbf{b} + c \\ &= -\frac{1}{2}\mathbf{b}^\top \mathbf{Q}^{-1}\mathbf{b} + c\end{aligned}$$

and

$$\begin{aligned}-\frac{1}{2}\mathbf{b}^\top \mathbf{Q}^{-1}\mathbf{b} + c \geq \underline{\psi} \quad &\Rightarrow \quad \psi\left(\mathbf{v}^*\right) \geq \underline{\psi} \\ &\Rightarrow \quad \left(\min_{\mathbf{v}\in\mathbb{R}^{d_u}} \psi(\mathbf{v})\right) \geq \underline{\psi},\end{aligned}$$

for any $\underline{\psi} \in \mathbb{R}$.

Now, according to Definitions 3.1 and 3.2, to prove the δ-passivity of Σ it suffices to show that

$$\nabla_{\mathbf{z}} S\left(\mathbf{z}, \mathbf{u}\right)^\top \mathcal{Z}(\mathbf{z}, \mathbf{u}) + \nabla_{\mathbf{u}} S(\mathbf{z}, \mathbf{u})^\top \mathbf{v} \leq -\sigma\left(\mathbf{z}, \mathbf{u}\right) + \mathbf{w}(\mathbf{z}, \mathbf{u}, \mathbf{v})^\top \mathbf{v}, \tag{3.8}$$

with $\mathbf{w}(\mathbf{z}, \mathbf{u}, \mathbf{v}) = \mathrm{D}_{\mathbf{z}}\, \mathcal{Y}(\mathbf{z}, \mathbf{u})\mathcal{Z}(\mathbf{z}, \mathbf{u}) + \mathrm{D}_{\mathbf{u}}\, \mathcal{Y}(\mathbf{z}, \mathbf{u})\mathbf{v}$, for all $\mathbf{z} \in \mathbb{R}^{d_z}$, $\mathbf{u} \in \mathbb{R}^{d_u}$, and $\mathbf{v} \in \mathbb{R}^{d_u}$. Such a condition can be equivalently written as

$$\begin{aligned}\sigma(\mathbf{z}, \mathbf{u}) \leq\ & \frac{1}{2}\mathbf{v}^\top \left(\mathrm{D}_{\mathbf{u}}\, \mathcal{Y}(\mathbf{z}, \mathbf{u}) + \mathrm{D}_{\mathbf{u}}\, \mathcal{Y}(\mathbf{z}, \mathbf{u})^\top\right)\mathbf{v} \\ & + \left(\mathrm{D}_{\mathbf{z}}\, \mathcal{Y}(\mathbf{z}, \mathbf{u})\mathcal{Z}(\mathbf{z}, \mathbf{u}) - \nabla_{\mathbf{u}} S(\mathbf{z}, \mathbf{u})\right)^\top \mathbf{v} \\ & + \left(-\nabla_{\mathbf{z}} S(\mathbf{z}, \mathbf{u})^\top \mathcal{Z}(\mathbf{z}, \mathbf{u})\right).\end{aligned} \tag{3.9}$$

Here, the right-hand side of the inequality is a quadratic form on $\mathbf{v}$ with

$$\begin{aligned}\mathbf{Q} &= \mathrm{D}_{\mathbf{u}}\,\mathcal{Y}(\mathbf{z},\mathbf{u}) + \mathrm{D}_{\mathbf{u}}\,\mathcal{Y}(\mathbf{z},\mathbf{u})^{\top}\\ \mathbf{b} &= \mathrm{D}_{\mathbf{z}}\,\mathcal{Y}(\mathbf{z},\mathbf{u})\mathcal{Z}(\mathbf{z},\mathbf{u}) - \nabla_{\mathbf{u}} S(\mathbf{z},\mathbf{u})\\ c &= -\nabla_{\mathbf{z}} S(\mathbf{z},\mathbf{u})^{\top}\mathcal{Z}(\mathbf{z},\mathbf{u}).\end{aligned}$$

Consequently, from Fact (3.7) it follows that Conditions (i) and (ii) imply that (3.9) holds; which in turn implies that (3.8) holds, leading to the desired result. □

Similarly, a simple modification in the proof of Lemma 3.2 yields an equivalent result for the case of δ-antipassivity, as outlined in Lemma 3.3.

Lemma 3.3 *Consider the system Σ in (3.1), and suppose that next conditions hold:*

(i) For every $(\mathbf{z},\mathbf{u}) \in \mathbb{R}^{d_z} \times \mathbb{R}^{d_u}$, the output map of Σ satisfies:

$$\mathrm{D}_{\mathbf{u}}\,\mathcal{Y}\,(\mathbf{z},\mathbf{u}) + \mathrm{D}_{\mathbf{u}}\,\mathcal{Y}\,(\mathbf{z},\mathbf{u})^{\top} \prec 0.$$

(ii) There exist some functions $S : \mathbb{R}^{d_z} \times \mathbb{R}^{d_u} \to \mathbb{R}_{\geq 0}$ and[3] $\sigma : \mathbb{R}^{d_z} \times \mathbb{R}^{d_u} \to \mathbb{R}_{\geq 0}$, with $S(\cdot,\cdot)$ continuously differentiable, such that: (3.2a) holds and

$$-\nabla_{\mathbf{z}} S(\mathbf{z},\mathbf{u})^{\top}\mathcal{Z}(\mathbf{z},\mathbf{u}) + \frac{1}{2}\boldsymbol{\Psi}^{\top}\left(\mathrm{D}_{\mathbf{u}}\,\mathcal{Y}(\mathbf{z},\mathbf{u}) + \mathrm{D}_{\mathbf{u}}\,\mathcal{Y}(\mathbf{z},\mathbf{u})^{\top}\right)^{-1}\boldsymbol{\Psi} \geq \sigma(\mathbf{z},\mathbf{u}),$$

for all $(\mathbf{z},\mathbf{u}) \in \mathbb{R}^{d} \times \Delta$. Here, $\boldsymbol{\Psi} := \mathrm{D}_{\mathbf{z}}\,\mathcal{Y}(\mathbf{z},\mathbf{u})\mathcal{Z}(\mathbf{z},\mathbf{u}) + \nabla_{\mathbf{u}} S(\mathbf{z},\mathbf{u})$.

Then, the system Σ is δ-antipassive characterized by σ. ◊

Lemmas 3.2 and 3.3 become particularly useful when applied to scalar systems. Examples 3.3 and 3.4 illustrate this remark.

Example 3.3 Consider a scalar dynamical system of the form

$$\begin{aligned}\dot{z}(t) &= \mathcal{Z}\,(z(t), u(t))\\ y(t) &= \mathcal{Y}_z\,(z(t)) + \mathcal{Y}_u\,(u(t))\,,\end{aligned}$$

with $\mathcal{Z} : \mathbb{R} \times \mathbb{R} \to \mathbb{R}$, $\mathcal{Y}_z : \mathbb{R} \to \mathbb{R}$, and $\mathcal{Y}_u : \mathbb{R} \to \mathbb{R}$. Moreover, let $\mathcal{Z}(\cdot,\cdot)$, $\mathcal{Y}_z(\cdot)$, and $\mathcal{Y}_u(\cdot)$ be continuously differentiable and define

$$S\,(z,u) = \frac{1}{2}\,(\mathcal{Z}\,(z,u))^2\,.$$

First, regard Lemma 3.2 and the notion of δ-passivity. Condition (i) requires

$$\frac{\partial \mathcal{Y}_u(u)}{\partial u} > 0, \quad \forall u \in \mathbb{R}, \tag{3.10}$$

while Condition (ii) requires

[3] Note that σ here is restricted to be a function only of $(\mathbf{z},\mathbf{u})$; not of $\mathbf{v}$.

$$-\left(\frac{\partial \mathcal{Z}(z,u)}{\partial z}+\frac{1}{4}\left(\frac{\partial \mathcal{Y}_u(u)}{\partial u}\right)^{-1}\left(\frac{\partial \mathcal{Y}_z(z)}{\partial z}-\frac{\partial \mathcal{Z}(z,u)}{\partial u}\right)^2\right)(\mathcal{Z}(z,u))^2 \geq \sigma(z,u).$$

Thus, if the dynamics of the system satisfy (3.10) and

$$-\frac{\partial \mathcal{Z}(z,u)}{\partial z} \geq \frac{1}{4}\left(\frac{\partial \mathcal{Y}_u(u)}{\partial u}\right)^{-1}\left(\frac{\partial \mathcal{Y}_z(z)}{\partial z}-\frac{\partial \mathcal{Z}(z,u)}{\partial u}\right)^2, \quad \forall (z,u) \in \mathbb{R}\times\mathbb{R}, \tag{3.11}$$

then σ can be taken as

$$\sigma(z,u) = -\left(\frac{\partial \mathcal{Z}(z,u)}{\partial z}+\frac{1}{4}\left(\frac{\partial \mathcal{Y}_u(u)}{\partial u}\right)^{-1}\left(\frac{\partial \mathcal{Y}_z(z)}{\partial z}-\frac{\partial \mathcal{Z}(z,u)}{\partial u}\right)^2\right)(\mathcal{Z}(z,u))^2,$$

which readily satisfies Condition (ii) and hence certifies the δ-passivity of the considered dynamics. Furthermore, if the inequality (3.11) holds strictly, then σ is informative in the sense of Definition 3.1.

Equivalent conditions as the above can be deduced to certify the δ-antipassivity of the considered dynamics. Specifically, repeating the previous analysis but under the light of Lemma 3.3 yields that, if the dynamics of the system satisfy

$$\frac{\partial \mathcal{Y}_u(u)}{\partial u} < 0, \quad \forall u \in \mathbb{R},$$

and

$$\frac{\partial \mathcal{Z}(z,u)}{\partial z} \leq \frac{1}{4}\left(\frac{\partial \mathcal{Y}_u(u)}{\partial u}\right)^{-1}\left(\frac{\partial \mathcal{Y}_z(z)}{\partial z}+\frac{\partial \mathcal{Z}(z,u)}{\partial u}\right)^2, \quad \forall (z,u) \in \mathbb{R}\times\mathbb{R}, \tag{3.12}$$

then the considered system is δ-antipassive characterized by

$$\sigma(z,u) = -\left(\frac{\partial \mathcal{Z}(z,u)}{\partial z}-\frac{1}{4}\left(\frac{\partial \mathcal{Y}_u(u)}{\partial u}\right)^{-1}\left(\frac{\partial \mathcal{Y}_z(z)}{\partial z}+\frac{\partial \mathcal{Z}(z,u)}{\partial u}\right)^2\right)(\mathcal{Z}(z,u))^2.$$

As before, if the inequality (3.12) holds strictly, then σ is informative in the sense of Definition 3.1.

Note that requirements (3.11) and (3.12) are further simplified in some particular cases. For instance, consider the dynamical system given by

$$\dot{z}(t) = \tau f(u(t)) - \tau z(t) \tag{3.13a}$$

$$y(t) = \alpha z(t) + \beta f(u(t)), \tag{3.13b}$$

with $\tau, \beta \in \mathbb{R}_{>0}$, $\alpha \in \mathbb{R}$, and $f : \mathbb{R} \to \mathbb{R}$ continuously differentiable. Under such a setup, (3.11) simplifies to

$$\tau \geq \frac{1}{4}\left(\beta\frac{\partial f(u)}{\partial u}\right)^{-1}\left(\alpha - \tau\frac{\partial f(u)}{\partial u}\right)^2, \quad \forall (z,u) \in \mathbb{R}\times\mathbb{R}, \tag{3.14}$$

while (3.12) reduces to

$$-\tau \leq \frac{1}{4}\left(\beta\frac{\partial f(u)}{\partial u}\right)^{-1}\left(\alpha + \tau\frac{\partial f(u)}{\partial u}\right)^2, \quad \forall (z,u) \in \mathbb{R}\times\mathbb{R}. \tag{3.15}$$

Consequently, the δ-passivity or δ-antipassivity of the system (3.13) can be easily asserted by appropriately designing the constants τ, β, α, and the map $f(\cdot)$. For instance, if $\alpha > 0$ and $f(u) = (\alpha/\tau)u$, then (3.14) is readily satisfied for every $\tau, \beta \in \mathbb{R}_{>0}$. Similarly, if $\alpha > 0$ and $f(u) = -(\alpha/\tau)u$, then (3.15) is readily satisfied for every $\tau, \beta \in \mathbb{R}_{>0}$. ◊

Example 3.4 Consider a scalar dynamical system of the form

$$\dot{z}(t) = [\overline{z} - z(t)]_+ [u(t)]_+ - z(t)[-u(t)]_+ \tag{3.16a}$$
$$y(t) = \alpha(z(t) + u(t)), \tag{3.16b}$$

where $[\cdot]_+ := \max\{\cdot, 0\}$, $\alpha \in \{-1, 1\}$, and $\overline{z} \in \mathbb{R}_{>0}$. Moreover, let the initial condition be subject to $z(0) \in [0, \overline{z}]$. It is straightforward to check that such a constraint on the initial condition implies that the interval $[0, \overline{z}]$ is positively invariant under the dynamics (3.16), i.e., $z(0) \in [0, \overline{z}] \Rightarrow z(t) \in [0, \overline{z}], \ \forall t \geq 0$. Thus, without loss of generality and keeping in mind the assumption $z(0) \in [0, \overline{z}]$, the dynamics (3.16) can be equivalently written as

$$\dot{z}(t) = (\overline{z} - z(t))[u(t)]_+ - z(t)[-u(t)]_+ \tag{3.17a}$$
$$y(t) = \alpha(z(t) + u(t)). \tag{3.17b}$$

Consequently, the δ-passivity or δ-antipassivity properties of (3.17) can be characterized by investigating these properties for (3.17) instead. As such, consider

$$S(z,u) = (\overline{z} - z)\int_0^u [\zeta]_+ \, d\zeta + z\int_0^{-u} [\zeta]_+ \, d\zeta$$

as a δ-storage function for (3.17).

For the considered system, it holds that

$$\mathrm{D}_{\mathbf{u}}\boldsymbol{\mathcal{Y}}(\mathbf{z},\mathbf{u}) + \mathrm{D}_{\mathbf{u}}\boldsymbol{\mathcal{Y}}(\mathbf{z},\mathbf{u})^\top = 2\alpha.$$

Therefore, if $\alpha = 1$, then Condition (i) of Lemma 3.2 is satisfied; and if $\alpha = -1$, then Condition (i) of Lemma 3.3 holds instead.

On the other hand, note that for the considered $S(\cdot,\cdot)$ it follows that

$$\nabla_{\mathbf{z}} S(\mathbf{z}, \mathbf{u}) = \frac{\partial S(z,u)}{\partial z} = -\int_0^u [\zeta]_+ \, d\zeta + \int_0^{-u} [\zeta]_+ \, d\zeta = -\frac{1}{2}[u]_+^2 + \frac{1}{2}[-u]_+^2$$
$$\nabla_{\mathbf{u}} S(\mathbf{z}, \mathbf{u}) = \frac{\partial S(z,u)}{\partial u} = (\bar{z} - z)\,[u]_+ - z\,[-u]_+ = \mathcal{Z}(z,u).$$

Then, setting $\alpha = 1$ or $\alpha = -1$ accordingly, Condition (ii) of Lemmas 3.2 and 3.3 reduces to

$$\left(\frac{1}{2}[u]_+^2 - \frac{1}{2}[-u]_+^2\right)\mathcal{Z}(z,u) \geq \sigma(z,u).$$

Since the left-hand side of this inequality is always non-negative, one can set

$$\sigma(z,u) = \left(\frac{1}{2}[u]_+^2 - \frac{1}{2}[-u]_+^2\right)\mathcal{Z}(z,u).$$

As a result, Condition (ii) is satisfied for both Lemmas 3.2 and 3.3.

In conclusion, when $\alpha = 1$, the system (3.16) is δ-passive characterized by σ. In contrast, when $\alpha = -1$, the same system becomes δ-antipassive, characterized by the same σ. This result suggests that simply changing the sign of the system's output is sufficient to switch its characterization from δ-passive to δ-antipassive, and vice versa. As will be proven in Lemma 3.5 of Sect. 3.3, this is indeed the case. ◇

3.2 Stability of Delta-Dissipative Systems

Consider the system (3.1) and let $\mathcal{E}_0$ be defined as

$$\mathcal{E}_0 = \left\{\mathbf{z} \in \mathbb{R}^{d_z} : \mathcal{Z}\left(\mathbf{z}, \mathbf{0}_{d_u}\right) = \mathbf{0}_{d_z}\right\}. \tag{3.18}$$

Namely, $\mathcal{E}_0$ is the set of equilibrium points of the system (3.1) under the zero input $\mathbf{u}(t) = \mathbf{0}_{d_u}$, for all $t \geq 0$. If the system (3.1) is shown to be δ-dissipative with a characteristic supply rate and $\mathcal{E}_0$ satisfies some regularity conditions, then it is possible to assert some stability properties of $\mathcal{E}_0$ under the dynamics (3.1).

Lemma 3.4 *Suppose that the system (3.1) is δ-dissipative characterized by $(\sigma, \mathbf{\Pi})$, and that the set $\mathcal{E}_0$ in (3.18) is nonempty and compact. Moreover, assume that the system (3.1) is subject to a constant input $\mathbf{u}(t) = \mathbf{0}_{d_u}$, for all $t \geq 0$. If it holds that*

$$\sigma(\mathbf{z}, \mathbf{0}_{d_u}, \mathbf{0}_{d_u}) \geq \left\|\mathrm{D}_{\mathbf{z}}\,\mathcal{Y}(\mathbf{z}, \mathbf{0}_{d_u})\mathcal{Z}(\mathbf{z}, \mathbf{0}_{d_u})\right\|_{\mathbf{\Pi}_{22}}^2, \quad \forall \mathbf{z} \in \mathbb{R}^{d_z}, \tag{3.19}$$

then the set $\mathcal{E}_0$ is stable in the sense of Lyapunov. If in addition, it holds that

$$\sigma(\mathbf{z}, \mathbf{0}_{d_u}, \mathbf{0}_{d_u}) > \left\|\mathrm{D}_{\mathbf{z}}\,\mathcal{Y}(\mathbf{z}, \mathbf{0}_{d_u})\mathcal{Z}(\mathbf{z}, \mathbf{0}_{d_u})\right\|_{\mathbf{\Pi}_{22}}^2, \quad \forall \mathbf{z} \in \mathbb{R}^{d_z} \setminus \mathcal{E}_0, \tag{3.20}$$

then the set $\mathcal{E}_0$ is asymptotically stable. For both claims, $\mathbf{\Pi}_{22} \in \mathbb{R}^{d_y \times d_y}$ is the lower-right block of $\mathbf{\Pi}$, i.e.,

$$\mathbf{\Pi} := \begin{bmatrix} \mathbf{\Pi}_{11} & \mathbf{\Pi}_{12} \\ \mathbf{\Pi}_{21} & \mathbf{\Pi}_{22} \end{bmatrix},$$

and $V(\mathbf{z}) = \mathbf{S}(\mathbf{z}, \mathbf{0}_{d_u})$ is a valid Lyapunov function. ◇

Proof From Definition 3.1, it holds that $S(\cdot,\cdot)$ is non-negative and $S(\mathbf{z}, \mathbf{0}_{d_u}) = 0 \Leftrightarrow \mathcal{Z}(\mathbf{z}, \mathbf{0}_{d_u}) = \mathbf{0}_{d_z} \Leftrightarrow \mathbf{z} \in \mathcal{E}_0$. Thus, take $V(\mathbf{z}) = S(\mathbf{z}, \mathbf{0}_{d_u})$ as a valid Lyapunov function candidate. Moreover, from (3.2b) it follows that

$$\begin{aligned}
\dot{V}(\mathbf{z}(t)) &= \dot{S}\left(\mathbf{z}(t), \mathbf{0}_{d_u}, \mathbf{0}_{d_u}\right) \\
&\le -\sigma\left(\mathbf{z}(t), \mathbf{0}_{d_u}, \mathbf{0}_{d_u}\right) + \left\| \begin{bmatrix} \mathbf{0}_{d_u} \\ \mathbf{w}(\mathbf{z}, \mathbf{0}_{d_u}, \mathbf{0}_{d_u}) \end{bmatrix} \right\|_{\mathbf{\Pi}}^2 \\
&= -\sigma\left(\mathbf{z}(t), \mathbf{0}_{d_u}, \mathbf{0}_{d_u}\right) + \left\| \mathrm{D}_{\mathbf{z}} \mathcal{Y}(\mathbf{z}, \mathbf{0}_{d_u}) \mathcal{Z}(\mathbf{z}, \mathbf{0}_{d_u}) \right\|_{\mathbf{\Pi}_{22}}^2 .
\end{aligned}$$

Using (3.19) together with the non-negativity of $\sigma(\cdot,\cdot,\cdot)$ (see Definition 3.1), it follows that $\dot{V}(\mathbf{z}(t)) \le 0$, for all $t \ge 0$, which implies that $\mathcal{E}_0$ is stable in the sense of Lyapunov [3, Corollary 4.7]. Similarly, from (3.20) and the non-negativity of $\sigma(\cdot,\cdot,\cdot)$, it holds that $\dot{V}(\mathbf{z}(t)) < 0$, for all $t \ge 0$ such that $\mathbf{z}(t) \notin \mathcal{E}_0$, which implies that $\mathcal{E}_0$ is asymptotically stable [3, Corollary 4.7]. □

Lemma 3.4 provides some sufficient conditions to certify the stability and asymptotic stability of the set of equilibria of an arbitrary δ-dissipative system under the constant zero input. Moreover, note that for δ-passive and δ-antipassive systems condition (3.19) is automatically satisfied (since $\mathbf{\Pi}_{22} = \mathbf{0}_{d_u \times d_u}$), and condition (3.19) is readily satisfied if the function $\sigma(\cdot,\cdot,\cdot)$ is informative in the sense of Definition 3.1. Thus, the informativity of $\sigma(\cdot,\cdot,\cdot)$ is particularly useful when characterizing the stability properties of δ-passive and δ-antipassive systems.

Example 3.5 Recall Example 3.2 and let $\gamma < 0$. Under such a setup, the considered system becomes Hurwitz, implying that $\mathcal{E}_0 = \{\mathbf{z} \in \mathbb{R}^{d_z} : \gamma\mathbf{z} = \mathbf{0}_{d_z}\}$ (which is a singleton) is asymptotically stable. Lemma 3.4 can be used to confirm this fact. For instance, take any $\beta \in \mathbb{R}_{\ge 0}$ and note that

$$\begin{aligned}
\sigma(\mathbf{z}, \mathbf{0}_{d_u}, \mathbf{0}_{d_u}) - \left\| \mathrm{D}_{\mathbf{z}} \mathcal{Y}(\mathbf{z}, \mathbf{0}_{d_u}) \mathcal{Z}(\mathbf{z}, \mathbf{0}_{d_u}) \right\|_{\mathbf{\Pi}_{22}}^2 &= \sigma_\beta(\mathbf{z}, \mathbf{0}_{d_u}) - \|\gamma\mathbf{z}\|_{\beta+\gamma}^2 \\
&= \beta\, \|\gamma\mathbf{z}\|_2^2 - (\beta+\gamma)\, \|\gamma\mathbf{z}\|_2^2 \\
&= -\gamma\, \|\gamma\mathbf{z}\|_2^2 \\
&> 0.
\end{aligned}$$

Thus, condition (3.20) holds and the asymptotic stability of $\mathcal{E}_0$ is verified. ◇

Example 3.6 Consider the system (3.13), with $\alpha > 0$ and $f(u) = (\alpha/\tau)u$. As discussed in Example 3.3, such a system is said to be δ-passive characterized by

$$\sigma(z, u) = \tau\left(\mathcal{Z}(z, u)\right)^2,$$

which is informative in the sense of Definition 3.1. On the other hand, for the considered system the set $\mathcal{E}_0$ in (3.18) is given by

$$\mathcal{E}_0 = \{z \in \mathbb{R} : z = f(0)\},$$

which is a singleton and hence is nonempty and compact. Therefore, from Lemma 3.4 it follows that $\mathcal{E}_0$ is asymptotically stable under the considered dynamics (3.13) with input $u(t) = 0$, for all $t \geq 0$. The same claim holds true for the δ-antipassive counterpart of (3.13) with $f(u) = -(\alpha/\tau)u$. ◊

3.3 Interconnected Delta-Dissipative Systems

As with other notions of dissipativity, the one of δ-dissipativity is useful for compositional analyses of interconnected systems. Namely, under certain conditions it is possible to show that a given interconnection of arbitrary δ-dissipative systems is itself a δ-dissipative system. This section considers some particular types of interconnections and characterizes their δ-dissipativity analytically.

Cascaded Gain Interconnection

Consider a system Σ of the form (3.1) and a non-zero scalar gain $\kappa \in \mathbb{R} \setminus \{0\}$ interconnected in cascade as depicted in Fig. 3.1. Let $\bar{\mathbf{u}} = \mathbf{u}$ and $\bar{\mathbf{y}} = \kappa\mathbf{y}$ be the input and output vectors of the interconnected system, respectively.

Lemma 3.5 *Consider the cascaded gain interconnection of Fig. 3.1 and suppose that Σ is δ-dissipative characterized by $(\sigma, \mathbf{\Pi})$. Then, the interconnected system (from the input $\bar{\mathbf{u}}$ to the output $\bar{\mathbf{y}}$) is δ-dissipative characterized by $(\sigma, \bar{\mathbf{\Pi}})$ with*

$$\bar{\mathbf{\Pi}} = \begin{bmatrix} \mathbf{\Pi}_{11} & \frac{1}{\kappa}\mathbf{\Pi}_{12} \\ \frac{1}{\kappa}\mathbf{\Pi}_{21} & \frac{1}{\kappa^2}\mathbf{\Pi}_{22} \end{bmatrix},$$

where $\mathbf{\Pi}_{ij}$ is the (i, j)th block of $\mathbf{\Pi}$, for all $i, j \in \{1, 2\}$. Furthermore, the δ-storage function $S(\cdot, \cdot)$ of Σ serves as a valid δ-storage function for the cascaded system. ◊

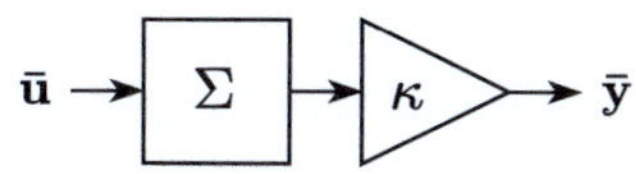

Fig. 3.1 Cascaded interconnection of a system and a non-zero scalar gain

Proof From the δ-dissipativity of Σ, it holds that

$$\dot{S}(\mathbf{z}, \mathbf{u}) \leq -\sigma(\mathbf{z}, \mathbf{u}, \mathbf{v}) + \left\| \begin{bmatrix} \mathbf{v} \\ \mathbf{w} \end{bmatrix} \right\|_{\Pi}^{2},$$

where $\mathbf{w} := \mathbf{w}(\mathbf{z}, \mathbf{u}, \mathbf{v})$. Now, let $\bar{\mathbf{v}}$ and $\bar{\mathbf{w}}$ play the roles of $\mathbf{v}$ and $\mathbf{w}(\cdot, \cdot, \cdot)$ for the interconnected system, respectively, and note that

$$\begin{aligned} \bar{\mathbf{u}} &= \mathbf{u} \\ \bar{\mathbf{v}} &= \mathbf{v} \\ \bar{\mathbf{w}} &= \kappa \mathbf{w}. \end{aligned}$$

Thus, $\mathbf{w} = (1/\kappa)\bar{\mathbf{w}}$, and it holds that

$$\begin{aligned} \dot{S}(\mathbf{z}, \bar{\mathbf{u}}) &\leq -\sigma(\mathbf{z}, \bar{\mathbf{u}}, \bar{\mathbf{v}}) + \left\| \begin{bmatrix} \bar{\mathbf{v}} \\ \frac{1}{\kappa}\bar{\mathbf{w}} \end{bmatrix} \right\|_{\Pi}^{2} \\ &= -\sigma(\mathbf{z}, \bar{\mathbf{u}}, \bar{\mathbf{v}}) + \left\| \begin{bmatrix} \bar{\mathbf{v}} \\ \bar{\mathbf{w}} \end{bmatrix} \right\|_{\bar{\Pi}}^{2}, \end{aligned}$$

which yields the desired result. □

Lemma 3.5 shows that multiplying the output of a δ-dissipative system by some non-zero scalar gain preserves the δ-dissipativity of the system but with a different supply rate. In particular, note that if $\kappa = -1$ and Σ is δ-passive characterized by some σ, then the cascaded system is δ-antipassive characterized by the same σ. Similarly, if $\kappa = -1$ and Σ is δ-antipassive characterized by some σ, then the cascaded system is δ-passive characterized by the same σ. This fact has been illustrated in Example 3.4.

Cascaded Memoryless Encapsulation of a Class of Scalar Systems

Consider the *cascaded encapsulation* depicted in Fig. 3.2, where Σ is a scalar instance of (3.1) given by

$$\Sigma : \begin{cases} \dot{z}(t) = \mathcal{Z}(z(t), u(t)) \\ y(t) = z(t), \end{cases} \tag{3.21}$$

with $z \in \mathbb{R}$, $u \in \mathbb{R}$, $y \in \mathbb{R}$; and $\phi : \mathbb{R}^{d_\phi} \to \mathbb{R}$ is a twice continuously differentiable function. The overall system takes the vector $\bar{\mathbf{u}} \in \mathbb{R}^{d_\phi}$ as input and yields the vector $\bar{\mathbf{y}} \in \mathbb{R}^{d_\phi}$ as output.

Lemma 3.6 *Consider the encapsulation of Fig. 3.2. Let Σ in (3.21) be δ-passive characterized by σ, and suppose that at least one of the following two conditions hold:*

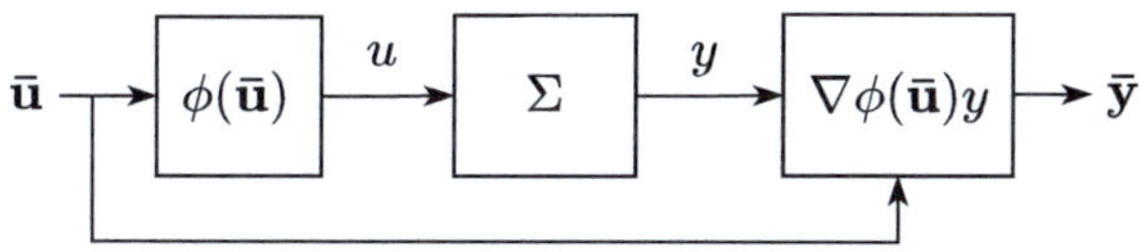

Fig. 3.2 Cascaded encapsulation of the scalar system (3.21) by the smooth memoryless function $\phi : \mathbb{R}^{d_\phi} \to \mathbb{R}$

(i) Σ *is a positive system,*[4] $z(0) \in \mathbb{R}_{\geq 0}$, *and* $\phi(\cdot)$ *is convex.*
(ii) $\phi(\cdot)$ *is affine.*

Then, the overall encapsulated system (from the input $\bar{\mathbf{u}}$ *to the output* $\bar{\mathbf{y}}$*) is also* δ*-passive characterized by* $\bar{\sigma}(z, \bar{\mathbf{u}}, \bar{\mathbf{v}}) = \sigma\left(z, \phi(\bar{\mathbf{u}}), \nabla\phi(\bar{\mathbf{u}})^\top \bar{\mathbf{v}}\right)$*. Furthermore,* $\bar{S}(z, \bar{\mathbf{u}}) = S(z, \phi(\bar{\mathbf{u}}))$ *is a valid* δ*-storage function for the encapsulated system. Here,* $S(\cdot, \cdot)$ *is the* δ*-storage function of* Σ*.* ◊

Proof The δ-passivity of Σ implies that

$$\frac{\partial S(z, u)}{\partial z}\mathcal{Z}(z, u) + \frac{\partial S(z, u)}{\partial u} v \leq -\sigma(z, u, v) + \mathcal{Z}(z, u)v, \tag{3.22}$$

for all $z, u, v \in \mathbb{R}$. Now, for the encapsulated system to be δ-passive it must hold that

$$\frac{\partial \bar{S}(z, \bar{\mathbf{u}})}{\partial z}\mathcal{Z}(z, \phi(\bar{\mathbf{u}})) + \nabla_{\bar{\mathbf{u}}}\bar{S}(z, \bar{\mathbf{u}})^\top \bar{\mathbf{v}} \leq -\bar{\sigma}(z, \bar{\mathbf{u}}, \bar{\mathbf{v}}) + \bar{\mathbf{w}}(z, \bar{\mathbf{u}}, \bar{\mathbf{v}})^\top \bar{\mathbf{v}},$$

for all $z \in \mathbb{R}, \bar{\mathbf{u}}, \bar{\mathbf{v}} \in \mathbb{R}^{d_\phi}$. Equivalently, replacing $\bar{\mathbf{w}}(z, \bar{\mathbf{u}}, \bar{\mathbf{v}}) = \nabla\phi(\bar{\mathbf{u}})\mathcal{Z}(z, \phi(\bar{\mathbf{u}})) + z\,\mathrm{D}\phi(\bar{\mathbf{u}})\,\bar{\mathbf{v}}$, $\bar{S}(z, \bar{\mathbf{u}}) = S(z, \phi(\bar{\mathbf{u}}))$, and $\bar{\sigma}(z, \bar{\mathbf{u}}, \bar{\mathbf{v}}) = \sigma\left(z, \phi(\bar{\mathbf{u}}), \nabla\phi(\bar{\mathbf{u}})^\top \bar{\mathbf{v}}\right)$, yields that for the entire system to be δ-passive it must hold that

$$\frac{\partial S(z, u)}{\partial z}\mathcal{Z}(z, u) + \frac{\partial S(z, u)}{\partial u} v \leq -\sigma(z, u, v) + \mathcal{Z}(z, u)v + \bar{\mathbf{v}}^\top (z\,\mathrm{D}\phi(\bar{\mathbf{u}}))\,\bar{\mathbf{v}}, \tag{3.23}$$

with $u := \phi(\bar{\mathbf{u}})$ and $v := \nabla\phi(\bar{\mathbf{u}})^\top \bar{\mathbf{v}}$. If Σ is a positive system, $z(0) \in \mathbb{R}_{\geq 0}$, and $\phi(\cdot)$ is convex, then $z\,\mathrm{D}\phi(\bar{\mathbf{u}}) \succeq 0$ for all $z \in \mathbb{R}_{\geq 0}$ and all $\bar{\mathbf{u}} \in \mathbb{R}^{d_\phi}$ (here, the analysis can be restricted to $z \in \mathbb{R}_{\geq 0}$ as those are the only relevant states granted the positivity of Σ and $z(0) \in \mathbb{R}_{\geq 0}$). On the other hand, if $\phi(\cdot)$ is affine, then $\mathrm{D}\phi(\bar{\mathbf{u}}) = \mathbf{0}_{d_\phi \times d_\phi}$, for all $\bar{\mathbf{u}} \in \mathbb{R}^{d_\phi}$. Overall, under the considered setup it always holds that $\bar{\mathbf{v}}^\top (z\,\mathrm{D}\phi(\bar{\mathbf{u}}))\,\bar{\mathbf{v}} \geq 0$. Therefore, (3.23) holds by means of (3.22), leading to the desired result. □

Lemma 3.6 can be easily adapted to δ-antipassive systems, as established next.

[4] A positive system is one in which all state variables remain non-negative for all time, provided the initial conditions are non-negative.

Lemma 3.7 *Consider the encapsulation of Fig. 3.2. Let*

$$\Sigma : \begin{cases} \dot{z}(t) = \mathcal{Z}(z(t), u(t)) \\ y(t) = -z(t), \end{cases}$$

be δ-antipassive characterized by σ, and suppose that at least one of Conditions (i)–(ii) of Lemma 3.6 holds. Then, the overall encapsulated system (from the input $\bar{\mathbf{u}}$ to the output $\bar{\mathbf{y}}$) is also δ-antipassive characterized by $\bar{\sigma}(z, \bar{\mathbf{u}}, \bar{\mathbf{v}}) = \sigma\left(z, \phi(\bar{\mathbf{u}}), \nabla\phi(\bar{\mathbf{u}})^\top \bar{\mathbf{v}}\right)$. Furthermore, $\bar{S}(z, \bar{\mathbf{u}}) = S(z, \phi(\bar{\mathbf{u}}))$ is a valid δ-storage function for the encapsulated system. Here, $S(\cdot, \cdot)$ is the δ-storage function of Σ. ◊

Proof The proof is virtually identical to that of Lemma 3.6. In fact, the result can be obtained by direct application of Lemmas 3.5 and 3.6, considering a gain interconnection with $\kappa = -1$. □

Example 3.7 Consider the dynamical system given by

$$\dot{z}(t) = [u(t)]_+ - z(t)\,[-u(t)]_+ \tag{3.24a}$$

$$y(t) = z(t), \tag{3.24b}$$

with $z(0) \in \mathbb{R}_{>0}$. It is not difficult to check that (3.24) is a positive system, since $z(0) \in \mathbb{R}_{\geq 0} \Rightarrow z(t) \in \mathbb{R}_{\geq 0}$, for all $t \geq 0$. Let

$$S(z, u) = \int_0^u [\zeta]_+\, d\zeta + z \int_0^{-u} [\zeta]_+\, d\zeta$$

$$\sigma(z, u) = -\frac{1}{2}[-u]_+^2 \left([u]_+ - z[-u]_+\right).$$

Observe that both $S(\cdot, \cdot)$ and $\sigma(\cdot, \cdot)$ are non-negative within the relevant domain $(z, u) \in \mathbb{R}_{\geq 0} \times \mathbb{R}$. Besides, while (3.2a) holds, (3.3) does not hold (as $\sigma(z, u) = 0$ for every $u \geq 0$). Hence, σ is not informative. Moreover, notice that

$$\frac{\partial S(z, u)}{\partial z} = \int_0^{-u} [\zeta]_+\, d\zeta = \frac{1}{2}[-u]_+^2$$

$$\frac{\partial S(z, u)}{\partial u} = [u]_+ - z[-u]_+ =: \mathcal{Z}(z, u).$$

Therefore,

$$\begin{aligned} \frac{\partial S(z, u)}{\partial z}\mathcal{Z}(z, u) + \frac{\partial S(z, u)}{\partial u} v &= \frac{1}{2}[-u]_+^2 \left([u]_+ - z[-u]_+\right) + \left([u]_+ - z[-u]_+\right) v \\ &= -\sigma(z, u) + \mathcal{Z}(z, u) v, \end{aligned}$$

which implies that (3.24) is δ-passive characterized by σ.

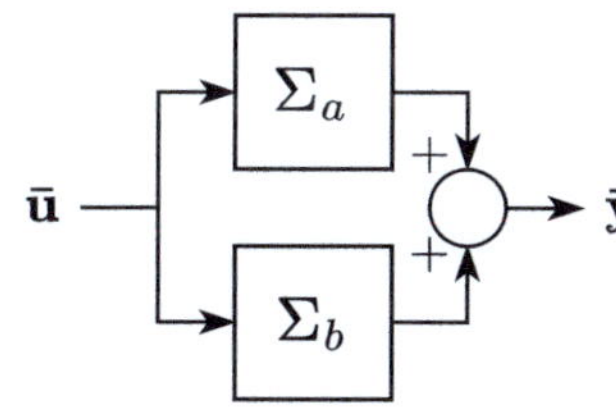

Fig. 3.3 Parallel interconnection of two systems

Now, let $\phi : \mathbb{R}^{d_\phi} \to \mathbb{R}$ be any twice continuously differentiable convex function, and consider the encapsulated system given by

$$\dot{z}(t) = [\phi(\bar{\mathbf{u}}(t))]_+ - z(t)[-\phi(\bar{\mathbf{u}}(t))]_+ \tag{3.25a}$$

$$\bar{\mathbf{y}}(t) = \nabla\phi(\bar{\mathbf{u}}(t))\, z(t). \tag{3.25b}$$

Lemma 3.6 and the δ-passivity of (3.24) imply that (3.25) is δ-passive characterized by $\bar{\sigma}(z, \bar{\mathbf{u}}) = -(1/2)[-\phi(\bar{\mathbf{u}})]_+^2([\phi(\bar{\mathbf{u}})]_+ - z[-\phi(\bar{\mathbf{u}})]_+)$. ◊

Example 3.8 Consider the simple scalar integrator dynamics given by

$$\dot{z}(t) = u(t) \tag{3.26a}$$

$$y(t) = z(t). \tag{3.26b}$$

From Example 3.2 it follows that (3.26) is δ-passive characterized by $\sigma = 0$ (non-informative). Then, by Lemma 3.6, it holds that, for every $\mathbf{a} \in \mathbb{R}^{d_{\bar{u}}}$ and $b \in \mathbb{R}$, the encapsulated system

$$\dot{z}(t) = \mathbf{a}^\top \bar{\mathbf{u}}(t) + b$$

$$\bar{\mathbf{y}}(t) = \mathbf{a}z(t)$$

is δ-passive characterized by $\bar{\sigma} = 0$. ◊

Parallel Interconnection

Consider two systems, Σ_a and Σ_b, interconnected in parallel as depicted in Fig. 3.3. For each $i \in \{a, b\}$, let the corresponding system Σ_i have dynamics of the form (3.1) and let $\mathbf{z}_i \in \mathbb{R}^{d_{z_i}}$, $\mathbf{u}_i \in \mathbb{R}^{d_{u_i}}$, and $\mathbf{y}_i \in \mathbb{R}^{d_{y_i}}$ denote its state, input, and output vectors, respectively. Clearly, it must hold that $d_{u_a} = d_{u_b}$ and $d_{y_a} = d_{y_b}$. On the other hand, $\bar{\mathbf{u}} \in \mathbb{R}^{d_{u_a}}$ and $\bar{\mathbf{y}} = \mathbf{y}_a + \mathbf{y}_b$ are the input and output vectors of the interconnected system, respectively, and $\mathbf{z} = \mathrm{col}\,(\mathbf{z}_a, \mathbf{z}_b)$ denotes the state vector of the interconnected system.

Lemma 3.8 *Consider the parallel interconnection of Fig. 3.3 and suppose that Σ_a and Σ_b are δ-dissipative characterized by $(\sigma_a, \boldsymbol{\Pi})$ and $(\sigma_b, \boldsymbol{\Pi})$, respectively. Moreover, let $\boldsymbol{\Pi}$ be of the form*

$$\mathbf{\Pi} = \begin{bmatrix} \mathbf{0}_{d_{ua}\times d_{ua}} & \hat{\mathbf{\Pi}} \\ \hat{\mathbf{\Pi}}^\top & \mathbf{0}_{d_{ya}\times d_{ya}} \end{bmatrix},$$

where $\hat{\mathbf{\Pi}} \in \mathbb{R}^{d_{ua}\times d_{ya}}$. Then, the interconnected system (from the input $\bar{\mathbf{u}}$ to the output $\bar{\mathbf{y}}$) is δ-dissipative characterized by $(\sigma, \mathbf{\Pi})$ with

$$\sigma\left(\mathbf{z}, \bar{\mathbf{u}}, \bar{\mathbf{v}}\right) = \sigma_a\left(\mathbf{z}_a, \bar{\mathbf{u}}, \bar{\mathbf{v}}\right) + \sigma_b\left(\mathbf{z}_b, \bar{\mathbf{u}}, \bar{\mathbf{v}}\right).$$

Furthermore, $S(\mathbf{z}, \bar{\mathbf{u}}) = S_a(\mathbf{z}_a, \bar{\mathbf{u}}) + S_b(\mathbf{z}_b, \bar{\mathbf{u}})$ is a valid δ-storage function for the interconnected system. Here, $S_a(\cdot, \cdot)$ and $S_b(\cdot, \cdot)$ are the δ-storage functions of Σ_a and Σ_b, respectively. ◊

Proof Using $S(\mathbf{z}, \bar{\mathbf{u}}) = S_a(\mathbf{z}_a, \bar{\mathbf{u}}) + S_b(\mathbf{z}_b, \bar{\mathbf{u}})$, it follows that

$$\begin{aligned}
\dot{S}(\mathbf{z}, \bar{\mathbf{u}}) &= \dot{S}_a(\mathbf{z}_a, \bar{\mathbf{u}}) + \dot{S}_b(\mathbf{z}_b, \bar{\mathbf{u}}) \\
&\leq -\sigma_a(\mathbf{z}_a, \bar{\mathbf{u}}, \bar{\mathbf{v}}) - \sigma_b(\mathbf{z}_b, \bar{\mathbf{u}}, \bar{\mathbf{v}}) + 2\bar{\mathbf{v}}^\top\hat{\mathbf{\Pi}}\mathbf{w}_a + 2\bar{\mathbf{v}}^\top\hat{\mathbf{\Pi}}\mathbf{w}_b \\
&= -\sigma_a(\mathbf{z}_a, \bar{\mathbf{u}}, \bar{\mathbf{v}}) - \sigma_b(\mathbf{z}_b, \bar{\mathbf{u}}, \bar{\mathbf{v}}) + 2\bar{\mathbf{v}}^\top\hat{\mathbf{\Pi}}\left(\mathbf{w}_a + \mathbf{w}_b\right) \\
&= -\sigma\left(\mathbf{z}, \bar{\mathbf{u}}, \bar{\mathbf{v}}\right) + \left\|\begin{bmatrix} \bar{\mathbf{v}} \\ \bar{\mathbf{w}} \end{bmatrix}\right\|_{\mathbf{\Pi}}^2,
\end{aligned}$$

which yields the desired result. Here, $\mathbf{w}_i := \mathbf{w}_i(\mathbf{z}_i, \mathbf{u}_i, \bar{\mathbf{v}})$, for all $i \in \{a, b\}$, and $\bar{\mathbf{v}}$ and $\bar{\mathbf{w}} = \mathbf{w}_a + \mathbf{w}_b$ play the roles of $\mathbf{v}$ and $\mathbf{w}(\cdot, \cdot, \cdot)$ for the interconnected system, respectively. □

Note that Lemma 3.8 can be easily generalized to the parallel interconnection of $N \in \mathbb{Z}_{\geq 2}$ δ-dissipative systems, all characterized by the same $\mathbf{\Pi}$ considered in Lemma 3.8. To see this fact it suffices to apply Lemma 3.8 recursively. Namely, pick two systems and apply Lemma 3.8 as it is. Then, group the two selected systems as a new (bigger) system leaving a total of $N-1$ systems to be considered. Repeat this process $N-1$ times so that all systems are included into the interconnection.

Example 3.9 Recall Examples 3.3 and 3.4, and consider the two systems given by

$$\Sigma_a : \begin{cases} \dot{z}_a(t) = \alpha u_a(t) - \tau z_a(t) \\ y_a(t) = \alpha z_a(t) + \frac{\alpha\beta}{\tau} u_a(t) \end{cases}$$
$$\Sigma_b : \begin{cases} \dot{z}_b(t) = \left[\gamma - z_b(t)\right]_+ [u_b(t)]_+ - z_b(t)\,[-u_b(t)]_+ \\ y_b(t) = z_b(t) + u_b(t), \end{cases}$$

where $\alpha, \beta, \tau, \gamma \in \mathbb{R}_{>0}$ are constant parameters. Moreover, consider the parallel interconnection of Fig. 3.3, which under the considered setup is given by

$$\Sigma : \begin{cases} \begin{bmatrix} \dot{z}_a(t) \\ \dot{z}_b(t) \end{bmatrix} = \begin{bmatrix} \alpha\bar{u}(t) - \tau z_a(t) \\ \left[\gamma - z_b(t)\right]_+ [\bar{u}(t)]_+ - z_b(t)\,[-\bar{u}(t)]_+ \end{bmatrix} \\ \bar{y}(t) = \alpha z_a(t) + z_b(t) + \left(\frac{\alpha\beta}{\tau} + 1\right)\bar{u}(t), \end{cases}$$

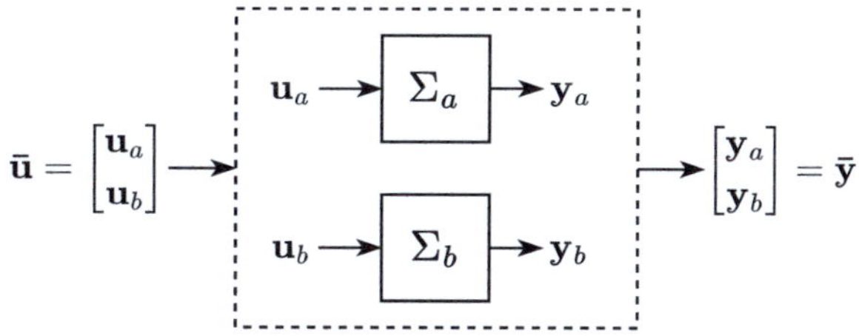

Fig. 3.4 Parallel concatenation of two systems

where $\overline{u}$ and $\overline{y}$ denote the input and output of the parallel interconnection.

From Example 3.3 it follows that Σ_a is δ-passive characterized by some σ_a. Similarly, from Example 3.4, it holds that Σ_b is δ-passive characterized by some σ_b. Therefore, the parallel interconnection Σ is also δ-passive characterized by $\sigma = \sigma_a + \sigma_b$ as established in Lemma 3.8. ◊

Parallel Concatenation

Consider two systems, Σ_a and Σ_b, concatenated in parallel as depicted in Fig. 3.4. For each $i \in \{a, b\}$, let the corresponding system Σ_i have dynamics of the form (3.1) and let $\mathbf{z}_i \in \mathbb{R}^{d_{z_i}}$, $\mathbf{u}_i \in \mathbb{R}^{d_{u_i}}$, and $\mathbf{y}_i \in \mathbb{R}^{d_{y_i}}$ denote its state, input, and output vectors, respectively. Under the considered setup, $\mathbf{z} = \mathrm{col}\,(\mathbf{z}_a, \mathbf{z}_b)$, $\bar{\mathbf{u}} = \mathrm{col}(\mathbf{u}_a, \mathbf{u}_b)$, and $\bar{\mathbf{y}} = \mathrm{col}(\mathbf{y}_a, \mathbf{y}_b)$ are the state, input and output vectors of the concatenated system, respectively.

Lemma 3.9 *Consider the parallel concatenation of Fig. 3.4 and suppose that Σ_a and Σ_b are δ-dissipative characterized by $(\sigma_a, \mathbf{\Pi}_a)$ and $(\sigma_b, \mathbf{\Pi}_b)$, respectively, with $\mathbf{\Pi}_a$ and $\mathbf{\Pi}_b$ defined as*

$$\mathbf{\Pi}_a = \begin{bmatrix} \mathbf{\Pi}_{a,11} & \mathbf{\Pi}_{a,12} \\ \mathbf{\Pi}_{a,21} & \mathbf{\Pi}_{a,22} \end{bmatrix}, \qquad \mathbf{\Pi}_b = \begin{bmatrix} \mathbf{\Pi}_{b,11} & \mathbf{\Pi}_{b,12} \\ \mathbf{\Pi}_{b,21} & \mathbf{\Pi}_{b,22} \end{bmatrix}.$$

Then, the concatenated system (from the input $\bar{\mathbf{u}}$ to the output $\bar{\mathbf{y}}$) is δ-dissipative characterized by $(\sigma, \mathbf{\Pi})$ with[5]

$$\mathbf{\Pi} = \begin{bmatrix} \mathbf{\Pi}_{a,11} & \mathbf{0} & \mathbf{\Pi}_{a,12} & \mathbf{0} \\ \mathbf{0} & \mathbf{\Pi}_{b,11} & \mathbf{0} & \mathbf{\Pi}_{b,12} \\ \mathbf{\Pi}_{a,21} & \mathbf{0} & \mathbf{\Pi}_{a,22} & \mathbf{0} \\ \mathbf{0} & \mathbf{\Pi}_{b,21} & \mathbf{0} & \mathbf{\Pi}_{b,22} \end{bmatrix}$$

and

$$\sigma\,(\mathbf{z}, \bar{\mathbf{u}}, \bar{\mathbf{v}}) = \sigma_a\,(\mathbf{z}_a, \mathbf{u}_a, \mathbf{v}_a) + \sigma_b\,(\mathbf{z}_b, \mathbf{u}_b, \mathbf{v}_b)\,.$$

Furthermore, $S(\mathbf{z}, \bar{\mathbf{u}}) = S_a(\mathbf{z}_a, \mathbf{u}_a) + S_b(\mathbf{z}_b, \mathbf{u}_b)$ is a valid δ-storage function for the concatenated system. Here, $S_a(\cdot, \cdot)$ and $S_b(\cdot, \cdot)$ are the δ-storage functions of Σ_a and Σ_b, respectively. ◊

[5] Here, each $\mathbf{0}$ denotes a zero-matrix of appropriate dimensions, which can be deduced by context.

Proof Using $S(\mathbf{z}, \bar{\mathbf{u}}) = S_a(\mathbf{z}_a, \mathbf{u}_a) + S_b(\mathbf{z}_b, \mathbf{u}_b)$, it follows that

$$\begin{aligned}
\dot{S}(\mathbf{z}, \bar{\mathbf{u}}) &= \dot{S}_a(\mathbf{z}_a, \mathbf{u}_a) + \dot{S}_b(\mathbf{z}_b, \mathbf{u}_b) \\
&\leq -\sigma_a(\mathbf{z}_a, \mathbf{u}_a, \mathbf{v}_a) - \sigma_b(\mathbf{z}_b, \mathbf{u}_b, \mathbf{v}_b) + \begin{bmatrix} \mathbf{v}_a \\ \mathbf{w}_a \end{bmatrix}^\top \mathbf{\Pi}_a \begin{bmatrix} \mathbf{v}_a \\ \mathbf{w}_a \end{bmatrix} + \begin{bmatrix} \mathbf{v}_b \\ \mathbf{w}_b \end{bmatrix}^\top \mathbf{\Pi}_b \begin{bmatrix} \mathbf{v}_b \\ \mathbf{w}_b \end{bmatrix} \\
&= -\sigma_a(\mathbf{z}_a, \mathbf{u}_a, \mathbf{v}_a) - \sigma_b(\mathbf{z}_b, \mathbf{u}_b, \mathbf{v}_b) + \begin{bmatrix} \mathbf{v}_a \\ \mathbf{v}_b \\ \mathbf{w}_a \\ \mathbf{w}_b \end{bmatrix}^\top \mathbf{\Pi} \begin{bmatrix} \mathbf{v}_a \\ \mathbf{v}_b \\ \mathbf{w}_a \\ \mathbf{w}_b \end{bmatrix} \\
&= -\sigma(\mathbf{z}, \bar{\mathbf{u}}, \bar{\mathbf{v}}) + \left\| \begin{bmatrix} \bar{\mathbf{v}} \\ \bar{\mathbf{w}} \end{bmatrix} \right\|_{\mathbf{\Pi}}^2,
\end{aligned}$$

which yields the desired result. Here, $\mathbf{w}_i := \mathbf{w}_i(\mathbf{z}_i, \mathbf{u}_i, \mathbf{v}_i)$, for all $i \in \{a, b\}$, and $\bar{\mathbf{v}} = \mathrm{col}(\mathbf{v}_a, \mathbf{v}_b)$ and $\bar{\mathbf{w}} = \mathrm{col}(\mathbf{w}_a, \mathbf{w}_b)$ play the roles of $\mathbf{v}$ and $\mathbf{w}(\cdot, \cdot, \cdot)$ for the concatenated system, respectively. □

Lemma 3.9 can be easily generalized to the parallel concatenation of $N \in \mathbb{Z}_{\geq 2}$ δ-dissipative systems, each characterized by its own (possibly distinct) $\mathbf{\Pi}_i$. To see this fact it suffices to apply Lemma 3.9 recursively, following the same procedure that was described to generalize Lemma 3.8. Specifically, pick two systems and apply Lemma 3.9 as it is. Then, view the concatenated pair as a new (bigger) system leaving a total of $N-1$ systems to be considered. Repeat this process $N-1$ times so that all systems are included into the parallel concatenation.

A trivial yet relevant simplification of Lemma 3.9 arises when both systems are either δ-passive or δ-antipassive. These results are highlighted in Corollaries 3.2 and 3.3, respectively.

Corollary 3.2 *Consider the parallel concatenation of Fig. 3.4 and suppose that both* Σ_a *and* Σ_b *are* δ*-passive characterized by* σ_a *and* σ_b*, respectively. Then, the concatenated system (from the input* $\bar{\mathbf{u}}$ *to the output* $\bar{\mathbf{y}}$*) is* δ*-passive characterized by* $\sigma(\mathbf{z}, \bar{\mathbf{u}}, \bar{\mathbf{v}}) = \sigma_a(\mathbf{z}_a, \mathbf{u}_a, \mathbf{v}_a) + \sigma_b(\mathbf{z}_b, \mathbf{u}_b, \mathbf{v}_b)$. ♢

Corollary 3.3 *Consider the parallel concatenation of Fig. 3.4 and suppose that both* Σ_a *and* Σ_b *are* δ*-antipassive characterized by* σ_a *and* σ_b*, respectively. Then, the concatenated system (from the input* $\bar{\mathbf{u}}$ *to the output* $\bar{\mathbf{y}}$*) is* δ*-antipassive characterized by* $\sigma(\mathbf{z}, \bar{\mathbf{u}}, \bar{\mathbf{v}}) = \sigma_a(\mathbf{z}_a, \mathbf{u}_a, \mathbf{v}_a) + \sigma_b(\mathbf{z}_b, \mathbf{u}_b, \mathbf{v}_b)$. ♢

Example 3.10 Recall again Examples 3.3 and 3.4, and consider the two systems:

$$\Sigma_a : \begin{cases} \dot{z}_a(t) = \alpha u_a(t) - \tau z_a(t) \\ y_a(t) = \alpha z_a(t) + \frac{\alpha\beta}{\tau} u_a(t) \end{cases}$$

$$\Sigma_b : \begin{cases} \dot{z}_b(t) = \left[\gamma - z_b(t)\right]_+ [u_b(t)]_+ - z_b(t)\,[-u_b(t)]_+ \\ y_b(t) = -z_b(t) - u_b(t), \end{cases}$$

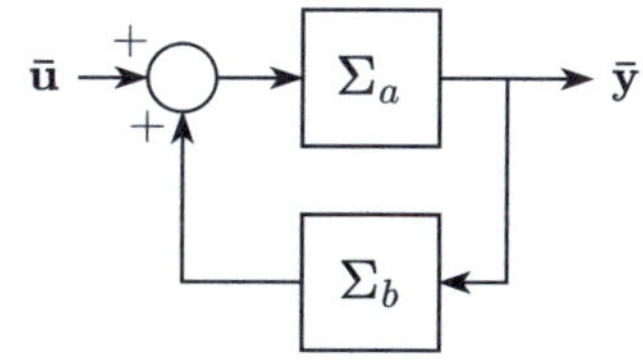

Fig. 3.5 Positive feedback interconnection of two systems

where $\alpha, \beta, \tau, \gamma \in \mathbb{R}_{>0}$ are constant parameters. Furthermore, consider the parallel concatenation of Fig. 3.4, which under the considered setup is given by

$$\Sigma : \begin{cases} \begin{bmatrix} \dot{z}_a(t) \\ \dot{z}_b(t) \end{bmatrix} = \begin{bmatrix} \alpha u_a(t) - \tau z_a(t) \\ \left[\gamma - z_b(t)\right]_+ [u_b(t)]_+ - z_b(t)\,[-u_b(t)]_+ \end{bmatrix} \\ \begin{bmatrix} y_a(t) \\ y_b(t) \end{bmatrix} = \begin{bmatrix} \alpha z_a(t) + \frac{\alpha\beta}{\tau} u_a(t) \\ -z_b(t) - u_b(t) \end{bmatrix}. \end{cases}$$

From Example 3.3, it follows that Σ_a is δ-passive characterized by some σ_a; while from Example 3.4, it holds that Σ_b is δ-antipassive characterized by some σ_b. Consequently, it follows from Lemma 3.9 that their parallel concatenation Σ is δ-dissipative characterized by $(\sigma, \mathbf{\Pi})$ with

$$\mathbf{\Pi} = \begin{bmatrix} 0 & 0 & \frac{1}{2} & 0 \\ 0 & 0 & 0 & -\frac{1}{2} \\ \frac{1}{2} & 0 & 0 & 0 \\ 0 & -\frac{1}{2} & 0 & 0 \end{bmatrix},$$

and $\sigma = \sigma_a + \sigma_b$ as established in Lemma 3.9. ◊

Positive Feedback Interconnection

Consider two systems, Σ_a and Σ_b, interconnected in positive feedback as depicted in Fig. 3.5. For each $i \in \{a, b\}$, let the corresponding system Σ_i have dynamics of the form (3.1) and let $\mathbf{z}_i \in \mathbb{R}^{d_{z_i}}$, $\mathbf{u}_i \in \mathbb{R}^{d_{u_i}}$, and $\mathbf{y}_i \in \mathbb{R}^{d_{y_i}}$ denote its state, input, and output vectors, respectively. Clearly, it must hold that $d_{u_a} = d_{y_b}$ and $d_{u_b} = d_{y_a}$. On the other hand, let $\bar{\mathbf{u}} \in \mathbb{R}^{d_{ua}}$ and $\bar{\mathbf{y}} = \mathbf{y}_a$ be the input and output vectors of the interconnected system, respectively, and let $\mathbf{z} = \operatorname{col}(\mathbf{z}_a, \mathbf{z}_b)$. Moreover, it is assumed that the interconnected system is well posed in the sense that $\mathbf{y}_a = \boldsymbol{\mathcal{Y}}_a(\mathbf{z}_a, \bar{\mathbf{u}} + \boldsymbol{\mathcal{Y}}_b(\mathbf{z}_b, \mathbf{y}_a))$ has the unique solution $\mathbf{y}_a = \bar{\boldsymbol{\mathcal{Y}}}(\mathbf{z}, \bar{\mathbf{u}})$, where $\bar{\boldsymbol{\mathcal{Y}}} : \mathbb{R}^{d_{xa}+d_{x_b}} \times \mathbb{R}^{d_{ua}} \to \mathbb{R}^{d_{ya}}$ is continuously differentiable.

Lemma 3.10 *Consider the positive feedback interconnection of Fig. 3.5 and suppose that Σ_a is δ-passive characterized by σ_a while Σ_b is δ-antipassive characterized by σ_b. Then, the interconnected system (from the input $\bar{\mathbf{u}}$ to the output $\bar{\mathbf{y}}$) is δ-passive characterized by σ with*

$$\sigma(\mathbf{z}, \bar{\mathbf{u}}, \bar{\mathbf{v}}) = \sigma_a(\mathbf{z}_a, \bar{\mathbf{u}} + \mathbf{y}_b, \bar{\mathbf{v}} + \mathbf{w}_b) + \sigma_b(\mathbf{z}_b, \mathbf{y}_a, \mathbf{w}_a),$$

where

$$
\begin{aligned}
\mathbf{y}_a &:= \mathcal{Y}_a\left(\mathbf{z}_a, \bar{\mathbf{u}} + \mathbf{y}_b\right) \\
\mathbf{y}_b &:= \mathcal{Y}_b\left(\mathbf{z}_b, \mathbf{y}_a\right) \\
\mathbf{w}_a &:= \mathrm{D}_{\mathbf{z}_a} \mathcal{Y}_a(\mathbf{z}_a, \bar{\mathbf{u}} + \mathbf{y}_b)\mathcal{Z}_a(\mathbf{z}_a, \bar{\mathbf{u}} + \mathbf{y}_b) + \mathrm{D}_{\mathbf{u}_a} \mathcal{Y}_a(\mathbf{z}_a, \bar{\mathbf{u}} + \mathbf{y}_b)\left(\bar{\mathbf{v}} + \mathbf{w}_b\right) \\
\mathbf{w}_b &:= \mathrm{D}_{\mathbf{z}_b} \mathcal{Y}_b(\mathbf{z}_b, \mathbf{y}_a)\mathcal{Z}_b(\mathbf{z}_b, \mathbf{y}_a) + \mathrm{D}_{\mathbf{u}_b} \mathcal{Y}_b(\mathbf{z}_b, \mathbf{y}_a)\mathbf{w}_a.
\end{aligned}
$$

Furthermore, $S(\mathbf{z}, \bar{\mathbf{u}}) = S_a(\mathbf{z}_a, \bar{\mathbf{u}} + \mathbf{y}_b) + S_b(\mathbf{z}_b, \mathbf{y}_a)$ *is a valid* δ*-storage function for the interconnected system. Here,* $S_a(\cdot, \cdot)$ *and* $S_b(\cdot, \cdot)$ *are the* δ*-storage functions of* Σ_a *and* Σ_b*, respectively.* ◊

Proof Using $S(\mathbf{z}, \bar{\mathbf{u}}) = S_a(\mathbf{z}_a, \bar{\mathbf{u}} + \mathbf{y}_b) + S_b(\mathbf{z}_b, \mathbf{y}_a)$, it follows that

$$
\begin{aligned}
\dot{S}(\mathbf{z}, \bar{\mathbf{u}}) &= \dot{S}_a(\mathbf{z}_a, \bar{\mathbf{u}} + \mathbf{y}_b) + \dot{S}_b(\mathbf{z}_b, \mathbf{y}_a) \\
&\le -\sigma_a(\mathbf{z}_a, \bar{\mathbf{u}} + \mathbf{y}_b, \mathbf{v}_a) - \sigma_b(\mathbf{z}_b, \mathbf{y}_a, \mathbf{v}_b) + \mathbf{v}_a^\top \mathbf{w}_a - \mathbf{v}_b^\top \mathbf{w}_b.
\end{aligned}
$$

Now, let $\bar{\mathbf{v}}$ and $\bar{\mathbf{w}}$ play the roles of $\mathbf{v}$ and $\mathbf{w}(\cdot, \cdot, \cdot)$ for the interconnected system, respectively. It then follows that

$$
\begin{aligned}
\mathbf{v}_a &:= \bar{\mathbf{v}} + \mathbf{w}_b \\
\mathbf{v}_b &:= \mathbf{w}_a \\
\mathbf{w}_a &:= \bar{\mathbf{w}}.
\end{aligned}
$$

Consequently, $\mathbf{v}_a^\top \mathbf{w}_a = \bar{\mathbf{v}}^\top \bar{\mathbf{w}} + \mathbf{w}_b^\top \mathbf{w}_a$, $\mathbf{v}_b^\top \mathbf{w}_b = \mathbf{w}_a^\top \mathbf{w}_b = \mathbf{w}_b^\top \mathbf{w}_a$, and

$$
\dot{S}(\mathbf{z}, \bar{\mathbf{u}}) \le -\sigma(\mathbf{z}, \bar{\mathbf{u}}, \bar{\mathbf{v}}) + \bar{\mathbf{v}}^\top \bar{\mathbf{w}},
$$

which yields the desired result. □

Lemma 3.10 characterizes perhaps the most important interconnection that would be considered throughout this book. As will be discussed in Chap. 4, the positive feedback interconnection of δ-passive and δ-antipassive systems is relevant to characterize the interconnection between several evolutionary and payoff dynamics models. Furthermore, when combined with Lemmas 3.4 and 3.10 yields a powerful tool to certify the stability properties of the closed-loop system.

3.4 Illustrative Examples

This section provides two examples illustrating the application of the various results derived in this chapter. The first example regards an unconstrained optimization problem, while the second shows how a seemingly complex system can be broken down into simpler subsystems to ease the analysis. While quite simple, these examples integrate the main ideas discussed throughout the chapter.

Unconstrained Optimization

Consider the unconstrained optimization problem given by

$$\max_{\mathbf{z}\in\mathbb{R}^n} \varphi(\mathbf{z}), \tag{3.27}$$

where $\varphi : \mathbb{R}^n \to \mathbb{R}$ is μ-strongly concave[6] and continuously differentiable. Assume that there exists a finite $\mathbf{z}^*$ (i.e., $\|\mathbf{z}^*\| < \infty$) such that

$$\mathbf{z}^* = \arg\max_{\mathbf{z}\in\mathbb{R}^n} \varphi(\mathbf{z}).$$

From strong concavity it follows that $\nabla_{\mathbf{z}}\varphi(\cdot)$ is μ-strongly contractive, i.e.,

$$\mathrm{D}_{\mathbf{z}} \left(\nabla_{\mathbf{z}}\varphi(\cdot)\right) \preceq -\mu\mathbf{I}_n.$$

Therefore, letting $\boldsymbol{\mathcal{Y}}(\cdot) = \nabla_{\mathbf{z}}\varphi(\cdot)$ and following the analysis in Example 3.1, one can conclude that the memoryless system $\mathbf{y}(t) = \boldsymbol{\mathcal{Y}}(\mathbf{u}(t))$ (with input $\mathbf{u}$ and output $\mathbf{y}$) is δ-antipassive characterized by $\sigma(\mathbf{v}) = \mu\mathbf{v}^\top\mathbf{v}$.

The δ-antipassivity of $\nabla_{\mathbf{z}}\varphi(\cdot)$ implies that to solve the optimization problem (3.27) it suffices to design a continuous-time maximization algorithm with δ-passive dynamics. Namely, since $\nabla_{\mathbf{z}}\varphi(\cdot)$ and the corresponding continuous-time maximization algorithm form a closed-loop system as in Fig. 3.5, the asymptotic stability of $\mathbf{z}^*$ follows from Lemmas 3.4 and 3.10.

Following these observations, consider the continuous-time gradient ascent dynamics given by

$$\dot{\mathbf{z}}(t) = \nabla_{\mathbf{z}}\varphi\left(\mathbf{z}(t)\right). \tag{3.28}$$

Note that the dynamics (3.28) are indeed equivalent to the feedback interconnection depicted in Fig. 3.5 (subject to $\bar{\mathbf{u}} = \mathbf{0}_n$) with:

$$\Sigma_a : \begin{cases} \dot{\mathbf{z}}_a(t) = \mathbf{u}_a(t) \\ \mathbf{y}_a(t) = \mathbf{z}_a(t) \end{cases}$$
$$\Sigma_b : \left\{ \mathbf{y}_b(t) = \nabla_{\mathbf{z}}\varphi\left(\mathbf{u}_b\right). \right.$$

Namely, Σ_a captures the optimization algorithm (the gradient ascent) while Σ_b represents the memoryless map specifying the maximization problem (the gradient of the objective function).

From the analysis in Example 3.1, it follows that the subsystem Σ_a is indeed δ-passive characterized by $\sigma_a = 0$. Consequently, from Lemma 3.10 it follows that the overall interconnected system (from input $\bar{\mathbf{u}}$ to output $\bar{\mathbf{y}}$) is δ-passive with

$$\sigma(\mathbf{z}_a, \bar{\mathbf{u}}) = \mu \left\|\bar{\mathbf{u}} + \nabla_{\mathbf{z}}\varphi(\mathbf{z}_a)\right\|_2^2.$$

[6] Here, $\varphi : \mathbb{R}^n \to \mathbb{R}$ is μ-strongly concave if and only if $\tilde{\varphi}(\mathbf{z}) = \varphi(\mathbf{z}) + (\mu/2)\mathbf{z}^\top\mathbf{z}$ is concave, for some $\mu \in \mathbb{R}_{>0}$.

Now, consider the set $\mathcal{E}_0 = \{\tilde{\mathbf{z}} \in \mathbb{R}^n : \nabla_{\mathbf{z}}\varphi(\tilde{\mathbf{z}}) = \mathbf{0}_n\}$. By the assumption of the existence of a finite solution to (3.27), it follows that $\mathcal{E}_0$ is nonempty and compact (in fact, $\mathcal{E}_0$ is a singleton due to strong concavity and boundedness). Finally, Lemma 3.4 implies that $\mathcal{E}_0$ is asymptotically stable under the considered interconnected system. Hence, the continuous-time gradient ascent dynamics (3.28) can indeed be used to solve the optimization problem (3.27).

Bounded Dual-Lagrangian-Based Dynamics

Motivated by the PDM introduced in Example 2.5, consider the dynamics given by

$$\dot{\mathbf{z}}_1(t) = [\overline{\mathbf{z}}_1 - \mathbf{z}_1(t)]_+ \odot \hat{\vartheta}\left(\mathbf{A}\mathbf{u}(t) - \mathbf{b}\right) - [\mathbf{z}_1(t) + \underline{\mathbf{z}}_1]_+ \odot \check{\vartheta}\left(\mathbf{b} - \mathbf{A}\mathbf{u}(t)\right), \tag{3.29a}$$

$$\dot{\mathbf{z}}_2(t) = [\overline{\mathbf{z}}_2 - \mathbf{z}_2(t)]_+ \odot \hat{\theta}\left(\boldsymbol{\phi}\left(\mathbf{u}(t)\right)\right) - \mathbf{z}_2(t) \odot \check{\theta}\left(-\boldsymbol{\phi}\left(\mathbf{u}(t)\right)\right), \tag{3.29b}$$

$$\mathbf{y}(t) = \boldsymbol{\mathcal{Y}}\left(\mathbf{u}(t)\right) - \mathbf{A}^\top \mathbf{z}_1(t) - \mathrm{D}_{\mathbf{u}}\, \boldsymbol{\phi}\left(\mathbf{u}(t)\right)^\top \mathbf{z}_2(t), \tag{3.29c}$$

where $\mathbf{z} := \left[\mathbf{z}_1^\top,\, \mathbf{z}_2^\top\right]^\top \in \mathbb{R}^{d_1+d_2}$; $\overline{\mathbf{z}}_1 \in \mathbb{R}^{d_1}_{>0}$, $\underline{\mathbf{z}}_1 \in \mathbb{R}^{d_1}_{>0}$ and $\overline{\mathbf{z}}_2 \in \mathbb{R}^{d_2}_{>0}$ are constant vectors with positive entries and $\underline{\mathbf{z}}_1 \preceq \overline{\mathbf{z}}_1$; the matrix $\mathbf{A} \in \mathbb{R}^{d_u \times d_u}$ and the vector $\mathbf{b} \in \mathbb{R}^{d_u}$ are arbitrary; the function $\boldsymbol{\phi} : \mathbb{R}^{d_u} \to \mathbb{R}^{d_u}$ is of the form $\boldsymbol{\phi}\left(\mathbf{u}\right) = \mathrm{col}\left(\phi_\ell(\mathbf{u})\right)_{\ell \in \{1,2,\ldots,d_2\}}$, with each $\phi_\ell : \mathbb{R}^{d_u} \to \mathbb{R}$ being a twice continuously differentiable convex function; the function $\boldsymbol{\mathcal{Y}} : \mathbb{R}^{d_u} \to \mathbb{R}^{d_u}$ is continuously differentiable and satisfies $\mathrm{D}_{\mathbf{u}}\, \boldsymbol{\mathcal{Y}}(\mathbf{u}) \preceq 0$, for all $\mathbf{u} \in \mathbb{R}^{d_u}$; the symbol $\odot$ denotes the Hadamard product; and each function

$$\xi(\cdot) \in \left\{\hat{\vartheta}(\cdot),\, \check{\vartheta}(\cdot),\, \hat{\theta}(\cdot),\, \check{\theta}(\cdot)\right\}$$

is a Lipschitz continuous scalar function $\xi : \mathbb{R} \to \mathbb{R}_{\geq 0}$, applied element-wise as

$$\xi(\boldsymbol{\zeta}) := \begin{bmatrix} \xi(\zeta_1) \\ \xi(\zeta_2) \\ \vdots \\ \xi(\zeta_{d_i}) \end{bmatrix}, \qquad \forall \boldsymbol{\zeta} = \begin{bmatrix} \zeta_1 \\ \zeta_2 \\ \vdots \\ \zeta_{d_i} \end{bmatrix} \in \mathbb{R}^{d_i},$$

and satisfies $\xi(\alpha) > 0$ if $\alpha > 0$, and $\xi(\alpha) = 0$ if $\alpha \leq 0$, for all $\alpha \in \mathbb{R}$. A canonical example satisfying these properties would be $\xi(\cdot) = [\cdot]_+$, applied element-wise as described.

The goal in this example is to investigate the δ-dissipativity properties of (3.29) by breaking it into smaller subsystems and applying the results of Sect. 3.3. Specifically, it will be shown that (3.29) is δ-antipassive with an informative σ.

Granted that the initial condition of (3.29) satisfies

$$-\underline{\mathbf{z}}_1 \preceq \mathbf{z}_1(0) \preceq \overline{\mathbf{z}}_1 \tag{3.30a}$$

$$\mathbf{0}_{d_2} \preceq \mathbf{z}_2(0) \preceq \overline{\mathbf{z}}_2, \tag{3.30b}$$

it then follows that

$$\begin{aligned} -\underline{\mathbf{z}}_1 &\preceq \mathbf{z}_1(t) \preceq \overline{\mathbf{z}}_1, \\ \mathbf{0}_{d_2} &\preceq \mathbf{z}_2(t) \preceq \overline{\mathbf{z}}_2, \end{aligned}$$

for all $t \geq 0$. To see this fact, note that, subject to appropriate initial conditions (3.30), the following implications hold for all $t \geq 0$:

$$\begin{aligned} \mathbf{z}_1(t) = -\underline{\mathbf{z}}_1 &\quad \Rightarrow \quad \dot{\mathbf{z}}_1(t) \succeq \mathbf{0}_{d_1} \\ \mathbf{z}_1(t) = \overline{\mathbf{z}}_1 &\quad \Rightarrow \quad \dot{\mathbf{z}}_1(t) \preceq \mathbf{0}_{d_1} \\ \mathbf{z}_2(t) = \mathbf{0}_d &\quad \Rightarrow \quad \dot{\mathbf{z}}_2(t) \succeq \mathbf{0}_{d_2} \\ \mathbf{z}_2(t) = \overline{\mathbf{z}}_2 &\quad \Rightarrow \quad \dot{\mathbf{z}}_2(t) \preceq \mathbf{0}_{d_2}. \end{aligned}$$

Therefore, every trajectory of the considered system is confined within the box-like constraints: $\left[-\underline{\mathbf{z}}_1, \overline{\mathbf{z}}_1\right] \times \left[\mathbf{0}_{d_2}, \overline{\mathbf{z}}_2\right]$. Thus, assuming that the initial condition satisfies (3.30), the relevant domain of analysis is restricted to $\left[-\underline{\mathbf{z}}_1, \overline{\mathbf{z}}_1\right] \times \left[\mathbf{0}_{d_2}, \overline{\mathbf{z}}_2\right]$.

Accordingly, subject to the initial condition constraint (3.30), note that (3.29) can be rewritten equivalently as

$$\dot{\mathbf{z}}_1(t) = (\overline{\mathbf{z}}_1 - \mathbf{z}_1(t)) \odot \hat{\vartheta}\,(\mathbf{A}\mathbf{u}(t) - \mathbf{b}) - (\mathbf{z}_1(t) + \underline{\mathbf{z}}_1) \odot \check{\vartheta}\,(\mathbf{b} - \mathbf{A}\mathbf{u}(t)), \tag{3.32a}$$

$$\dot{\mathbf{z}}_2(t) = (\overline{\mathbf{z}}_2 - \mathbf{z}_2(t)) \odot \hat{\theta}\,(\boldsymbol{\phi}\,(\mathbf{u}(t))) - \mathbf{z}_2(t) \odot \check{\theta}\,(-\boldsymbol{\phi}\,(\mathbf{u}(t))), \tag{3.32b}$$

$$\mathbf{y}(t) = \boldsymbol{\mathcal{Y}}\,(\mathbf{u}(t)) - \mathbf{A}^\top \mathbf{z}_1(t) - \mathrm{D}_{\mathbf{u}}\,\boldsymbol{\phi}\,(\mathbf{u}(t))^\top\, \mathbf{z}_2(t). \tag{3.32c}$$

Furthermore, note that (3.32) is equivalent to the parallel interconnection of the three systems:

$$\Sigma_0 : \left\{\mathbf{y}_0(t) = \boldsymbol{\mathcal{Y}}(\mathbf{u})\right.$$

$$\Sigma_1 : \begin{cases} \dot{\mathbf{z}}_1(t) = (\overline{\mathbf{z}}_1 - \mathbf{z}_1(t)) \odot \hat{\vartheta}\,(\mathbf{A}\mathbf{u}(t) - \mathbf{b}) - (\mathbf{z}_1(t) + \underline{\mathbf{z}}_1) \odot \check{\vartheta}\,(\mathbf{b} - \mathbf{A}\mathbf{u}(t)) \\ \mathbf{y}_1(t) = -\mathbf{A}^\top \mathbf{z}_1(t) \end{cases}$$

$$\Sigma_2 : \begin{cases} \dot{\mathbf{z}}_2(t) = (\overline{\mathbf{z}}_2 - \mathbf{z}_2(t)) \odot \hat{\theta}\,(\boldsymbol{\phi}\,(\mathbf{u}(t))) - \mathbf{z}_2(t) \odot \check{\theta}\,(-\boldsymbol{\phi}\,(\mathbf{u}(t))) \\ \mathbf{y}_2(t) = -\,\mathrm{D}_{\mathbf{u}}\,\boldsymbol{\phi}\,(\mathbf{u}(t))^\top\, \mathbf{z}_2(t). \end{cases}$$

Specifically, such a parallel interconnected system takes $\mathbf{u}(t)$ as input and yields $\mathbf{y}(t) = \mathbf{y}_0(t) + \mathbf{y}_1(t) + \mathbf{y}_2(t)$ as output. According to Lemma 3.8, the interconnected system is δ-antipassive granted that each of its components is δ-antipassive. As such, one may investigate the δ-dissipativity properties of each subsystem independently.

Subsystem Σ_0: Subsystem Σ_0 is a memoryless system characterized by the function $\boldsymbol{\mathcal{Y}}(\cdot)$. Following Example 3.1, the property $\mathrm{D}_{\mathbf{u}}\boldsymbol{\mathcal{Y}}(\mathbf{u}) \preceq 0$ implies that Σ_0 is δ-antipassive with an informative σ (recall that σ is always informative for every memoryless δ-dissipative system). ◊

Subsystem Σ_1*:* Regarding Σ_1, note that it is itself the parallel interconnection of d_1 subsystems of the form

$$\Sigma_{1,\ell}: \begin{cases} \dot{z}_{1,\ell}(t) = \left(\overline{z}_{1,\ell} - z_{1,\ell}(t)\right) \hat{\vartheta}\left(\mathbf{a}_\ell^\top \mathbf{u}(t) - b_\ell\right) - (z_{1,\ell}(t) + \underline{z}_{1,\ell})\check{\vartheta}\left(b_\ell - \mathbf{a}_\ell^\top \mathbf{u}(t)\right) \\ \mathbf{y}_{1,\ell}(t) = -\mathbf{a}_\ell z_{1,\ell}(t), \end{cases}$$

where $\mathbf{a}_\ell^\top$ denotes the ℓth row of $\mathbf{A}$, for all $\ell \in \{1, 2, \ldots, d_1\}$. Furthermore, each subsystem $\Sigma_{1,\ell}$ can be seen as the *cascaded encapsulation* (as in Fig. 3.2) of the scalar system

$$\tilde{\Sigma}_{1,\ell}: \begin{cases} \dot{z}_{1,\ell}(t) = \left(\overline{z}_{1,\ell} - z_{1,\ell}(t)\right) \hat{\vartheta}\left(u_{1,\ell}(t)\right) - (z_{1,\ell}(t) + \underline{z}_{1,\ell})\check{\vartheta}\left(-u_{1,\ell}(t)\right) \\ y_{1,\ell}(t) = -z_{1,\ell}(t), \end{cases}$$

by the affine function $\phi_\ell(\mathbf{u}) = \mathbf{a}_\ell^\top \mathbf{u} - b_\ell$. Here, $u_{1,\ell} \in \mathbb{R}$ is the input of $\tilde{\Sigma}_{1,\ell}$. Consequently, by applying Lemmas 3.7 and 3.8, it suffices to establish the δ-antipassivity of $\tilde{\Sigma}_{1,\ell}$ in order to conclude the δ-antipassivity of Σ_1.

Accordingly, let $z_{1,\ell}(0) \in [-\underline{z}_{1,\ell}, \overline{z}_{1,\ell}]$ and take

$$\begin{aligned} S(z_{1,\ell}, u_{1,\ell}) &= (\overline{z}_{1,\ell} - z_{1,\ell}) \int_0^{u_{1,\ell}} \hat{\vartheta}(\zeta)\, d\zeta + (z_{1,\ell} + \underline{z}_{1,\ell}) \int_0^{-u_{1,\ell}} \check{\vartheta}(\zeta)\, d\zeta \\ \sigma(z_{1,\ell}, u_{1,\ell}) &= \left(\int_0^{u_{1,\ell}} \hat{\vartheta}(\zeta)\, d\zeta - \int_0^{-u_{1,\ell}} \check{\vartheta}(\zeta)\, d\zeta\right) \mathcal{Z}_{1,\ell}(z_{1,\ell}, u_{1,\ell}) \end{aligned}$$

as the δ-storage and auxiliary σ functions for $\tilde{\Sigma}_{1,\ell}$, respectively. Here,

$$\mathcal{Z}_{1,\ell}(z_{1,\ell}, u_{1,\ell}) = (\overline{z}_{1,\ell} - z_{1,\ell})\hat{\vartheta}(u_{1,\ell}) - (z_{1,\ell} + \underline{z}_{1,\ell})\check{\vartheta}(-u_{1,\ell}).$$

Note that both $S(\cdot,\cdot)$ and $\sigma(\cdot,\cdot)$ are always non-negative over the relevant domain $[-\underline{z}_{1,\ell}, \overline{z}_{1,\ell}]$, and they satisfy (3.2a) and (3.3), respectively. Also, it follows that

$$\begin{aligned} \frac{\partial S(z_{1,\ell}, u_{1,\ell})}{\partial z_{1,\ell}} &= -\int_0^{u_{1,\ell}} \hat{\vartheta}(\zeta)\, d\zeta + \int_0^{-u_{1,\ell}} \check{\vartheta}(\zeta)\, d\zeta \\ \frac{\partial S(z_{1,\ell}, u_{1,\ell})}{\partial u_{1,\ell}} &= (\overline{z}_{1,\ell} - z_{1,\ell})\hat{\vartheta}(u_{1,\ell}) - (z_{1,\ell} + \underline{z}_{1,\ell})\check{\vartheta}(-u_{1,\ell}) \\ &= \mathcal{Z}_{1,\ell}(z_{1,\ell}, u_{1,\ell}). \end{aligned}$$

Thus, for $\tilde{\Sigma}_{1,\ell}$ to be δ-antipassive it must hold that

$$\frac{\partial S(z_{1,\ell}, u_{1,\ell})}{\partial z_{1,\ell}} \mathcal{Z}_{1,\ell} + \frac{\partial S(z_{1,\ell}, u_{1,\ell})}{\partial u_{1,\ell}} v \leq -\sigma(z_{1,\ell}, u_{1,\ell}) - \left(-\mathcal{Z}_{1,\ell}\right) v,$$

where $\mathcal{Z}_{1,\ell} := \mathcal{Z}_{1,\ell}(z_{1,\ell}, u_{1,\ell})$. Replacing the partial derivatives and simplifying yields the equivalent condition:

$$-\left(\int_0^{u_{1,\ell}} \hat{\vartheta}(\zeta)\,d\zeta - \int_0^{-u_{1,\ell}} \check{\vartheta}(\zeta)\,d\zeta\right) \mathcal{Z}_{1,\ell} \le -\sigma(z_{1,\ell}, u_{1,\ell}),$$

which holds with equality by construction. ◊

Subsystem Σ_2*:* Similar to Σ_1, subsystem Σ_2 is itself the parallel interconnection of $c_{\le}$ subsystems of the form

$$\Sigma_{2,\ell}:\begin{cases}\dot{z}_{2,\ell}(t) = \left(\overline{z}_{2,\ell} - z_{2,\ell}(t)\right)\hat{\theta}\left(\phi_\ell\left(\mathbf{u}(t)\right)\right) - z_{2,\ell}(t)\check{\theta}\left(-\phi_\ell\left(\mathbf{u}(t)\right)\right)\\ \mathbf{y}_{2,\ell}(t) = -\nabla_{\mathbf{u}}\phi_\ell\left(\mathbf{u}(t)\right)z_{2,\ell}(t),\end{cases}$$

where $\phi_\ell(\cdot)$ is the ℓth element of $\boldsymbol{\phi}(\cdot)$, for all $\ell \in \{1, 2, \ldots, d_2\}$. Furthermore, each subsystem $\Sigma_{2,\ell}$ can be seen as the *cascaded encapsulation* (as in Fig. 3.2) of the scalar system

$$\tilde{\Sigma}_{2,\ell}:\begin{cases}\dot{z}_{2,\ell}(t) = \left(\overline{z}_{2,\ell} - z_{2,\ell}(t)\right)\hat{\theta}\left(u_{2,\ell}(t)\right) - z_{2,\ell}(t)\check{\theta}\left(-u_{2,\ell}(t)\right)\\ y_{2,\ell}(t) = -z_{2,\ell}(t),\end{cases}$$

by the convex function $\phi_\ell(\mathbf{u}) = \phi_\ell(\mathbf{u})$. Here, $u_{2,\ell} \in \mathbb{R}$ is the input of $\tilde{\Sigma}_{2,\ell}$. Consequently, by applying Lemmas 3.7 and 3.8, and using the fact that $\tilde{\Sigma}_{2,\ell}$ is positive system, it suffices to establish the δ-antipassivity of $\tilde{\Sigma}_{2,\ell}$ in order to conclude the δ-antipassivity of Σ_2.

Letting $z_{2,\ell}(0) \in [0, \overline{z}_{2,\ell}]$ and focusing only on the relevant domain $[0, \overline{z}_{2,\ell}]$, the δ-antipassivity of $\tilde{\Sigma}_{2,\ell}$ can be proven by taking

$$S(z_{2,\ell}, u_{2,\ell}) = (\overline{z}_{2,\ell} - z_{2,\ell})\int_0^{u_{2,\ell}} \hat{\theta}(\zeta)\,d\zeta + z_{2,\ell}\int_0^{-u_{2,\ell}} \check{\theta}(\zeta)\,d\zeta$$
$$\sigma(z_{2,\ell}, u_{2,\ell}) = \left(\int_0^{u_{2,\ell}} \hat{\theta}(\zeta)\,d\zeta - \int_0^{-u_{2,\ell}} \check{\theta}(\zeta)\,d\zeta\right)\mathcal{Z}_{2,\ell}(z_{2,\ell}, u_{2,\ell}),$$

with

$$\mathcal{Z}_{2,\ell}(z_{2,\ell}, u_{2,\ell}) = (\overline{z}_{2,\ell} - z_{2,\ell})\hat{\theta}(u_{2,\ell}) - z_{2,\ell}\check{\theta}(-u_{2,\ell}).$$

The proof is virtually identical to that of $\tilde{\Sigma}_{1,\ell}$ and is thus omitted. ◊

Based on the previous analysis, one can conclude that the overall system (3.29) is indeed δ-antipassive.

Notes

The concept of δ-dissipativity was first introduced in [2], and the term was later formalized in [1, 6]. It serves as a generalization of the classical dissipativity theory developed in [11, 12], with a focus on the time derivatives of input–output signals.

Several related notions have been proposed in the literature, including Krasovskii passivity [5], differential passivity [10], and incremental passivity [8], among others. For comparative discussions of these concepts, see [7, Sect. VI-D] and [4, Sect. II].

Applications of δ-dissipativity to the analysis of interconnected systems are further explored in [9]. In fact, a quite general result on such a matter is given by [9, Theorem 1], which considers arbitrary (well posed) interconnections of δ-dissipative systems and the δ-dissipativity of the interconnection is asserted by (numerically) solving a linear matrix inequality.

References

1. Arcak M, Martins NC (2021) Dissipativity tools for convergence to Nash equilibria in population games. IEEE Trans Control Netw Syst 8(1):39–50. https://doi.org/10.1109/TCNS.2020.3029990
2. Fox MJ, Shamma JS (2013) Population games, stable games, and passivity. Games 4(4):561–583. https://doi.org/10.3390/g4040561
3. Haddad WM, Chellaboina V (2008) Nonlinear dynamical systems and control: a Lyapunov-based approach. Princeton University Press
4. Kawano Y, Kosaraju KC, Scherpen JMA (2021) Krasovskii and shifted passivity-based control. IEEE Trans Autom Control 66(10):4926–4932. https://doi.org/10.1109/TAC.2020.3040252
5. Kosaraju KC, Kawano Y, Scherpen JM (2019) Krasovskii's passivity. IFAC-PapersOnLine 52(16):466–471. Proceedings of the 11th IFAC symposium on nonlinear control systems (NOLCOS)
6. Park S, Martins NC, Shamma JS (2019) From population games to payoff dynamics models: a passivity-based approach. In: Proceedings of the 58th IEEE conference on decision and control (CDC), pp 6584–6601. https://doi.org/10.1109/CDC40024.2019.9029756
7. Park S, Martins NC, Shamma JS (2019) Payoff dynamics model and evolutionary dynamics model: feedback and convergence to equilibria. https://doi.org/10.48550/ARXIV.1903.02018. arXiv:1903.02018
8. Pavlov A, Marconi L (2006) Incremental passivity and output regulation. In: Proceedings of the 45th IEEE conference on decision and control, pp 5263–5268. https://doi.org/10.1109/CDC.2006.377803
9. Schweidel KS, Arcak M (2022) Compositional analysis of interconnected systems using Delta dissipativity. IEEE Control Syst Lett 6:662–667. https://doi.org/10.1109/LCSYS.2021.3084974
10. van der Schaft A (2013) On differential passivity. IFAC Proc Vol 46(23):21–25. 9th IFAC symposium on nonlinear control systems
11. Willems JC (1972) Dissipative dynamical systems part i: general theory. Arch Ration Mech Anal 45(5):321–351
12. Willems JC (1972) Dissipative dynamical systems part ii: linear systems with quadratic supply rates. Arch Ration Mech Anal 45:352–393

Chapter 4
Convergence of Deterministic Dynamics Models

This chapter builds on the modeling framework introduced in Chap. 2 and the theoretical tools developed in Chap. 3 to investigate the convergence properties of EDM–PDM systems.

More precisely, building on the developments of Chap. 2, the present chapter rigorously establishes sufficient conditions under which the equilibrium set of an EDM–PDM system:

(i) coincides with the set of generalized Nash equilibria of the underlying population game; and
(ii) is asymptotically stable.

Together, these conditions certify the convergence of EDM–PDM dynamics to a GNE, thereby providing a predictor of the society's long-term strategic behavior.

Furthermore, leveraging δ-dissipativity as a key analytical concept for stability analysis in interconnected systems, the results of Chap. 3 are employed to derive convergence guarantees for broad families of EDM–PDM interconnections. These results enable the unified analysis of diverse revision processes and payoff mechanisms.

As an illustration, the representative classes of EDMs and the smoothing-anticipatory PDM introduced in Chap. 2 are revisited. It is shown that, under suitable sufficient conditions (which are fully characterized), every interconnection of these models converges to an NE. That is, to a GNE in the unconstrained setting with $\Omega = \mathbb{R}^n$. These results serve as a motivation for the more general constrained case that will be addressed in Chap. 5.

The remainder of this chapter is structured as follows. Section 4.1 characterizes the equilibrium set of EDM–PDM systems and formalizes its relationship with the GNE set of the underlying population game. Section 4.2 establishes sufficient conditions to ensure Lyapunov and asymptotic stability of this equilibrium set. Section 4.3 presents simplified sufficient conditions tailored to unconstrained game settings. Finally, Sect. 4.4 offers simulation-based validations of the theoretical results.

J. Martinez-Piazuelo et al., *Generalized Nash Equilibrium Seeking in Population Games*, Communications and Control Engineering,
https://doi.org/10.1007/978-3-032-06081-5_4

4.1 Characterization of the Equilibrium Set

This section revisits the relationship between the equilibrium set $\mathcal{E}$ of EDM–PDM systems and the set GNE($\mathbf{f}$) of the underlying population game $G = (\mathcal{P}, \mathcal{X}, \mathbf{f}(\cdot))$. In particular, the formal connections established in Lemmas 2.8, 2.9 and Corollary 2.2 are reformulated in a more technical form, tailored to the developments of this chapter. The discussion relies on Definitions 2.4, 2.5, 2.10, 2.11, and 2.12; and thus the reader is encouraged to review these definitions before proceeding.

To introduce the theoretical results of this chapter, some necessary and sufficient conditions are first established for a strategic distribution $\mathbf{x}^*$ to qualify as a GNE.[1] To this end, recall that

$$\mathrm{GNE}(\mathbf{f}) = \mathrm{SOL}\,(\mathcal{X}, -\mathbf{f})\,,$$

where SOL($\mathcal{X}, -\mathbf{f}$) denotes the solution set of the variational inequality VI $(\mathcal{X}, -\mathbf{f})$, defined as:

$$\text{find } \mathbf{x}^* \in \mathcal{X} \quad : \quad \left(\mathbf{x} - \mathbf{x}^*\right)^\top \left(-\mathbf{f}(\mathbf{x}^*)\right) \geq 0, \quad \text{for all } \mathbf{x} \in \mathcal{X}.$$

Accordingly, $\mathbf{x}^* \in$ GNE($\mathbf{f}$) if and only if[2] there exist multipliers $\boldsymbol{\mu}^* \in \mathbb{R}^{c_=}$ and $\boldsymbol{\lambda}^* \in \mathbb{R}^{c_\leq}$ such that

$$\mathbf{x}^* \in \underset{\mathbf{x} \in \Delta}{\operatorname{argmax}} \left(\mathbf{x}^\top \mathbf{f}\left(\mathbf{x}^*\right) - \mathbf{x}^\top \mathbf{A}^\top \boldsymbol{\mu}^* - \mathbf{g}\left(\mathbf{x}\right)^\top \boldsymbol{\lambda}^*\right) \tag{4.1a}$$

$$\mathbf{x}^* \in \Omega \tag{4.1b}$$

$$\boldsymbol{\lambda}^* \odot \mathbf{g}\left(\mathbf{x}^*\right) = \mathbf{0}_{c_\leq} \tag{4.1c}$$

$$\boldsymbol{\lambda}^* \succeq \mathbf{0}_{c_\leq}. \tag{4.1d}$$

Expressions in (4.1) correspond to the Karush–Kuhn–Tucker (KKT) conditions associated with the solution set SOL($\mathcal{X}, -\mathbf{f}$), formulated with respect to the constraint set Ω. The remaining constraints, related to Δ, are addressed separately by the stationarity condition in (4.1a). Accordingly, a tuple $(\mathbf{x}^*, \boldsymbol{\mu}^*, \boldsymbol{\lambda}^*)$ is referred to as a KKT point associated with GNE($\mathbf{f}$) if and only if it satisfies (4.1).

Building on this characterization, sufficient conditions on the EDM–PDM pair are now provided to establish the connection between the sets $\mathcal{E}$ and GNE($\mathbf{f}$), as stated next.

---

[1] Remember that, as established in Definition 2.12 and Remark 2.9, the notion of GNE adopted in this book corresponds to that of a variational GNE.

[2] A proof of this result can be found in [2, Proposition 1.3.4].

Proposition 4.1 *Consider a population game* $G = (\mathcal{P}, \mathcal{X}, \mathbf{f}(\cdot))$ *and an EDM–PDM pair subject to the following conditions:*

(C1) *The EDM is Nash stationary, i.e., it satisfies the Nash stationarity property:*

$$\mathcal{V}\left(\mathbf{x}^*, \mathbf{p}^*\right) = \mathbf{0}_n \Leftrightarrow \mathbf{x}^* \in \arg\max_{\mathbf{x}\in\Delta} \mathbf{x}^\top \mathbf{p}^*.$$

(C2) *For every* $(\mathbf{q}^*, \mathbf{x}^*) \in \mathbb{R}^d \times \Delta$*, the PDM satisfies that* $\mathcal{W}(\mathbf{q}^*, \mathbf{x}^*) = \mathbf{0}_d$ *implies:*

$$\begin{aligned}\mathcal{H}\left(\mathbf{q}^*, \mathbf{x}^*\right) &= \mathbf{f}\left(\mathbf{x}^*\right) - \mathbf{A}^\top \mathbf{M}_\mu \mathbf{q}^* - \mathrm{D}\mathbf{g}\left(\mathbf{x}^*\right)^\top \mathbf{M}_\lambda \mathbf{q}^* \\ \left(\mathbf{M}_\lambda \mathbf{q}^*\right) \odot \mathbf{g}\left(\mathbf{x}^*\right) &= \mathbf{0}_{c_\leq} \\ \mathbf{M}_\lambda \mathbf{q}^* &\succeq \mathbf{0}_{c_\leq} \\ \mathbf{x}^* &\in \Omega,\end{aligned}$$

for some matrices $\mathbf{M}_\mu \in \mathbb{R}^{c_= \times d}$ *and* $\mathbf{M}_\lambda \in \mathbb{R}^{c_\leq \times d}$.

Then, every $(\mathbf{x}^*, \mathbf{q}^*) \in \mathcal{E}$ *implies that* $\left(\mathbf{x}^*, \mathbf{M}_\mu \mathbf{q}^*, \mathbf{M}_\lambda \mathbf{q}^*\right)$ *is a KKT point associated with* GNE($\mathbf{f}$)*. Specifically,* $\left(\mathbf{x}^*, \mathbf{M}_\mu \mathbf{q}^*, \mathbf{M}_\lambda \mathbf{q}^*\right)$ *satisfies (4.1), with* $\boldsymbol{\mu}^* = \mathbf{M}_\mu \mathbf{q}^*$ *and* $\boldsymbol{\lambda}^* = \mathbf{M}_\lambda \mathbf{q}^*$*. Consequently,* $(\mathbf{x}^*, \mathbf{q}^*) \in \mathcal{E}$ *implies that* $\mathbf{x}^* \in$ GNE($\mathbf{f}$). ◊

Proof Let $(\mathbf{x}^*, \mathbf{q}^*) \in \mathcal{E}$. From (C1) it follows that

$$\mathbf{x}^* \in \operatorname*{argmax}_{\mathbf{x}\in\Delta} \left(\mathbf{x}^\top \mathcal{H}\left(\mathbf{q}^*, \mathbf{x}^*\right)\right).$$

This fact in conjunction with (C2) yield that the KKT conditions in (4.1) are satisfied with $\boldsymbol{\mu}^* = \mathbf{M}_\mu \mathbf{q}^*$ and $\boldsymbol{\lambda}^* = \mathbf{M}_\lambda \mathbf{q}^*$, which implies that $\left(\mathbf{x}^*, \mathbf{M}_\mu \mathbf{q}^*, \mathbf{M}_\lambda \mathbf{q}^*\right)$ is a KKT point associated with GNE($\mathbf{f}$) and $\mathbf{x}^* \in$ GNE($\mathbf{f}$). □

Proposition 4.1 establishes sufficient conditions under which the $\mathbf{x}$-component of the equilibrium set $\mathcal{E}$ of a given EDM–PDM pair coincides exactly with the set GNE($\mathbf{f}$). As specified in Condition (C1), this correspondence relies on the Nash stationarity property of the EDM.

Following Lemmas 2.3, 2.4 and Corollary 2.1, the Nash stationarity property in Condition (C1) is satisfied by EDMs corresponding to societies where all populations behave according to either SEPT or IPC revision protocols, but does not hold for EDMs describing societies where at least one population uses a PBR revision protocol. Consequently, the result of Proposition 4.1 does not apply to the PBR mixed scenario.

To allow the consideration of this latter setting, one should extend Proposition 4.1 to the notion of ϖ-perturbed Nash stationarity of Definition 2.10, which jointly addresses scenarios with mixed societies comprised of populations using different classes of revision protocols (recall Corollary 2.1). To this end, regard the variational inequality VI($\mathcal{X}, -\tilde{\mathbf{f}}$), defined as:

$$\text{find } \mathbf{x}^* \in \mathcal{X} \quad : \quad \left(\mathbf{x} - \mathbf{x}^*\right)^\top \left(-\tilde{\mathbf{f}}(\mathbf{x}^*)\right) \geq 0, \quad \text{for all } \mathbf{x} \in \mathcal{X},$$

where

$$\tilde{\mathbf{f}}(\mathbf{x}) = \mathbf{f}(\mathbf{x}) - \nabla_{\mathbf{x}} \varpi(\mathbf{x}), \tag{4.2}$$

with $\varpi(\cdot)$ taken as in Definition 2.10. Then, $\mathbf{x}^* \in \text{GNE}(\tilde{\mathbf{f}}) = \text{SOL}(\mathcal{X}, -\tilde{\mathbf{f}})$ if and only if there exist multipliers $\boldsymbol{\mu}^*$ and $\boldsymbol{\lambda}^*$ such that

$$\mathbf{x}^* \in \arg \max_{\mathbf{x} \in \text{int}_{\varpi}(\Delta)} \left(\mathbf{x}^\top \tilde{\mathbf{f}}(\mathbf{x}^*) - \mathbf{x}^\top \mathbf{A}^\top \boldsymbol{\mu}^* - \mathbf{g}(\mathbf{x})^\top \boldsymbol{\lambda}^* \right) \tag{4.3a}$$

$$\mathbf{x}^* \in \Omega \tag{4.3b}$$

$$\boldsymbol{\lambda}^* \odot \mathbf{g}(\mathbf{x}^*) = \mathbf{0}_{c_\leq} \tag{4.3c}$$

$$\boldsymbol{\lambda}^* \succeq \mathbf{0}_{c_\leq}, \tag{4.3d}$$

where $\text{int}_{\varpi}(\Delta)$ is implicitly enforced by the domain of $\varpi(\cdot)$ and is defined according to Definition 2.10. As before, a tuple $(\mathbf{x}^*, \boldsymbol{\mu}^*, \boldsymbol{\lambda}^*)$ is referred to as a KKT point associated with $\text{GNE}(\tilde{\mathbf{f}})$ if and only if it satisfies (4.3).

Building on this reformulation, a generalization of Proposition 4.1 for societies with mixed classes of revision protocols is presented next in Proposition 4.2.

Proposition 4.2 *Consider a population game $G = (\mathcal{P}, \mathcal{X}, \mathbf{f}(\cdot))$ and an EDM–PDM pair where:*

(A) *The EDM satisfies the ϖ-perturbed Nash stationarity property:*

$$\mathcal{V}(\mathbf{x}^*, \mathbf{p}^*) = \mathbf{0}_n \Leftrightarrow \mathbf{x}^* \in \arg\max_{\mathbf{x} \in \text{int}_{\varpi}(\Delta)} \left(\mathbf{x}^\top \mathbf{p}^* - \varpi(\mathbf{x}) \right),$$

with $\varpi(\cdot)$ and $\text{int}_{\varpi}(\Delta)$ defined as in Definition 2.10;

and the PDM satisfies Condition (C2). Then, every $(\mathbf{x}^*, \mathbf{q}^*) \in \mathcal{E}$ implies that $(\mathbf{x}^*, \mathbf{M}_\mu \mathbf{q}^*, \mathbf{M}_\lambda \mathbf{q}^*)$ is a KKT point associated with $\text{GNE}(\tilde{\mathbf{f}})$, where $\tilde{\mathbf{f}}(\cdot)$ is given in (4.2). More formally, the point $(\mathbf{x}^*, \mathbf{M}_\mu \mathbf{q}^*, \mathbf{M}_\lambda \mathbf{q}^*)$ satisfies (4.3), with $\boldsymbol{\mu}^* = \mathbf{M}_\mu \mathbf{q}^*$ and $\boldsymbol{\lambda}^* = \mathbf{M}_\lambda \mathbf{q}^*$. Consequently, $(\mathbf{x}^*, \mathbf{q}^*) \in \mathcal{E}$ implies that $\mathbf{x}^* \in \text{GNE}(\tilde{\mathbf{f}})$. ◊

Proof Let $(\mathbf{x}^*, \mathbf{q}^*) \in \mathcal{E}$. From Corollary 2.1, it follows that

$$\mathbf{x}^* \in \arg\max_{\mathbf{x} \in \text{int}_{\varpi}(\Delta)} \left(\mathbf{x}^\top \mathcal{H}(\mathbf{q}^*, \mathbf{x}^*) - \varpi(\mathbf{x}) \right).$$

This fact, together with Condition (C2), imply that the KKT conditions in (4.3) are satisfied with $\boldsymbol{\mu}^* = \mathbf{M}_\mu \mathbf{q}^*$ and $\boldsymbol{\lambda}^* = \mathbf{M}_\lambda \mathbf{q}^*$, which means that $(\mathbf{x}^*, \mathbf{M}_\mu \mathbf{q}^*, \mathbf{M}_\lambda \mathbf{q}^*)$ is a KKT point associated with $\text{GNE}(\tilde{\mathbf{f}})$ and $\mathbf{x}^* \in \text{GNE}(\tilde{\mathbf{f}})$. □

Remark 4.1 Based on Remark 2.7, these instances of Proposition 4.2 can be deduced:

- If each population of the society uses either a SEPT or IPC revision protocol, then the resulting EDM is 0-perturbed Nash stationary, which is equivalent to being Nash stationary; and Condition (A) reduces to Condition (C1). In such case, Proposition 4.2 becomes exactly Proposition 4.1.

- If all populations of the society use PBR revision protocols, then the resulting EDM is ω-perturbed Nash stationary, with ω defined as in Lemma 2.4 and Remark 2.7; and Condition (A) reduces to the perturbed Nash stationarity property of Lemma 2.4.
- If some populations of the society use PBR revision protocols and all the remaining ones use either SEPT or IPC revision protocols, then Condition (A) is satisfied by means of Corollary 2.1.

Consequently, Proposition 4.2 fully generalizes Proposition 4.1. $\diamondsuit$

Propositions 4.1 and 4.2 can be regarded as reformulations of Lemma 2.8 and Corollary 2.2, respectively, in which Assumption 2.3 is encapsulated in Condition (C2). In contrast to Lemma 2.8 and Corollary 2.2, however, Propositions 4.1 and 4.2 explicitly emphasize the role of the PDM state $\mathbf{q}$ in the KKT optimality conditions. In particular, for the $\mathbf{x}$-component of $\mathcal{E}$ to coincide with the set GNE($\mathbf{f}$) (or GNE($\tilde{\mathbf{f}}$)), the PDM's steady-state $\mathbf{q}^*$ must encode the information corresponding to the optimal Ω-related dual-Lagrange multipliers $\boldsymbol{\mu}^*$ and $\boldsymbol{\lambda}^*$. For Nash stationary EDMs (e.g., the SEPT and IPC EDMs), this correspondence is exact. In contrast, for non-Nash stationary EDMs satisfying Condition (A) (e.g., PBR EDMs or EDMs corresponding to mixed societies with PBR, SEPT, or IPC revision protocols), the correspondence holds with respect to the perturbed set GNE($\tilde{\mathbf{f}}$).

Remark 4.2 Notice that in the unconstrained scenario, where no explicit coupling constraints are considered (i.e., $\Omega = \mathbb{R}^n$), the feasible set $\mathcal{X}$ becomes exactly Δ. As a consequence, the sets GNE($\mathbf{f}$) and GNE($\tilde{\mathbf{f}}$) reduce to NE($\mathbf{f}$) and NE($\tilde{\mathbf{f}}$), respectively (see Definition 2.13); and Condition (C2) simplifies to:

(S) For every $(\mathbf{q}^*, \mathbf{x}^*) \in \mathbb{R}^d \times \Delta$, the PDM satisfies that $\mathcal{W}(\mathbf{q}^*, \mathbf{x}^*) = \mathbf{0}_d$ implies:

$$\mathcal{H}\left(\mathbf{q}^*, \mathbf{x}^*\right) = \mathbf{f}\left(\mathbf{x}^*\right).$$

Under such a setup, Propositions 4.1 and 4.2 guarantee the correspondence between the $\mathbf{x}$-component of $\mathcal{E}$ and the sets NE($\mathbf{f}$) and NE($\tilde{\mathbf{f}}$), respectively. $\diamondsuit$

Example 4.1 Consider a population game $G = (\mathcal{P}, \mathcal{X}, \mathbf{f}(\cdot))$ under the unconstrained scenario with $\Omega = \mathbb{R}^n$ and $\mathcal{X} = \Delta$. Moreover, consider an EDM–PDM pair comprised of a Nash stationary EDM (e.g., a SEPT or IPC EDM) and the smoothing-anticipatory PDM introduced in Example 2.4, which is of the form

$$\begin{aligned}\dot{\mathbf{q}}(t) &= \tau\left(\mathbf{f}\left(\mathbf{x}(t)\right) - \mathbf{q}(t)\right),\\ \mathbf{p}(t) &= \alpha_0 \mathbf{f}\left(\mathbf{x}(t)\right) + \alpha_1 \mathbf{q}(t) + \alpha_2 \dot{\mathbf{q}}(t),\end{aligned}$$

where $\tau \in \mathbb{R}_{>0}$ and $\alpha_0, \alpha_1, \alpha_2 \in \mathbb{R}_{\geq 0}$ are tuning parameters subject to $\alpha_0 + \alpha_1 = 1$.

From Lemma 2.3 it follows that the considered EDM is Nash stationary and therefore satisfies Condition (C1). On the other hand, note for every $(\mathbf{q}^*, \mathbf{x}^*) \in \mathbb{R}^d \times \Delta$ it holds that

$$\mathbf{f}\left(\mathbf{x}^*\right) - \mathbf{q}^* = \mathbf{0}_d \quad \Rightarrow \quad \mathcal{H}\left(\mathbf{q}^*, \mathbf{x}^*\right) = \mathbf{f}\left(\mathbf{x}^*\right).$$

That is, every point $(\mathbf{q}^*, \mathbf{x}^*) \in \mathbb{R}^d \times \Delta$ that renders an steady state in the smoothing-anticipatory PDM implies that the output of the PDM at that point is exactly $\mathbf{f}(\mathbf{x}^*)$. Therefore, the smoothing-anticipatory PDM satisfies Condition (S).

It then follows from Proposition 4.1 and Remark 4.2 that every $(\mathbf{x}^*, \mathbf{q}^*) \in \mathcal{E}$ implies that $\mathbf{x}^* \in \mathrm{NE}(\mathbf{f})$. In words: every equilibrium point $(\mathbf{x}^*, \mathbf{q}^*) \in \mathcal{E}$ of the considered EDM–PDM satisfies that the associated society state $\mathbf{x}^* \in \Delta$ is an NE of the underlying population game G. ◊

Example 4.2 Consider a population game $G = (\mathcal{P}, \mathcal{X}, \mathbf{f}(\cdot))$ with $\mathcal{X} \subset \Delta$, and let the associated EDM–PDM pair be comprised of the PBR EDM and an instance of the dual-Lagrangian-based PDM of Example 2.5 given by

$$\begin{aligned}
\dot{\mathbf{q}}_1(t) &= \mathbf{A}\mathbf{x} - \mathbf{b} \\
\dot{\mathbf{q}}_2(t) &= \left[\mathbf{g}\left(\mathbf{x}(t)\right)\right]_+ - \mathbf{q}_2(t) \odot \left[-\mathbf{g}\left(\mathbf{x}(t)\right)\right]_+ \\
\mathbf{p}(t) &= \mathbf{f}\left(\mathbf{x}(t)\right) - \mathbf{A}^\top \mathbf{q}_1(t) - \mathrm{D}_{\mathbf{x}}\,\mathbf{g}\left(\mathbf{x}(t)\right)^\top \mathbf{q}_2(t),
\end{aligned}$$

where $[\cdot]_+ := \max\{\cdot, 0\}$ and $\odot$ is the Hadamard product. Denote $\mathbf{q} = \mathrm{col}\,(\mathbf{q}_1, \mathbf{q}_2)$ and let $\mathbf{M}_1 \in \mathbb{R}^{c_= \times d}$ and $\mathbf{M}_2 \in \mathbb{R}^{c_\leq \times d}$ be defined as

$$\begin{aligned}
\mathbf{M}_1 &= \left[\mathbf{I}_{c_=}\ \mathbf{0}_{c_= \times c_\leq}\right] \\
\mathbf{M}_2 &= \left[\mathbf{0}_{c_\leq \times c_=}\ \mathbf{I}_{c_\leq}\right].
\end{aligned}$$

Namely, $d = c_= + c_\leq$ and

$$\begin{aligned}
\mathbf{q}_1 &= \mathbf{M}_1 \mathbf{q} \\
\mathbf{q}_2 &= \mathbf{M}_2 \mathbf{q}.
\end{aligned}$$

In addition, let the initial condition $\mathbf{q}_2(0)$ satisfy $\mathbf{q}_2(0) \in \mathbb{R}^{c_\leq}_{\geq 0}$. It can be shown that[3] this requirement is enough to ensure that $\mathbf{q}_2(t) \succeq \mathbf{0}_{c_\leq}$ for all $t \geq 0$.

From Lemma 2.4 and Remark 4.1, it follows that the PBR EDM satisfies Condition (A). Moreover, observe that for every $(\mathbf{q}^*, \mathbf{x}^*) \in \mathbb{R}^d \times \Delta$ it holds that

$$\left.\begin{aligned}
\mathbf{A}\mathbf{x}^* - \mathbf{b} = \mathbf{0}_{c_=} \\
\left[\mathbf{g}\left(\mathbf{x}^*\right)\right]_+ - \mathbf{q}_2^* \odot \left[-\mathbf{g}\left(\mathbf{x}^*\right)\right]_+ = \mathbf{0}_{\leq}
\end{aligned}\right\} \Rightarrow \left\{\begin{aligned}
\mathcal{H}\left(\mathbf{q}^*, \mathbf{x}^*\right) = \mathbf{f}(\mathbf{x}^*) - \mathbf{A}^\top \mathbf{q}_1^* - \mathrm{D}_{\mathbf{x}}\,\mathbf{g}(\mathbf{x}^*)^\top \mathbf{q}_2^* \\
\mathbf{q}_2^* \odot \mathbf{g}\left(\mathbf{x}^*\right) = \mathbf{0}_{c_\leq} \\
\mathbf{q}_2^* \succ \mathbf{0}_{c_\leq} \\
\mathbf{x}^* \in \Omega.
\end{aligned}\right.$$

Hence, the considered PDM satisfies Condition (C2) with $\mathbf{M}_\mu = \mathbf{M}_1$ and $\mathbf{M}_\lambda = \mathbf{M}_2$.

Consequently, using Proposition 4.2 and Remark 4.1 one can conclude that every $(\mathbf{x}^*, \mathbf{q}^*) \in \mathcal{E}$ implies that $\mathbf{x}^* \in \mathrm{GNE}(\tilde{\mathbf{f}})$. Equivalently: every equilibrium point $(\mathbf{x}^*, \mathbf{q}^*) \in \mathcal{E}$ of the considered EDM–PDM satisfies that the associated society state

[3] This claim is formally proven later in Sect. 4.3.1.

$\mathbf{x}^* \in \Delta$ is a GNE of the perturbed population game $\tilde{G} = \left(\mathcal{P}, \mathcal{X}, \tilde{\mathbf{f}}(\cdot)\right)$, where $\tilde{\mathbf{f}}(\cdot)$ is set as in (4.2), with $\varpi(\cdot) = \omega(\cdot)$ and $\omega(\cdot)$ defined as in Lemma 2.4.

Alternatively, if the PBR EDM is replaced by any Nash stationary EDM (e.g., SEPT EDM, IPC EDM), then every equilibrium point $(\mathbf{x}^*, \mathbf{q}^*) \in \mathcal{E}$ of the resulting EDM–PDM satisfies that the associated society state $\mathbf{x}^* \in \Delta$ is a GNE of the underlying population game G. This claim follows from Proposition 4.1. ◊

4.2 Sufficient Conditions for Asymptotic Stability

While Propositions 4.1 and 4.2 address only the steady-state behavior of EDM–PDM pairs, the present section establishes sufficient conditions to ensure both Lyapunov and asymptotic stability of the equilibrium set $\mathcal{E}$. The results presented here rely on the δ-dissipativity notions developed in Chap. 3. The reader is encouraged to review them before proceeding.

Proposition 4.3 *Consider an EDM–PDM pair subject to the following conditions:*

(C3) *The equilibrium set $\mathcal{E}$ is nonempty and bounded.*

(C4) *The EDM and PDM are δ-dissipative characterized by $(\sigma_{EDM}, \mathbf{\Pi}_{EDM})$ and $(\sigma_{PDM}, \mathbf{\Pi}_{PDM})$, respectively, and for every $(\mathbf{x}, \mathbf{q}) \in \Delta \times \mathbb{R}^d$ it holds that*

$$\left\| \begin{matrix} \dot{\mathcal{H}}(\mathbf{q}, \mathbf{x}) \\ \overline{\mathcal{V}}(\mathbf{x}, \mathbf{q}) \end{matrix} \right\|^2_{\mathbf{\Pi}_{EDM}} + \left\| \begin{matrix} \overline{\mathcal{V}}(\mathbf{x}, \mathbf{q}) \\ \dot{\mathcal{H}}(\mathbf{q}, \mathbf{x}) \end{matrix} \right\|^2_{\mathbf{\Pi}_{PDM}} \leq 0,$$

where

$$\overline{\mathcal{V}}(\mathbf{x}, \mathbf{q}) = \mathcal{V}\left(\mathbf{x}, \mathcal{H}(\mathbf{q}, \mathbf{x})\right)$$
$$\dot{\mathcal{H}}(\mathbf{q}, \mathbf{x}) = \mathrm{D}_{\mathbf{q}} \mathcal{H}(\mathbf{q}, \mathbf{x}) \mathcal{W}(\mathbf{q}, \mathbf{x}) + \mathrm{D}_{\mathbf{x}} \mathcal{H}(\mathbf{q}, \mathbf{x}) \overline{\mathcal{V}}(\mathbf{x}, \mathbf{q}).$$

Then, $\mathcal{E}$ is stable in the sense of Lyapunov. Moreover, the largest invariant set of the EDM–PDM pair contained in $\mathcal{R}$, with

$$\mathcal{R} = \left\{ (\mathbf{x}, \mathbf{q}) \in \Delta \times \mathbb{R}^d : \begin{matrix} \sigma_{EDM}\left(\mathbf{x}, \mathcal{H}(\mathbf{q}, \mathbf{x}), \mathbf{0}_n\right) = 0 \\ \sigma_{PDM}\left(\mathbf{q}, \mathbf{x}, \mathbf{0}_n\right) = 0 \end{matrix} \right\},$$

is asymptotically stable. ◊

Proof Consider the Lyapunov function candidate given by

$$V(\mathbf{x}, \mathbf{q}) = S_{EDM}\left(\mathbf{x}, \mathcal{H}(\mathbf{q}, \mathbf{x})\right) + S_{PDM}(\mathbf{q}, \mathbf{x}),$$

where $S_{EDM}(\cdot, \cdot)$ and $S_{PDM}(\cdot, \cdot)$ are the storage functions of the EDM and the PDM, respectively (see Definition 3.1). Note that $\mathcal{E}$ is the preimage of the singleton $\{0\}$

under the continuous map $V(\cdot,\cdot)$. Thus, $\mathcal{E}$ is a closed set, and from (C3) one can conclude that $\mathcal{E}$ is nonempty and compact. On the other hand, let

$$\dot{V}(\mathbf{x},\mathbf{q}) = \nabla_{\mathbf{x}} V(\mathbf{x},\mathbf{q})^\top \overline{\boldsymbol{\mathcal{V}}}(\mathbf{x},\mathbf{q}) + \nabla_{\mathbf{q}} V(\mathbf{x},\mathbf{q})^\top \boldsymbol{\mathcal{W}}(\mathbf{q},\mathbf{x})$$

denote the time derivative of $V(\cdot,\cdot)$ over the trajectories of the (closed loop) EDM–PDM pair. Then,

$$\begin{aligned}
\dot{V}(\mathbf{x},\mathbf{q}) &= \nabla_{\mathbf{x}} S_{EDM}\left(\mathbf{x},\boldsymbol{\mathcal{H}}(\mathbf{q},\mathbf{x})\right)^\top \overline{\boldsymbol{\mathcal{V}}}(\mathbf{x},\mathbf{q}) \\
&\quad + \nabla_{\boldsymbol{\mathcal{H}}} S_{EDM}\left(\mathbf{x},\boldsymbol{\mathcal{H}}(\mathbf{q},\mathbf{x})\right)^\top \dot{\boldsymbol{\mathcal{H}}}(\mathbf{q},\mathbf{x}) \\
&\quad + \nabla_{\mathbf{q}} S_{PDM}\left(\mathbf{q},\mathbf{x}\right)^\top \boldsymbol{\mathcal{W}}(\mathbf{q},\mathbf{x}) \\
&\quad + \nabla_{\mathbf{x}} S_{PDM}\left(\mathbf{q},\mathbf{x}\right)^\top \overline{\boldsymbol{\mathcal{V}}}(\mathbf{x},\mathbf{q}) \\
&\le -\sigma_{EDM}(\mathbf{x},\mathbf{q},\dot{\boldsymbol{\mathcal{H}}}(\mathbf{q},\mathbf{x})) - \sigma_{PDM}(\mathbf{q},\mathbf{x},\overline{\boldsymbol{\mathcal{V}}}(\mathbf{x},\mathbf{q})) \\
&\quad + \left\| \begin{matrix} \dot{\boldsymbol{\mathcal{H}}}(\mathbf{q},\mathbf{x}) \\ \overline{\boldsymbol{\mathcal{V}}}(\mathbf{x},\mathbf{q}) \end{matrix} \right\|_{\Pi_{EDM}}^2 + \left\| \begin{matrix} \overline{\boldsymbol{\mathcal{V}}}(\mathbf{x},\mathbf{q}) \\ \dot{\boldsymbol{\mathcal{H}}}(\mathbf{q},\mathbf{x}) \end{matrix} \right\|_{\Pi_{PDM}}^2 .
\end{aligned}$$

Thus, by Condition (C4), it follows that

$$\dot{V}(\mathbf{x},\mathbf{q}) \le -\sigma_{EDM}(\mathbf{x},\mathbf{q},\dot{\boldsymbol{\mathcal{H}}}(\mathbf{q},\mathbf{x})) - \sigma_{PDM}(\mathbf{q},\mathbf{x},\overline{\boldsymbol{\mathcal{V}}}(\mathbf{x},\mathbf{q})).$$

Consequently, from the non-negativity of $\sigma_{EDM}(\cdot,\cdot,\cdot)$ and $\sigma_{PDM}(\cdot,\cdot,\cdot)$, it holds that $\dot{V}(\cdot,\cdot)$ is non-positive, which implies that $\mathcal{E}$ is stable in the sense of Lyapunov [3, Corollary 4.7]. Furthermore, it follows that $\dot{V}(\mathbf{x},\mathbf{q}) = 0$ only if $(\mathbf{x},\mathbf{q}) \in \mathcal{R}$. Hence, the asymptotic stability of $\mathcal{E}$ follows from LaSalle's Theorem [3, Theorem 3.3]. □

Building on Propositions 4.1, 4.2, and 4.3, the main supporting results concerning the asymptotic stability of the sets $\mathrm{GNE}(\mathbf{f})$ and $\mathrm{GNE}(\tilde{\mathbf{f}})$ are now established.

Theorem 4.1 *Consider a population game $G = (\mathcal{P}, \mathcal{X}, \mathbf{f}(\cdot))$ and an EDM–PDM pair subject to Conditions (C1)–(C4). Besides, suppose that the next condition is satisfied:*

(C5) *The equilibrium set $\mathcal{E}$ is the largest invariant set of the EDM–PDM pair contained in $\mathcal{R}$.*

Then, the set $\mathrm{GNE}(\mathbf{f})$ *is asymptotically stable.* ◊

Proof The proof is immediate from Propositions 4.1 and 4.3. □

Theorem 4.2 *Consider a population game $G = (\mathcal{P}, \mathcal{X}, \mathbf{f}(\cdot))$ and an EDM–PDM pair subject to Conditions (A) and (C2)–(C5). Then, the set* $\mathrm{GNE}(\tilde{\mathbf{f}})$ *is asymptotically stable. Here, $\tilde{\mathbf{f}}(\cdot)$ is given in (4.2).* ◊

Proof The proof is immediate from Propositions 4.2 and 4.3. □

Theorems 4.1 and 4.2 establish that, under appropriate assumptions, the dynamics of EDM–PDM pairs exhibit desirable stability properties consistent with the GNE concept (or with its perturbed variant in the case of suitable non-Nash stationary EDMs). Importantly, the developed theoretical framework does not depend on specific EDM or PDM instances. Instead, the sufficient conditions are formulated in terms of general equilibrium-related and dissipativity properties, thereby ensuring broad applicability across a wide range of models.

4.3 Decoupled Sufficient Conditions for Unconstrained Games

While Conditions (C1)–(C5) and (A) do not rely on specific instances of EDM–PDM pairs and are therefore applicable to a broad class of models, they can be challenging to verify in practice. To facilitate analysis, one can derive simplified sufficient conditions that ensure the asymptotic stability of the associated set of Nash equilibria in the unconstrained scenario with $\Omega = \mathbb{R}^n$. Although these simplified conditions apply to a more restricted class of EDM–PDM pairs and focus only on the unconstrained case, they are easier to verify in practical scenarios. Furthermore, these simplified conditions are fully decoupled between the EDM and the PDM, meaning that they can be verified independently for each subsystem while still ensuring the stability properties of the interconnected EDM–PDM pair.

Corollary 4.1 *Consider a population game $G = (\mathcal{P}, \Delta, \mathbf{f}(\cdot))$ without explicit coupling constraints, i.e., $\Omega = \mathbb{R}^n$, and an EDM–PDM pair subject to the following conditions:*

(D1) *The EDM is Nash stationary and δ-passive characterized by an informative σ_{EDM}.*

(D2) *The PDM is bounded, δ-antipassive characterized by an informative σ_{PDM}, satisfies Condition (S), and the set $\mathcal{E}_{\mathcal{W}}(\mathbf{x}^*) = \left\{\mathbf{q}^* \in \mathbb{R}^d : \mathcal{W}(\mathbf{q}^*, \mathbf{x}^*) = \mathbf{0}_d\right\}$ is nonempty for every $\mathbf{x}^* \in \mathrm{NE}(\mathbf{f})$.*

Then, the set $\mathrm{NE}(\mathbf{f})$ *is asymptotically stable.* $\diamondsuit$

Proof Condition (C1) is readily satisfied, while Condition (C2) is satisfied for the unconstrained case via Condition (S) (see Remark 4.2). The nonemptiness of $\mathcal{E}$ is guaranteed by the nonemptiness of NE($\mathbf{f}$) (recall Lemma 2.5) together with Condition (D1), the Nash stationarity of the EDM, and the nonemptiness of $\mathcal{E}_{\mathcal{W}}(\mathbf{x}^*)$, for all $\mathbf{x}^* \in \mathrm{NE}(\mathbf{f})$. The boundedness of $\mathcal{E}$ is guaranteed by the boundedness of the PDM and the fact that the trajectories of every EDM are bounded within Δ (see Definition 2.5). Therefore, Condition (C3) holds. Next, Condition (C4) is satisfied by the δ-passivity and δ-antipassivity of the EDM and PDM, respectively. Finally, Condition (C5) is guaranteed by the informativity of σ_{EDM} and σ_{PDM}, which implies that $\mathcal{R} = \mathcal{E}$. Consequently, the claim follows from Theorem 4.1. $\square$

Corollary 4.2 *Consider a population game* $G = (\mathcal{P}, \Delta, \mathbf{f}(\cdot))$ *without explicit coupling constraints, i.e.,* $\Omega = \mathbb{R}^n$*, and an EDM–PDM pair subject to Condition (D2) and:*

(D3) *The EDM satisfies Condition (A) and is* δ*-passive characterized by an informative* σ_{EDM}.

Then, the set $\mathrm{NE}(\tilde{\mathbf{f}})$ *is asymptotically stable. Here,* $\tilde{\mathbf{f}}(\cdot)$ *is given in (4.2).* ◊

Proof The proof is virtually identical to that of Corollary 4.1 but regarding Theorem 4.2 instead of Theorem 4.1. □

Corollaries 4.1 and 4.2 provide decoupled sufficient conditions, formulated separately in terms of the EDM and the PDM, that certify the asymptotic stability of $\mathrm{GNE}(\mathbf{f})$ under the resulting interconnected EDM–PDM pairs. Accordingly, by independently designing EDM and PDM systems that satisfy their respective decoupled assumptions, the asymptotic stability of the overall interconnection can be guaranteed.

For the sake of illustration, Sects. 4.3.1 and 4.3.2 provide concrete EDM and PDM instances that readily satisfy the decoupled conditions stated in Corollaries 4.1 and 4.2. Specifically, Sect. 4.3.1 shows that all of the representative EDMs of Sect. 2.4 (e.g., the PBR, SEPT, IPC, and mixed EDMs) readily satisfy Conditions (D1) or (D3). On the other hand, Sect. 4.3.2 focuses on the smoothing-anticipatory PDM introduced in Example 2.4, which is suitable for the unconstrained case[4] of Corollaries 4.1 and 4.2. It is shown that, subject to some sufficient conditions, such a PDM readily satisfies Condition (D2). The resulting implication is that the interconnection of any representative EDM and the smoothing-anticipatory PDM yields a closed-loop system for which the underlying set of Nash equilibria is asymptotically stable.

4.3.1 Representative Evolutionary Dynamics Models

As established in Chap. 2, the PBR EDM satisfies Condition (A) (see Lemma 2.4 and Remark 2.7), while every SEPT EDM and every IPC EDM satisfy Condition (C1) (see Lemma 2.3). Moreover, every EDM that represents a mixed society of populations implementing either PBR, SEPT, or IPC protocols satisfies Condition (A) (recall Corollary 2.1). Therefore, in order for these classes of EDMs to fulfill Conditions (D1) or (D3), it only remains to show that they are δ-passive with informative σ functions. The rest of this subsection is devoted to establishing that these classes of EDMs indeed satisfy this required δ-passivity condition.

On the δ-Passivity of the PBR EDM

To prove the δ-passivity of the PBR EDM one may consider the δ-storage function given by

[4] The design of PDMs suitable for the more general constrained case (with $\Omega \cap \Delta \subset \Delta$) is addressed in Chap. 5.

$$S_{PBR}(\mathbf{x},\mathbf{p}) = \max_{\tilde{\mathbf{x}}\in\mathrm{int}(\Delta)} \left(\sum_{k\in\mathcal{P}} R^k \left(\tilde{\mathbf{x}}^{k\top}\mathbf{p}^k - \omega^k\left(\tilde{\mathbf{x}}^k\right)\right)\right) - \sum_{k\in\mathcal{P}} R^k \left(\mathbf{x}^{k\top}\mathbf{p}^k - \omega^k\left(\mathbf{x}^k\right)\right)$$

in conjunction with the auxiliary function

$$\sigma_{PBR}(\mathbf{x},\mathbf{p}) = -\nabla_{\mathbf{x}} S_{PBR}(\mathbf{x},\mathbf{p})^\top \boldsymbol{\mathcal{V}}(\mathbf{x},\mathbf{p}). \tag{4.4}$$

Here, $\omega^k : \mathbb{R}^{n^k}_{>0} \to \mathbb{R}$ is the function characterizing the PBR revision protocol of population k. Recall from Sect. 2.3.2 that these functions are assumed to be twice continuously differentiable, strictly convex, and steep near $\mathrm{bd}\left(\Delta^k\right)$. From Lemma 2.4 and the strict convexity of the functions $\omega^k(\cdot)$ (see Remark 4.3 below), it follows that both $S_{PBR}(\cdot,\cdot)$ and $\sigma_{PBR}(\cdot,\cdot)$ are non-negative and that

$$\begin{aligned} S_{PBR}(\mathbf{x},\mathbf{p}) = 0 \quad &\Leftrightarrow \quad \boldsymbol{\mathcal{V}}(\mathbf{x},\mathbf{p}) = \mathbf{0}_n \\ \sigma_{PBR}(\mathbf{x},\mathbf{p}) = 0 \quad &\Leftrightarrow \quad \boldsymbol{\mathcal{V}}(\mathbf{x},\mathbf{p}) = \mathbf{0}_n. \end{aligned}$$

Thus, $S_{PBR}(\cdot,\cdot)$ can be taken as a valid δ-storage function candidate and σ_{PBR} is informative in the sense of Definition 3.1.

To show that the PBR EDM is δ-passive with informative σ_{PBR} it then suffices to prove that

$$\nabla_{\mathbf{x}} S_{PBR}(\mathbf{x},\mathbf{p})^\top \boldsymbol{\mathcal{V}}(\mathbf{x},\mathbf{p}) + \nabla_{\mathbf{p}} S_{PBR}(\mathbf{x},\mathbf{p})^\top \mathbf{v} \le -\sigma_{PBR}(\mathbf{x},\mathbf{p}) + \mathbf{v}^\top \boldsymbol{\mathcal{V}}(\mathbf{x},\mathbf{p}),$$

for every $\mathbf{v} \in \mathbb{R}^n$; which, by the definition of σ_{PBR}, simplifies to proving that

$$\nabla_{\mathbf{p}} S_{PBR}(\mathbf{x},\mathbf{p})^\top \mathbf{v} \le \mathbf{v}^\top \boldsymbol{\mathcal{V}}(\mathbf{x},\mathbf{p}), \tag{4.5}$$

for every $\mathbf{v} \in \mathbb{R}^n$.

Now, denote

$$\overline{\omega}(\mathbf{x}) = \sum_{k\in\mathcal{P}} R^k \omega^k\left(\mathbf{x}^k\right),$$

and

$$\overline{\omega}^\star(\boldsymbol{\pi}) = \max_{\tilde{\mathbf{x}}\in\mathrm{int}(\Delta)} \left(\tilde{\mathbf{x}}^\top\boldsymbol{\pi} - \overline{\omega}(\tilde{\mathbf{x}})\right).$$

The domain of $\overline{\omega}^\star(\cdot)$ is defined as

$$\mathbb{D}_{\overline{\omega}^\star} = \left\{ \boldsymbol{\pi} \in \mathbb{R}^n : \max_{\tilde{\mathbf{x}}\in\mathrm{int}(\Delta)} \left(\tilde{\mathbf{x}}^\top\boldsymbol{\pi} - \overline{\omega}(\tilde{\mathbf{x}})\right) < \infty \right\}.$$

From the strict convexity and steepness properties of $\overline{\omega}(\cdot)$, it follows that $\mathbb{D}_{\overline{\omega}^\star} = \mathbb{R}^n$. In addition, these properties of $\overline{\omega}(\cdot)$, together with Danskin's Theorem, yield that

$$\nabla_{\boldsymbol{\pi}} \overline{\omega}^\star(\boldsymbol{\pi}) = \arg\max_{\tilde{\mathbf{x}}\in\mathrm{int}(\Delta)} \left(\tilde{\mathbf{x}}^\top\boldsymbol{\pi} - \overline{\omega}(\tilde{\mathbf{x}})\right).$$

Besides, from the KKT optimality conditions, it follows that

$$\mathbf{x}^* = \nabla_{\pi}\overline{\omega}^{\star}(\boldsymbol{\pi}) \quad \Leftrightarrow \quad \begin{cases} \boldsymbol{\pi} = \nabla_{\mathbf{x}}\overline{\omega}(\mathbf{x}^*) + \mathrm{col}\left(\mu^{k*}\mathbf{1}_{n^k}\right)_{k\in\mathcal{P}} \\ \mathbf{x}^* \in \mathrm{int}(\Delta), \end{cases} \tag{4.6}$$

for some $\mu^{k*} \in \mathbb{R}$, for all $k \in \mathcal{P}$. Finally, following the notation

$$\boldsymbol{C}^{\omega}(\mathbf{p}) = \arg\max_{\tilde{\mathbf{x}}\in\mathrm{int}(\Delta)} \left(\tilde{\mathbf{x}}^{\top}\mathbf{p} - \omega(\tilde{\mathbf{x}})\right),$$

introduced in the proof of Lemma 2.4, denote

$$\boldsymbol{C}^{\overline{\omega}}(\boldsymbol{\pi}) = \arg\max_{\tilde{\mathbf{x}}\in\mathrm{int}(\Delta)} \left(\tilde{\mathbf{x}}^{\top}\boldsymbol{\pi} - \overline{\omega}(\tilde{\mathbf{x}})\right) \tag{4.7a}$$

$$= \nabla_{\pi}\overline{\omega}^{\star}(\boldsymbol{\pi}). \tag{4.7b}$$

It is straightforward to check that the following identity holds:

$$\boldsymbol{C}^{\overline{\omega}}(\mathbf{R}\odot\mathbf{p}) = \boldsymbol{C}^{\omega}(\mathbf{p}),$$

where $\mathbf{R} = \mathrm{col}\left(R^k\mathbf{1}_{n^k}\right)_{k\in\mathcal{P}}$ and $\odot$ is the Hadamard product.

By construction,

$$S_{PBR}(\mathbf{x},\mathbf{p}) = \overline{\omega}^{\star}(\mathbf{R}\odot\mathbf{p}) - \left(\mathbf{x}^{\top}(\mathbf{R}\odot\mathbf{p}) - \overline{\omega}(\mathbf{x})\right).$$

Therefore, it follows that

$$\begin{aligned} \nabla_{\mathbf{p}}S_{PBR}(\mathbf{x},\mathbf{p}) &= \mathbf{R}\odot\arg\max_{\tilde{\mathbf{x}}\in\mathrm{int}(\Delta)}\left(\tilde{\mathbf{x}}^{\top}(\mathbf{R}\odot\mathbf{p}) - \overline{\omega}(\tilde{\mathbf{x}})\right) - (\mathbf{R}\odot\mathbf{x}) \\ &= \mathbf{R}\odot\boldsymbol{C}^{\overline{\omega}}(\mathbf{R}\odot\mathbf{p}) - (\mathbf{R}\odot\mathbf{x}) \\ &= \mathbf{R}\odot(\boldsymbol{C}^{\omega}(\mathbf{p}) - \mathbf{x}) \\ &= \boldsymbol{\mathcal{V}}(\mathbf{x},\mathbf{p}). \end{aligned}$$

Consequently, the requirement (4.5) is satisfied and the PBR EDM is indeed δ-passive characterized by the informative σ_{PBR}. This fact is established in Lemma 4.1.

Lemma 4.1 *The PBR EDM of Definition 2.7 is δ-passive characterized by the informative σ_{PBR} function given in (4.4).* ◊

Remark 4.3 Notice that, under the considered framework,

$$\nabla_{\mathbf{x}}S_{PBR}(\mathbf{x},\mathbf{p}) = -\left(\mathbf{R}\odot\mathbf{p} - \nabla_{\mathbf{x}}\overline{\omega}(\mathbf{x})\right),$$

and thus

$$
\begin{aligned}
-\sigma_{PBR}(\mathbf{x}, \mathbf{p}) &= \nabla_{\mathbf{x}} S_{PBR}(\mathbf{x}, \mathbf{p})^\top \boldsymbol{\mathcal{V}}(\mathbf{x}, \mathbf{p}) \\
&= -\left(\mathbf{R} \odot \mathbf{p} - \nabla_{\mathbf{x}} \overline{\omega}(\mathbf{x})\right)^\top \left(\mathbf{R} \odot \boldsymbol{C}^{\overline{\omega}}\left(\mathbf{R} \odot \mathbf{p}\right) - (\mathbf{R} \odot \mathbf{x})\right) \\
&= -\left(\nabla_{\mathbf{x}} \overline{\omega}\left(\hat{\mathbf{x}}\right) + \hat{\boldsymbol{\mu}} - \nabla_{\mathbf{x}} \overline{\omega}(\mathbf{x})\right)^\top \left(\mathbf{R} \odot \left(\hat{\mathbf{x}} - \mathbf{x}\right)\right) \\
&= -\left(\nabla_{\mathbf{x}} \overline{\omega}\left(\hat{\mathbf{x}}\right) - \nabla_{\mathbf{x}} \overline{\omega}(\mathbf{x})\right)^\top \left(\mathbf{R} \odot \left(\hat{\mathbf{x}} - \mathbf{x}\right)\right) \\
&= -\sum_{k \in \mathcal{P}} \left(R^k\right)^2 \left(\nabla_{\mathbf{x}^k} \omega^k\left(\hat{\mathbf{x}}^k\right) - \nabla_{\mathbf{x}^k} \omega^k\left(\mathbf{x}^k\right)\right)^\top \left(\hat{\mathbf{x}}^k - \mathbf{x}^k\right) \\
&< 0, \quad \forall \hat{\mathbf{x}} \neq \mathbf{x}.
\end{aligned}
$$

Here, the third equality uses (4.6), with $\hat{\mathbf{x}} \in \operatorname{int}(\Delta)$ satisfying $\mathbf{R} \odot \mathbf{p} = \nabla_{\mathbf{x}} \overline{\omega}\left(\hat{\mathbf{x}}\right) + \hat{\boldsymbol{\mu}}$, where $\hat{\boldsymbol{\mu}} = \operatorname{col}\left(\hat{\mu}^k \mathbf{1}_{n^k}\right)_{k \in \mathcal{P}}$, for some $\hat{\mu}^k \in \mathbb{R}$, for all $k \in \mathcal{P}$; the fourth equality uses the fact $\hat{\boldsymbol{\mu}}^\top \left(\mathbf{R} \odot \left(\hat{\mathbf{x}} - \mathbf{x}\right)\right) = \sum_{k \in \mathcal{P}} \hat{\mu}^k R^k \mathbf{1}_{n^k}^\top \left(\hat{\mathbf{x}}^k - \mathbf{x}^k\right) = 0$, since both $\mathbf{1}_{n^k}^\top \hat{\mathbf{x}}^k = m^k$ and $\mathbf{1}_{n^k}^\top \mathbf{x}^k = m^k$, for all $k \in \mathcal{P}$; and the final inequality follows from the strict convexity of the $\omega^k(\cdot)$ functions. This observation confirms the informativity of σ_{PBR} given in (4.4). ◊

On the δ-Passivity of the SEPT EDM

The δ-passivity of the SEPT EDM (Definition 2.8) can be shown by considering the δ-storage function

$$
S_{SEPT}\left(\mathbf{x}, \mathbf{p}\right) = \sum_{k \in \mathcal{P}} R^k m^k \sum_{i \in \mathcal{S}^k} \int_0^{p_i^k - \frac{1}{m^k} \mathbf{x}^{k\top} \mathbf{p}^k} \phi_i^k(\zeta)\, d\zeta,
$$

and the auxiliary function

$$
\sigma_{SEPT}\left(\mathbf{x}, \mathbf{p}\right) = \sum_{k \in \mathcal{P}} R^k \sum_{i \in \mathcal{S}^k} \mathcal{V}_i^k\left(\mathbf{x}^k, \mathbf{p}^k\right) p_i^k \sum_{j \in \mathcal{S}^k} \phi_j^k\left(p_j^k - \frac{1}{m^k} \mathbf{x}^{k\top} \mathbf{p}^k\right), \tag{4.8}
$$

where

$$
\mathcal{V}_i^k\left(\mathbf{x}^k, \mathbf{p}^k\right) = R^\kappa \left(m^\kappa \phi_\iota^\kappa\left(p_\iota^\kappa - \frac{1}{m^\kappa} \mathbf{x}^{\kappa\top} \mathbf{p}^\kappa\right) - x_\iota^\kappa \sum_{i \in \mathcal{S}^\kappa} \phi_i^\kappa\left(p_i^\kappa - \frac{1}{m^\kappa} \mathbf{x}^{\kappa\top} \mathbf{p}^\kappa\right)\right),
$$

for all $i \in \mathcal{S}^k$ and all $k \in \mathcal{P}$. It is possible to show that both $S_{SEPT}(\cdot, \cdot)$ and $\sigma_{SEPT}(\cdot, \cdot)$ are non-negative and that[5]

$$
\begin{aligned}
S_{SEPT}\left(\mathbf{x}, \mathbf{p}\right) = 0 \quad &\Leftrightarrow \quad \boldsymbol{\mathcal{V}}\left(\mathbf{x}, \mathbf{p}\right) = \mathbf{0}_n \\
\sigma_{SEPT}\left(\mathbf{x}, \mathbf{p}\right) = 0 \quad &\Leftrightarrow \quad \boldsymbol{\mathcal{V}}\left(\mathbf{x}, \mathbf{p}\right) = \mathbf{0}_n.
\end{aligned}
$$

[5] See for instance the discussion in [4, Appendix A.2].

Thus, $S_{SEPT}(\cdot, \cdot)$ can be taken as a valid δ-storage function candidate and σ_{SEPT} is informative in the sense of Definition 3.1.

Based on these formulations, to show that the SEPT EDM is δ-passive with informative σ_{SEPT} it suffices to prove that

$$\nabla_{\mathbf{x}} S_{SEPT}(\mathbf{x}, \mathbf{p})^\top \mathcal{V}(\mathbf{x}, \mathbf{p}) + \nabla_{\mathbf{p}} S_{SEPT}(\mathbf{x}, \mathbf{p})^\top \mathbf{v} \le -\sigma_{SEPT}(\mathbf{x}, \mathbf{p}) + \mathbf{v}^\top \mathcal{V}(\mathbf{x}, \mathbf{p}), \tag{4.9}$$

for every $\mathbf{v} \in \mathbb{R}^n$.

Accordingly, note that[6]

$$\begin{aligned}
\frac{\partial S_{SEPT}(\mathbf{x}, \mathbf{p})}{\partial x_\iota^\kappa} &= -R^\kappa p_\iota^\kappa \sum_{i \in \mathcal{S}^\kappa} \phi_i^\kappa \left(p_i^\kappa - \frac{1}{m^\kappa} \mathbf{x}^{\kappa\top} \mathbf{p}^\kappa \right) \\
\frac{\partial S_{SEPT}(\mathbf{x}, \mathbf{p})}{\partial p_\iota^\kappa} &= R^\kappa m^\kappa \sum_{i \in \mathcal{S}^\kappa} \phi_i^\kappa \left(p_i^\kappa - \frac{1}{m^\kappa} \mathbf{x}^{\kappa\top} \mathbf{p}^\kappa \right) \left(\frac{\partial p_i^\kappa}{\partial p_\iota^\kappa} - \frac{1}{m^\kappa} x_\iota^\kappa \right) \\
&= R^\kappa \left(m^\kappa \phi_\iota^\kappa \left(p_\iota^\kappa - \frac{1}{m^\kappa} \mathbf{x}^{\kappa\top} \mathbf{p}^\kappa \right) - x_\iota^\kappa \sum_{i \in \mathcal{S}^\kappa} \phi_i^\kappa \left(p_i^\kappa - \frac{1}{m^\kappa} \mathbf{x}^{\kappa\top} \mathbf{p}^\kappa \right) \right) \\
&= \mathcal{V}_\iota^\kappa (\mathbf{x}^\kappa, \mathbf{p}^\kappa).
\end{aligned}$$

Then,

$$\begin{aligned}
\nabla_{\mathbf{x}} S_{SEPT}(\mathbf{x}, \mathbf{p})^\top \mathcal{V}(\mathbf{x}, \mathbf{p}) &= -\sum_{k \in \mathcal{P}} R^k \sum_{i \in \mathcal{S}^k} \mathcal{V}_i^k \left(\mathbf{x}^k, \mathbf{p}^k \right) p_i^k \sum_{j \in \mathcal{S}^k} \phi_j^k \left(p_j^k - \frac{1}{m^k} \mathbf{x}^{k\top} \mathbf{p}^k \right) \\
&= -\sigma_{SEPT}(\mathbf{x}, \mathbf{p}),
\end{aligned}$$

and

$$\nabla_{\mathbf{p}} S_{SEPT}(\mathbf{x}, \mathbf{p})^\top \mathbf{v} = \mathbf{v}^\top \mathcal{V}(\mathbf{x}, \mathbf{p}).$$

Therefore, (4.9) holds with equality and the SEPT EDM is indeed δ-passive with informative σ_{SEPT}. This fact is formalized in Lemma 4.2.

Lemma 4.2 *The SEPT EDM of Definition 2.8 is δ-passive characterized by the informative σ_{SEPT} function given in (4.8).* ◊

On the δ-Passivity of the IPC EDM

To show the δ-passivity of the IPC EDM (Definition 2.9), consider the δ-storage function defined as

$$S_{IPC}(\mathbf{x}, \mathbf{p}) = \sum_{k \in \mathcal{P}} R^k \sum_{i \in \mathcal{S}^k} x_i^k \sum_{\ell \in \mathcal{S}^k} \int_0^{p_\ell^k - p_i^k} \phi_\ell^k(\zeta)\, d\zeta,$$

and the σ function given by

[6] Observe that the indices in the partial derivatives are κ and ι rather than k and i.

$$\sigma_{IPC}(\mathbf{x},\mathbf{p}) = -\sum_{k\in\mathcal{P}} R^k \sum_{i\in\mathcal{S}^k} \mathcal{V}_i^k\left(\mathbf{x}^k,\mathbf{p}^k\right) \sum_{\ell\in\mathcal{S}^k} \int_0^{p_\ell^k - p_i^k} \phi_\ell^k(\zeta)\, d\zeta, \tag{4.10}$$

where

$$\mathcal{V}_i^k\left(\mathbf{x}^k,\mathbf{p}^k\right) = R^k \sum_{j\in\mathcal{S}^k} \left(x_j^k \phi_i^k\left(p_i^k - p_j^k\right) - x_i^k \phi_j^k\left(p_j^k - p_i^k\right)\right),$$

for all $i \in \mathcal{S}^k$ and all $k \in \mathcal{P}$. It is not too difficult to show that[7] $S_{IPC}(\cdot,\cdot)$ and $\sigma_{IPC}(\cdot,\cdot)$ are always non-negative, and that

$$\begin{aligned} S_{IPC}(\mathbf{x},\mathbf{p}) = 0 &\quad\Leftrightarrow\quad \mathcal{V}(\mathbf{x},\mathbf{p}) = \mathbf{0}_n \\ \sigma_{IPC}(\mathbf{x},\mathbf{p}) = 0 &\quad\Leftrightarrow\quad \mathcal{V}(\mathbf{x},\mathbf{p}) = \mathbf{0}_n. \end{aligned}$$

Implying that $S_{IPC}(\cdot,\cdot)$ can be taken as a valid δ-storage function candidate, and that σ_{IPC} is informative in the sense of Definition 3.1.

Consequently, to show that the IPC EDM is δ-passive with informative σ_{IPC} it suffices to prove that

$$\nabla_{\mathbf{x}} S_{IPC}(\mathbf{x},\mathbf{p})^\top \mathcal{V}(\mathbf{x},\mathbf{p}) + \nabla_{\mathbf{p}} S_{IPC}(\mathbf{x},\mathbf{p})^\top \mathbf{v} \le -\sigma_{IPC}(\mathbf{x},\mathbf{p}) + \mathbf{v}^\top \mathcal{V}(\mathbf{x},\mathbf{p}), \tag{4.11}$$

for every $\mathbf{v} \in \mathbb{R}^n$.

To prove the claim, note that[8]

$$\begin{aligned} \frac{\partial S_{IPC}(\mathbf{x},\mathbf{p})}{\partial x_\iota^\kappa} &= R^\kappa \sum_{\ell\in\mathcal{S}^\kappa} \int_0^{p_\ell^\kappa - p_\iota^\kappa} \phi_\ell^\kappa(\zeta)\, d\zeta \\ \frac{\partial S_{IPC}(\mathbf{x},\mathbf{p})}{\partial p_\iota^\kappa} &= R^\kappa \sum_{i\in\mathcal{S}^\kappa} x_i^\kappa \sum_{\ell\in\mathcal{S}^\kappa} \phi_\ell^\kappa\left(p_\ell^\kappa - p_i^\kappa\right)\left(\frac{\partial p_\ell^\kappa}{\partial p_\iota^\kappa} - \frac{\partial p_i^\kappa}{\partial p_\iota^\kappa}\right) \\ &= R^\kappa \sum_{i\in\mathcal{S}^\kappa} \sum_{\ell\in\mathcal{S}^\kappa} \left(x_i^\kappa \phi_\ell^\kappa\left(p_\ell^\kappa - p_i^\kappa\right) - x_\ell^\kappa \phi_i^\kappa\left(p_i^\kappa - p_\ell^\kappa\right)\right) \frac{\partial p_\ell^\kappa}{\partial p_\iota^\kappa} \\ &= \mathcal{V}_\iota^\kappa\left(\mathbf{x}^\kappa,\mathbf{p}^\kappa\right), \end{aligned}$$

which implies that

$$\begin{aligned} \nabla_{\mathbf{x}} S_{IPC}(\mathbf{x},\mathbf{p})^\top \mathcal{V}(\mathbf{x},\mathbf{p}) &= \sum_{k\in\mathcal{P}} R^k \sum_{i\in\mathcal{S}^k} \mathcal{V}_i^k\left(\mathbf{x}^k,\mathbf{p}^k\right) \sum_{\ell\in\mathcal{S}^k} \int_0^{p_\ell^k - p_i^k} \phi_\ell^k(\zeta)\, d\zeta \\ &= -\sigma_{IPC}(\mathbf{x},\mathbf{p}), \end{aligned}$$

and

$$\nabla_{\mathbf{p}} S_{IPC}(\mathbf{x},\mathbf{p})^\top \mathbf{v} = \mathbf{v}^\top \mathcal{V}(\mathbf{x},\mathbf{p}).$$

---

[7] See for instance the discussion in [4, Appendix A.4].

[8] Observe that the indices in the partial derivatives are κ and ι rather than k and i.

Therefore, (4.11) holds with equality and the IPC EDM is indeed δ-passive with informative σ_{IPC}. This fact is summarized in Lemma 4.3.

Lemma 4.3 *The IPC EDM of Definition 2.9 is δ-passive characterized by the informative σ_{IPC} function given in (4.10).* ◊

On the δ-Passivity of PBR-SEPT-IPC-Mixed EDMs

Lemmas 4.1, 4.2, and 4.3 assume that all populations in the society implement the same class of revision protocol: PBR, SEPT, and IPC, respectively. In practice, however, it is interesting to consider societies where each population might use a different class of revision protocol. As such, Lemma 4.4 characterizes the δ-passivity properties of the resulting *mixed* EDM.

Lemma 4.4 *Consider a society where each population uses either a PBR, SEPT, or IPC revision protocol. Then, the resulting EDM is δ-passive with informative σ function given by $\sigma(\mathbf{x}, \mathbf{p}) = \sum_{k \in \mathcal{P}} \sigma^k\left(\mathbf{x}^k, \mathbf{p}^k\right)$, where*

$$\sigma^k\left(\mathbf{x}^k, \mathbf{p}^k\right) = \begin{cases} \sigma^k_{PBR}\left(\mathbf{x}^k, \mathbf{p}^k\right), & \text{if population } k \text{ uses a PBR revision protocol} \\ \sigma^k_{SEPT}\left(\mathbf{x}^k, \mathbf{p}^k\right), & \text{if population } k \text{ uses a SEPT revision protocol} \\ \sigma^k_{IPC}\left(\mathbf{x}^k, \mathbf{p}^k\right), & \text{if population } k \text{ uses an IPC revision protocol,} \end{cases}$$

with

$$\begin{aligned} \sigma^k_{PBR}\left(\mathbf{x}^k, \mathbf{p}^k\right) &= -R^k\left(\mathbf{p}^k - \nabla_{\mathbf{x}^k}\omega^k\left(\mathbf{x}^k\right)\right)^\top \mathcal{V}^k\left(\mathbf{x}^k, \mathbf{p}^k\right) \\ \sigma^k_{SEPT}\left(\mathbf{x}^k, \mathbf{p}^k\right) &= R^k \sum_{i \in S^k} \mathcal{V}^k_i\left(\mathbf{x}^k, \mathbf{p}^k\right) p^k_i \sum_{j \in S^k} \phi^k_j\left(p^k_j - \frac{1}{m^k}\mathbf{x}^{k\top}\mathbf{p}^k\right) \\ \sigma^k_{IPC}\left(\mathbf{x}^k, \mathbf{p}^k\right) &= -R^k \sum_{i \in S^k} \mathcal{V}^k_i\left(\mathbf{x}^k, \mathbf{p}^k\right) \sum_{\ell \in S^k} \int_0^{p^k_\ell - p^k_i} \phi^k_\ell(\zeta)\, d\zeta. \end{aligned}$$

◊

Proof The proof follows readily from Lemmas 4.1, 4.2, and 4.3; viewing the overall EDM as the parallel concatenation of δ-passive systems as in Lemma 3.9. □

4.3.2 Smoothing-Anticipatory Payoff Dynamics Model

According to Corollaries 4.1 and 4.2, once the EDM satisfies Conditions (D1) or (D3), the asymptotic stability of the corresponding sets NE($\mathbf{f}$) or NE($\tilde{\mathbf{f}}$) is guaranteed by ensuring that the PDM satisfies Condition (D2). Specifically, Condition (D2) asks for the following requirements:

(i) The PDM is bounded.
(ii) The PDM is δ-antipassive characterized by an informative σ_{PDM}.

(iii) The PDM satisfies Condition (S).
(iv) The set

$$\mathcal{E}_{\mathcal{W}}\left(\mathbf{x}^*\right) = \left\{\mathbf{q}^* \in \mathbb{R}^d : \mathcal{W}\left(\mathbf{q}^*, \mathbf{x}^*\right) = \mathbf{0}_d\right\}$$

is nonempty for every $\mathbf{x}^* \in \mathrm{NE}(\mathbf{f})$.

This section shows that the so-called smoothing-anticipatory PDM, introduced in Example 2.4, readily satisfies Requirements (i), (iii), and (iv) and satisfies Requirement (ii) granted some sufficient conditions.

As introduced in Example 2.4 and revisited in Example 4.1, the so-called smoothing-anticipatory PDM is of the form

$$\dot{\mathbf{q}}(t) = \tau\left(\mathbf{f}\left(\mathbf{x}(t)\right) - \mathbf{q}(t)\right), \tag{4.12a}$$

$$\mathbf{p}(t) = \alpha_0 \mathbf{f}\left(\mathbf{x}(t)\right) + \alpha_1 \mathbf{q}(t) + \alpha_2 \dot{\mathbf{q}}(t), \tag{4.12b}$$

where $\tau \in \mathbb{R}_{>0}$ and $\alpha_0, \alpha_1, \alpha_2 \in \mathbb{R}_{\geq 0}$ are tuning parameters subject to $\alpha_0 + \alpha_1 = 1$.

It is straightforward to check that every smoothing-anticipatory PDM indeed satisfies Requirements (iii) and (iv). Namely, $\mathcal{W}(\mathbf{x}^*, \mathbf{q}^*) = \tau\left(\mathbf{f}(\mathbf{x}^*) - \mathbf{q}^*\right) = \mathbf{0}_n$ implies that $\mathbf{q}^* = \mathbf{f}(\mathbf{x}^*)$, which in turn implies that $\mathcal{H}(\mathbf{q}^*, \mathbf{x}^*) = \alpha_0 \mathbf{f}(\mathbf{x}^*) + \alpha_1 \mathbf{q}^* + \alpha_2 \mathbf{0}_n = \mathbf{f}\left(\mathbf{x}^*\right)$ (recall that $\alpha_0 + \alpha_1 = 1$ by assumption). Therefore, Requirement (iii) holds. On the other hand, note that for every $\mathbf{x} \in \Delta$, setting $\mathbf{q} = \mathbf{f}(\mathbf{x})$ yields $\mathcal{W}(\mathbf{q}, \mathbf{x}) = \tau\left(\mathbf{f}(\mathbf{x}) - \mathbf{f}(\mathbf{x})\right) = \mathbf{0}_n$, meaning that the set $\mathcal{E}_{\mathcal{W}}(\mathbf{x})$ is nonempty for every $\mathbf{x} \in \Delta$. Thus, Requirement (iv) holds. As such, it is only left to investigate whether the smoothing-anticipatory PDM satisfies Requirements (i) and (ii).

To show that (i) holds, notice that the explicit solution for $\mathbf{q}(t)$ is given by

$$\mathbf{q}(t) = \tau \int_0^t e^{-\tau(t-\zeta)} \mathbf{f}\left(\mathbf{x}(\zeta)\right) d\zeta + e^{-\tau t} \mathbf{q}(0),$$

for all $t \geq 0$. Since $\mathbf{f}(\cdot)$ is continuous (Assumption 2.2) and $\mathbf{x}(t) \in \Delta$, for all $t \geq 0$ (Definition 2.4), it follows that

$$\begin{aligned}\|\mathbf{q}(t)\|_\infty &\leq \max_{\tilde{\mathbf{x}} \in \Delta} \|\mathbf{f}(\tilde{\mathbf{x}})\|_\infty + \|\mathbf{q}(0)\|_\infty \\ &< \infty,\end{aligned}$$

implying that the PDM is bounded. Therefore, every smoothing-anticipatory PDM satisfies Requirement (i).

Finally, to investigate Requirement (ii), observe that every smoothing-anticipatory PDM can be rewritten in the equivalent form

$$\dot{\mathbf{q}}(t) = \tau\left(\mathbf{f}\left(\mathbf{x}(t)\right) - \mathbf{q}(t)\right), \tag{4.13a}$$

$$\mathbf{p}(t) = \beta_0 \mathbf{f}\left(\mathbf{x}(t)\right) + \beta_1 \mathbf{q}(t), \tag{4.13b}$$

with $\beta_0 = \alpha_0 + \alpha_2\tau$ and $\beta_1 = \alpha_1 - \alpha_2\tau$. Subject to the assumption that $\mathbf{f}(\cdot)$ is continuously differentiable, applying Lemma 3.3 to (4.13) reveals that the δ-antipassivity of the smoothing-anticipatory PDM is guaranteed if

$$\mathrm{D}_{\mathbf{x}}\,\mathbf{f}(\mathbf{x}) + \mathrm{D}_{\mathbf{x}}\,\mathbf{f}(\mathbf{x})^\top \prec 0, \tag{4.14}$$

and

$$-\tau\nabla_{\mathbf{q}}S(\mathbf{q},\mathbf{x})^\top\left(\mathbf{f}(\mathbf{x})-\mathbf{q}\right) + \frac{1}{2\beta_0}\boldsymbol{\Psi}^\top\left(\mathrm{D}_{\mathbf{x}}\,\mathbf{f}(\mathbf{x}) + \mathrm{D}_{\mathbf{x}}\,\mathbf{f}(\mathbf{x})^\top\right)^{-1}\boldsymbol{\Psi} \geq \sigma(\mathbf{q},\mathbf{x}), \tag{4.15}$$

for all $(\mathbf{q},\mathbf{x}) \in \mathbb{R}^n \times \Delta$, for some non-negative continuously differentiable $S(\cdot,\cdot)$ and some non-negative $\sigma(\cdot,\cdot)$. Here,

$$\boldsymbol{\Psi} := \beta_1\tau\left(\mathbf{f}(\mathbf{x})-\mathbf{q}\right) + \nabla_{\mathbf{x}}S(\mathbf{q},\mathbf{x}).$$

Namely, (4.14) asks for the underlying game $G = (\mathcal{P}, \Delta, \mathbf{f}(\cdot))$ to be strictly contractive in the sense of Definition 2.1; and (4.15) requires that its left-hand side can be lower-bounded by a non-negative scalar-valued function $\sigma(\cdot,\cdot)$. In fact, the informativity requirement in (ii) further implies for $\sigma(\cdot,\mathbf{x})$ to be positive definite with respect to $\mathcal{E}_{\mathcal{W}}(\mathbf{x})$, for all $\mathbf{x} \in \Delta$. Both of these conditions can be satisfied if the underlying game G satisfies some additional properties, as outlined next.

Lemma 4.5 *Let $G = (\mathcal{P}, \Delta, \mathbf{f}(\cdot))$ be a μ-strongly contractive population game (Definition 2.1), whose fitness vector $\mathbf{f} : \mathbb{R}^n_{\geq 0} \to \mathbb{R}^n$ is continuously differentiable and θ-Lipschitz continuous within Δ, i.e., $\|\mathbf{f}(\tilde{\mathbf{x}}) - \mathbf{f}(\mathbf{x})\|_2 \leq \theta\,\|\tilde{\mathbf{x}} - \mathbf{x}\|_2$, for all $\tilde{\mathbf{x}}, \mathbf{x} \in \Delta$, for some $\theta \in \mathbb{R}_{>0}$. If $\beta_1 > 0$ and*

$$4\beta_0\mu > (\beta_1 + \theta)^2, \tag{4.16}$$

then, the associated smoothing-anticipatory PDM (4.13) is δ-antipassive characterized by

$$\sigma(\mathbf{q},\mathbf{x}) = \tau^2\left(1 - \frac{(\beta_1+\theta)^2}{4\beta_0\mu}\right)\|\mathbf{f}(\mathbf{x}) - \mathbf{q}\|_2^2, \tag{4.17}$$

which is informative in the sense of Definition 3.1. ◊

Proof From the strong contractivity of G it follows that

$$\mathrm{D}_{\mathbf{x}}\,\mathbf{f}(\mathbf{x}) + \mathrm{D}_{\mathbf{x}}\,\mathbf{f}(\mathbf{x})^\top \prec -2\mu\mathbf{I}_n,$$

with $\mu > 0$. Thus, (4.14) holds and the δ-antipassivity of the PDM can be assessed via (4.15).

Consider the δ-storage function given by

$$S(\mathbf{q},\mathbf{x}) = \frac{1}{2\tau}\left\|\tau\left(\mathbf{f}(\mathbf{x}) - \mathbf{q}\right)\right\|_2^2.$$

It then follows that

$$\nabla_{\mathbf{q}} S(\mathbf{q}, \mathbf{x}) = -\tau \left(\mathbf{f}(\mathbf{x}) - \mathbf{q}\right)$$
$$\nabla_{\mathbf{x}} S(\mathbf{q}, \mathbf{x}) = \tau \left(\mathrm{D}_{\mathbf{x}} \mathbf{f}(\mathbf{x})\right)^\top \left(\mathbf{f}(\mathbf{x}) - \mathbf{q}\right)$$
$$\boldsymbol{\Psi} := \tau \left(\beta_1 \mathbf{I}_n + \mathrm{D}_{\mathbf{x}} \mathbf{f}(\mathbf{x})\right)^\top \left(\mathbf{f}(\mathbf{x}) - \mathbf{q}\right).$$

Thus, the left-hand side of (4.15) is equal to

$$\begin{aligned}
&= \tau^2 \|\mathbf{f}(\mathbf{x}) - \mathbf{q}\|_2^2 + \frac{1}{2\beta_0} \boldsymbol{\Psi}^\top \left(\mathrm{D}_{\mathbf{x}} \mathbf{f}(\mathbf{x}) + \mathrm{D}_{\mathbf{x}} \mathbf{f}(\mathbf{x})^\top\right)^{-1} \boldsymbol{\Psi} \\
&\geq \tau^2 \|\mathbf{f}(\mathbf{x}) - \mathbf{q}\|_2^2 - \frac{1}{4\beta_0 \mu} \boldsymbol{\Psi}^\top \boldsymbol{\Psi} \\
&\geq \tau^2 \|\mathbf{f}(\mathbf{x}) - \mathbf{q}\|_2^2 - \frac{\tau^2}{4\beta_0 \mu} \|\beta_1 \mathbf{I}_n + \mathrm{D}_{\mathbf{x}} \mathbf{f}(\mathbf{x})\|_2^2 \|\mathbf{f}(\mathbf{x}) - \mathbf{q}\|_2^2 \\
&\geq \tau^2 \|\mathbf{f}(\mathbf{x}) - \mathbf{q}\|_2^2 - \frac{\tau^2}{4\beta_0 \mu} (\beta_1 + \theta)^2 \|\mathbf{f}(\mathbf{x}) - \mathbf{q}\|_2^2 \\
&= \tau^2 \left(1 - \frac{(\beta_1 + \theta)^2}{4\beta_0 \mu}\right) \|\mathbf{f}(\mathbf{x}) - \mathbf{q}\|_2^2 .
\end{aligned}$$

Consequently, due to (4.16) one can take σ as in (4.17) to satisfy (4.15). □

Lemma 4.5 provides sufficient conditions on the underlying population game G and the parameters of the PDM that guarantee that δ-antipassivity with informative σ of the PDM. Consequently, by means of Lemma 4.5 the smoothing-anticipatory PDM satisfies Requirement (ii). Consequently, granted that the sufficient conditions in Lemma 4.5 hold, the smoothing-anticipatory PDM satisfies Condition (D2).

4.4 Simulation Examples

This section provides some numerical simulation examples to illustrate the main theoretical results of this chapter. The *Python* codes to replicate these simulations are available in Listings 4.1 and 4.2 at the end of the chapter.

Scientific Research Topic Selection

Recall the *research topic selection* problem introduced in Sect. 1.2.1 and revisited in Example 2.3. As in Example 2.3, consider the simplified scenario involving a single population of researchers, i.e., $P = 1$, and let the fitness functions be given by

$$f_i(x_i) = a_i \exp\left(-\left(\frac{x_i - b_i}{c_i}\right)^2\right),$$

with parameters $a_i, b_i, c_i \in \mathbb{R}_{>0}$ for all $i \in \mathcal{S}$. Since these fitness functions are decoupled across strategies, the Jacobian matrix $\mathrm{D}_{\mathbf{x}}\,\mathbf{f}(\cdot)$ is diagonal, with entries given by

$$\frac{df_i(x_i)}{dx_i} = -2a_i\left(\frac{x_i - b_i}{(c_i)^2}\right)\exp\left(-\left(\frac{x_i - b_i}{c_i}\right)^2\right), \quad \forall i \in \mathcal{S}.$$

It is evident that, for each i, it holds that

$$\frac{df_i(x_i)}{dx_i} > 0, \quad \forall x_i \in [0, b_i),$$
$$\frac{df_i(x_i)}{dx_i} \leq 0, \quad \forall x_i \in [b_i, m].$$

Hence, the corresponding population game $G = (\{1\}, \Delta, \mathbf{f}(\cdot))$ is not contractive for any $b_i > 0$, and its contractivity deficit is given by

$$\nu^* = \max_{i \in \mathcal{S}} \max_{x_i \in [0, b_i)} -2a_i\left(\frac{x_i - b_i}{(c_i)^2}\right)\exp\left(-\left(\frac{x_i - b_i}{c_i}\right)^2\right). \tag{4.18}$$

In contrast, consider the modified population game $\hat{G} = \left(\{1\}, \Delta, \hat{\mathbf{f}}(\cdot)\right)$, with

$$\hat{\mathbf{f}}(\mathbf{x}) = \mathbf{f}(\mathbf{x}) - (\nu^* + \mu)\mathbf{x}, \tag{4.19}$$

where $\mu \in \mathbb{R}_{>0}$ is an arbitrarily small positive scalar. This modified game is indeed μ-strongly contractive.

The contractivity properties of G and $\hat{G}$ play a critical role in certifying whether the researchers' decision-making process converges to an NE. According to Lemma 2.5, while $\hat{G}$ admits a unique NE, the original game G does not necessarily exhibit this uniqueness. In fact, G may have multiple isolated Nash equilibria that do not form a convex set, implying that convergence to any specific NE of G cannot be guaranteed from arbitrary initial conditions in Δ.

To illustrate this, consider a scenario in which 700 researchers choose among three available strategies, i.e., $\mathcal{S} = \{1, 2, 3\}$, and let $a_i = 1$, $b_i = 0.4$, and $c_i = 0.5$ for all $i \in \mathcal{S}$. A plot of the resulting $f_i(\cdot)$, $df_i(\cdot)/dx_i$, and the corresponding ν^* is shown in Fig. 4.1. Additionally, assume that the researchers revise their strategy selection using IPC-type revision protocols of the form

$$\rho_{ij}(\mathbf{p}, \mathbf{x}) = \frac{1}{\varsigma}\max\left\{p_j - p_i, 0\right\}, \quad \forall i, j \in \mathcal{S},$$

where the parameter $\varsigma \in \mathbb{R}_{>0}$ is chosen large enough to ensure satisfaction of the IPC condition in (2.14). Let revision times be distributed according to Poisson alarm clocks with rate $R = 1$ (in arbitrary time units), and suppose the underlying payoff

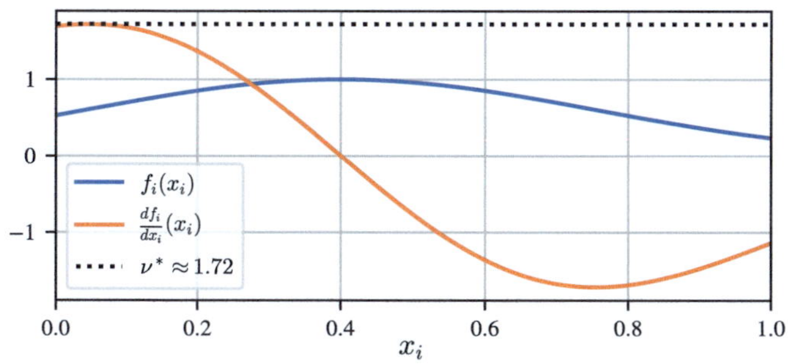

Fig. 4.1 Fitness function $f_i(\cdot)$, its derivative with respect to x_i, and the associated value of ν^* for the game G when $a_i = 1$, $b_i = 0.4$, and $c_i = 0.5$, for all $i \in S$

mechanism is characterized by the smoothing-anticipatory PDM (4.12) with $\alpha_0 = 1$ and $\alpha_1 = \alpha_2 = 0$. That is, the researchers' payoff mechanism reduces to a memoryless map of the form

$$\mathbf{p}(t) = \mathbf{f}(\mathbf{x}(t)).$$

Following Example 3.1, it can be shown that this simplified PDM is not δ-antipassive. However, the modified memoryless PDM

$$\mathbf{p}(t) = \hat{\mathbf{f}}(\mathbf{x}(t)),$$

with $\hat{\mathbf{f}}(\cdot)$ defined as in (4.19), is in fact δ-antipassive. Therefore, while convergence to a unique NE can be certified for the game $\hat{G}$ (via Corollary 4.1), such certification is not possible for the original game G.

Due to the homogeneity of the fitness function parameters across all strategies, and assuming a unitary population mass $m = 1$, it follows that

$$\mathbf{x}^* = \left[\tfrac{1}{3}, \tfrac{1}{3}, \tfrac{1}{3}\right]^\top$$

is an NE of both population games G and $\hat{G}$, with associated fitness values:

$$f_1(\mathbf{x}^*) \approx 0.98,$$
$$f_2(\mathbf{x}^*) \approx 0.98,$$
$$f_3(\mathbf{x}^*) \approx 0.98.$$

However, since G is not contractive, $\mathbf{x}^*$ may not be the unique NE. For instance,

$$\tilde{\mathbf{x}}^* = \left[\tfrac{1}{2}, 0, \tfrac{1}{2}\right]^\top$$

is also an NE of G (but not of $\hat{G}$), with corresponding fitness values:

$$f_1(\tilde{\mathbf{x}}^*) \approx 0.96,$$
$$f_2(\tilde{\mathbf{x}}^*) \approx 0.53,$$
$$f_3(\tilde{\mathbf{x}}^*) \approx 0.96.$$

It is therefore of interest to examine the behavior of the population near the shared equilibrium $\mathbf{x}^*$ under both G and $\hat{G}$. To this end, the following numerical simulations assume the initial condition

$$\mathbf{x}(0) = \left[\tfrac{1.1}{3}, \tfrac{0.9}{3}, \tfrac{1}{3}\right]^\top .$$

Figure 4.2 reveals that, for the specified parameters and when involved in game G, the population does not converge to $\mathbf{x}^*$. Instead, it converges to the alternative NE $\tilde{\mathbf{x}}^*$. Although both $\mathbf{x}^*$ and $\tilde{\mathbf{x}}^*$ are Nash equilibria, the former yields higher fitness values and is therefore deemed as more desirable. This result highlights how the lack of contractivity in G can significantly affect the population's long-term performance.

For comparison, the same simulation is repeated under the modified game $\hat{G} = \left(\{1\}, \Delta, \hat{\mathbf{f}}(\cdot)\right)$, where

$$\hat{\mathbf{f}}(\mathbf{x}) = \mathbf{f}(\mathbf{x}) - 1.73\mathbf{x}.$$

This instance corresponds to (4.19), with $\nu^* = 1.72$ and $\mu = 0.01$. As shown in Fig. 4.3, the population of researchers now does converge to $\mathbf{x}^*$.

This example remarks the importance of establishing sufficient conditions to certify the asymptotic stability of Nash equilibria. In the absence of such guarantees, equilibria with favorable outcomes may become unstable or unreachable under population dynamics, thereby compromising the long-term effectiveness and predictability of the system. The results presented in this chapter thus provide essential tools for ensuring the reliability of equilibrium-seeking processes in practical applications.

Load-Balancing in a Computing Center

Consider the *load-balancing in a computing center* example introduced in Sect. 1.2.2. The objective here is to validate convergence to Nash equilibria under the various classes of revision protocols, and their associated evolutionary dynamics, discussed throughout this chapter.

For simplicity, let the fitness functions be defined as

$$f_i(x_i) = s_i - \alpha_i x_i, \quad \forall i \in \mathcal{S},$$

where $s_i \in \mathbb{R}_{>0}$ denotes the nominal processing speed of server i, and $\alpha_i \in \mathbb{R}_{>0}$ is a strictly positive weighting parameter. Evidently, provided that $\alpha_i > 0$ for all $i \in \mathcal{S}$, the resulting population game is μ-strongly contractive, with $\mu = \min_{i \in \mathcal{S}} \alpha_i$. In fact, the game corresponds to a full-potential game with a (strictly concave) potential function given by

$$\varphi(\mathbf{x}) = \sum_{i \in \mathcal{S}} \left(s_i x_i - \frac{\alpha_i}{2} x_i^2 \right).$$

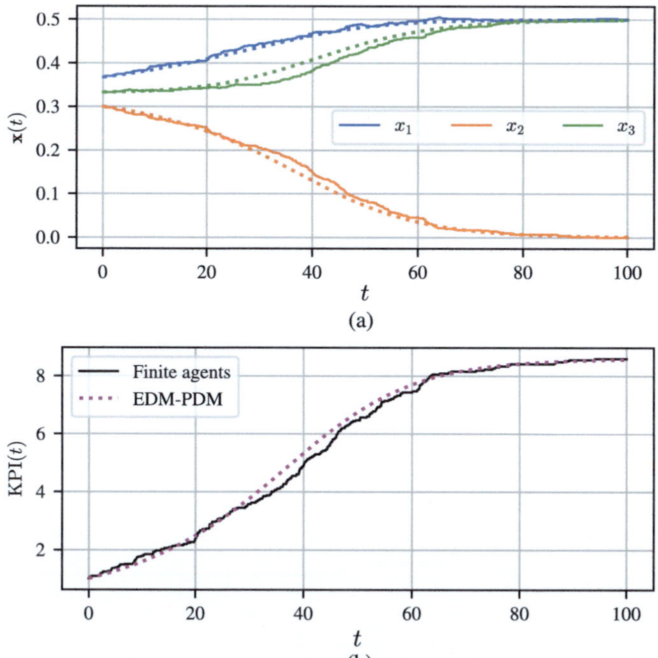

Fig. 4.2 Temporal evolution of **a** the strategic distribution $\mathbf{x}$; **b** the selected metric $\text{KPI}(t) = \|\mathbf{x}(t) - \mathbf{x}^*\|_2 \,/\, \|\mathbf{x}(0) - \mathbf{x}^*\|_2$, with $\mathbf{x}^* = \left[1/3,\ 1/3,\ 1/3\right]^\top$, for both the finite-agent simulation and the EDM–PDM approximation, under the original (not contractive) population game G. The dotted lines in (**a**) depict the EDM–PDM deterministic approximation

Assume the payoff signals are given by $p_i(t) = f_i(x_i(t))$ for all $i \in \mathcal{S}$, corresponding to a memoryless payoff mechanism. As shown in Example 3.1, the associated PDM is δ-antipassive. Consequently, by Corollary 4.1, it follows that $\mathbf{x}^* \in \text{NE}(\mathbf{f}) = \arg\max_{\mathbf{x}\in\Delta} \varphi(\mathbf{x})$ is asymptotically stable under the SEPT and IPC EDMs, as well as under any mixture of these EDMs. Similarly, by Corollary 4.2, it follows that $\mathbf{x}^* \in \text{NE}(\tilde{\mathbf{f}}) = \arg\max_{\mathbf{x}\in\Delta} \varphi(\mathbf{x}) - \varpi(\mathbf{x})$ is asymptotically stable under the PBR EDM, or under any combination of PBR with SEPT and IPC EDMs.

As an illustration, consider a scenario where 400 job schedulers choose among three servers, $\mathcal{S} = \{1, 2, 3\}$, with parameters $\alpha_1 = 1$, $\alpha_2 = 5$, $\alpha_3 = 2$, and $s_i = 10$ for all $i \in \mathcal{S}$. Let the initial conditions be $x_i(0) = 1/3$ for all $i \in \mathcal{S}$, and assume Poisson clocks with rate $R = 1$ (in arbitrary time units). Figures 4.4, 4.5, and 4.6 depict the temporal evolution of the population under SEPT-type, IPC-type, and PBR-type revision protocols, respectively. In all cases, convergence toward the expected outcome is observed, in accordance with Corollaries 4.1 and 4.2.

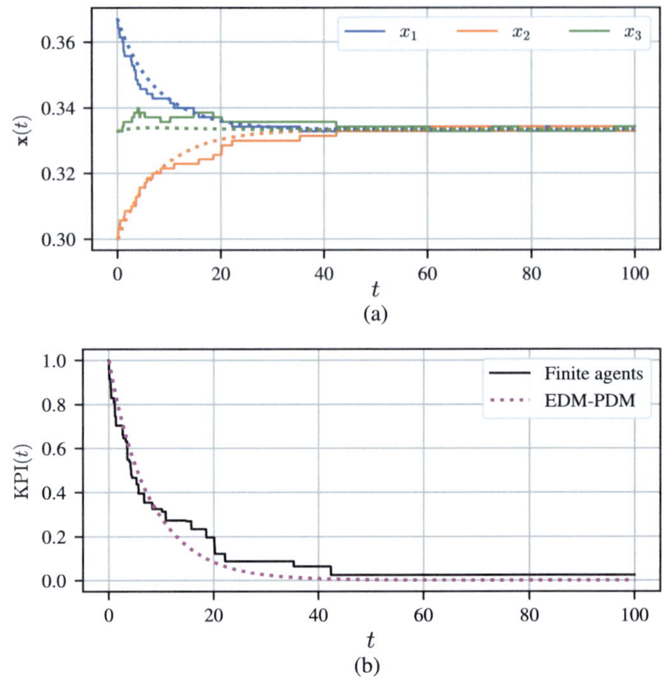

Fig. 4.3 Temporal evolution of **a** the strategic distribution **x**; **b** the selected metric $\text{KPI}(t) = \|\mathbf{x}(t) - \mathbf{x}^*\|_2 \,/\, \|\mathbf{x}(0) - \mathbf{x}^*\|_2$, with $\mathbf{x}^* = \left[1/3,\ 1/3,\ 1/3\right]^\top$, for both the finite-agent simulation and the EDM–PDM approximation, under the modified (strongly contractive) population game $\hat{G}$. The dotted lines in (**a**) depict the EDM–PDM deterministic approximation

Fig. 4.4 Evolution of the population under the BNN revision protocol

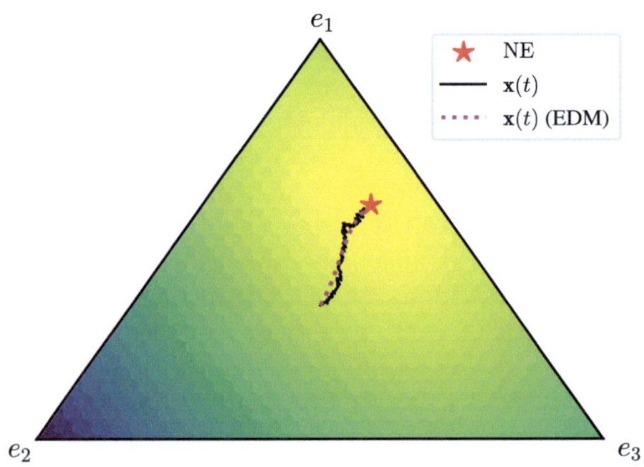

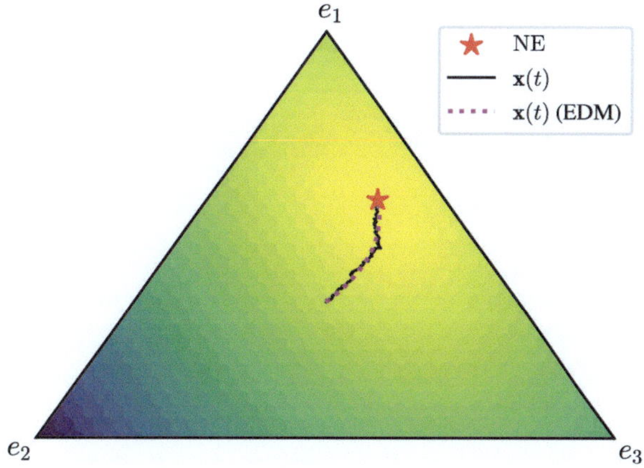

Fig. 4.5 Evolution of the population under the Smith revision protocol

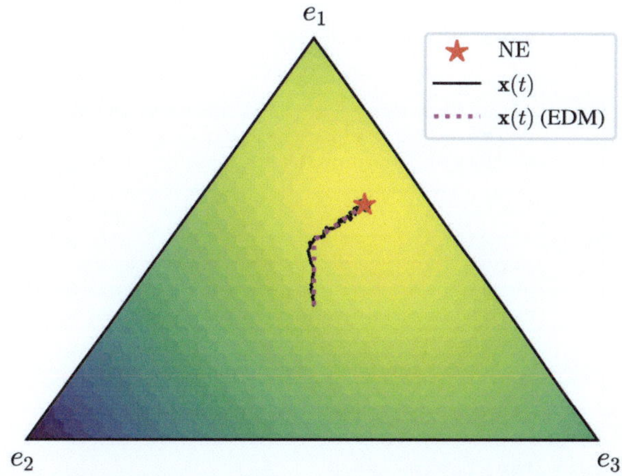

Fig. 4.6 Evolution of the population under the *logit-choice* revision protocol

Python Scripts

```
import numpy as np
import popgames as pg

a, b, c = 1, 0.4, 0.5 # Parameter values
modify_game = False # Set to True to simulate \hat{G}

def f(x):
    return a*np.exp(-np.power((x-b)/c, 2)) \
        - 1.73*x*modify_game

population_game = pg.SinglePopulationGame(
    fitness_function=f,
    num_strategies=3
)

payoff_mechanism = pg.PayoffMechanism(
    h_map=f, n=3,
)

revision_process = pg.PoissonRevisionProcess(
    Poisson_clock_rate=1.0,
    revision_protocol=pg.protocol.Smith(scale=0.1)
)

sim = pg.Simulator(
    population_game=population_game,
    payoff_mechanism=payoff_mechanism,
    revision_processes=revision_process,
    num_agents=700
)

x0 = np.array([[1.1/3], [0.9/3], [1/3]])
sim.reset(x0=x0)
sim.run(T_sim=100)
sim.plot(
    plot_type='univariate',
    plot_deterministic_approximation=True
)
sim.plot(
    plot_type='kpi',
    plot_deterministic_approximation=True
)
```

Listing 4.1 Python code to replicate the simulation of Figs. 4.2 and 4.3

```python
import numpy as np
import popgames as pg

# Parameters of the game
a, s = np.array([[1], [5], [2]]), 10

# Selected revision protocols to choose from
rev_protocols = [
    pg.protocol.BNN(scale=0.1),
    pg.protocol.Smith(scale=0.1),
    pg.protocol.Softmax(eta=0.01),
]

def f(x):
    return s - a*x

population_game = pg.SinglePopulationGame(
    num_strategies=3,
    fitness_function=f
)

payoff_mechanism = pg.PayoffMechanism(
    h_map=f, n=3
)

revision_process = pg.PoissonRevisionProcess(
    Poisson_clock_rate=1,
    revision_protocol=rev_protocols[0], # Set to 0, 1, or 2
                                        # depending on the
                                        # desired protocol to use
)

sim = pg.Simulator(
    population_game=population_game,
    payoff_mechanism=payoff_mechanism,
    revision_processes=revision_process,
    num_agents=400
)

sim.reset(x0=np.array([1/3, 1/3, 1/3]).reshape(3,1))
sim.run(T_sim=100)

def varphi(x):
    return (s*x).sum() - 0.5*np.dot(x.T, a*x)

sim.plot(
    plot_type='ternary',
    potential_function=varphi,
    plot_deterministic_approximation=True,
)
```

Listing 4.2 Python code to replicate the simulation of Figs. 4.4, 4.5 and 4.6

Notes

The results presented in this chapter synthesize several research contributions from recent literature. Key references include [1, 6, 8, 10], among others.

In particular, [10] establishes convergence properties of various evolutionary dynamics when coupled with memoryless payoff mechanisms of different classes. Building on these results, [1, 8] extend the analysis to dynamic payoff mechanisms, thereby fully generalizing the case studied in [10]. This extension leverages the δ-dissipativity framework introduced in Chap. 3. While these works provide important convergence guarantees for both static and dynamic payoff mechanisms, they primarily focus on convergence to conventional Nash equilibrium (NE) points. In contrast, [6] broadens the scope by addressing generalized Nash equilibrium (GNE) seeking, which is the focus of this book.

Several evolutionary dynamics models (EDMs) considered in these works satisfy Assumption 5.1. For example, EDMs derived from excess payoff targets or from impartial pairwise comparison revision protocols, such as the Brown–von Neumann–Nash (BNN) and Smith dynamics, comply with this assumption [1, 8]. The regularized logit EDM proposed in [7] also satisfies Assumption 5.1.

Multiple classes of payoff dynamics models (PDMs) also satisfy Assumption 5.2. One example is the family of systems introduced in [5, Sect. III], which includes smoothed learning dynamics and consensus-based distributed estimation schemes. These systems meet the assumption under suitable conditions, as established in [5, Theorem 1 and Lemma 6]. Further results and discussions on the smoothing-anticipatory PDM introduced in Sect. 4.3.2 can be found in [9]. The interested reader is encouraged to consult these works for additional insights.

References

1. Arcak M, Martins NC (2021) Dissipativity tools for convergence to Nash equilibria in population games. IEEE Trans Control of Netw Syst 8(1):39–50. https://doi.org/10.1109/TCNS.2020.3029990
2. Facchinei F, Pang JS (2003) Finite-dimensional variational inequalities and complementarity problems (vol 1). Springer
3. Haddad WM, Chellaboina V (2008) Nonlinear dynamical systems and control: a Lyapunov-based approach. Princeton University Press
4. Hofbauer J, Sandholm WH (2009) Stable games and their dynamics. J Econ Theory 144(4):1665–1693
5. Martinez-Piazuelo J, Ocampo-Martinez C, Quijano N (2024) Distributed Nash equilibrium seeking in strongly contractive aggregative population games. IEEE Trans Autom Control 69(7): 4427–4442. https://doi.org/10.1109/TAC.2023.3321208
6. Martinez-Piazuelo J, Ocampo-Martinez C, Quijano N (2024) Generalized Nash equilibrium seeking in a class of contractive population games over networks. In: Proceedings of the 63rd IEEE conference on decision and control (CDC), Milan, Italy, 2319–2324. https://doi.org/10.1109/CDC56724.2024.10886517

7. Park S, Certorio J, Martins NC, La RJ (2025) Epidemic population games and perturbed best response dynamics. IEEE Trans Autom Control 1–13. https://doi.org/10.1109/TAC.2025.3554601
8. Park S, Martins NC, Shamma JS (2019) From population games to payoff dynamics models: a passivity-based approach. In: Proceedings of the 58th IEEE conference on decision and control (CDC), pp 6584–6601. https://doi.org/10.1109/CDC40024.2019.9029756
9. Park S, Martins NC, Shamma JS (2019) Payoff dynamics model and evolutionary dynamics model Feedback and convergence to equilibria. https://doi.org/10.48550/ARXIV.1903.02018. arXiv:1903.02018
10. Sandholm WH (2010) Population games and evolutionary dynamics. MIT Press

Part III
Design

Chapter 5
Controlling Decision-Making in Large Populations

Chapter 4 has formulated sufficient conditions to certify the asymptotic convergence of EDM–PDM pairs toward a GNE of the underlying population game. Specifically, it is shown in Theorem 4.2 that the set GNE($\mathbf{f}$), associated to a population game $G = (\mathcal{P}, \mathcal{X}, \mathbf{f}(\cdot))$, is asymptotically stable under the EDM–PDM dynamics, granted that the EDM–PDM pair satisfies the sufficient Conditions (C1)–(C5), which are summarized next for convenience:

(C1) The EDM is Nash stationary, i.e., it satisfies the Nash stationarity property:

$$\boldsymbol{\mathcal{V}}\left(\mathbf{x}^*, \mathbf{p}^*\right) = \mathbf{0}_n \Leftrightarrow \mathbf{x}^* \in \arg\max_{\mathbf{x} \in \Delta} \mathbf{x}^\top \mathbf{p}^*.$$

(C2) For every $(\mathbf{q}^*, \mathbf{x}^*) \in \mathbb{R}^d \times \Delta$, the PDM satisfies that $\boldsymbol{\mathcal{W}}(\mathbf{q}^*, \mathbf{x}^*) = \mathbf{0}_d$ implies

$$\begin{aligned} \boldsymbol{\mathcal{H}}\left(\mathbf{q}^*, \mathbf{x}^*\right) &= \mathbf{f}\left(\mathbf{x}^*\right) - \mathbf{A}^\top \mathbf{M}_\mu \mathbf{q}^* - \mathrm{D}\,\mathbf{g}\left(\mathbf{x}^*\right)^\top \mathbf{M}_\lambda \mathbf{q}^* \\ \left(\mathbf{M}_\lambda \mathbf{q}^*\right) \odot \mathbf{g}\left(\mathbf{x}^*\right) &= \mathbf{0}_{c_\leq} \\ \mathbf{M}_\lambda \mathbf{q}^* &\succeq \mathbf{0}_{c_\leq} \\ \mathbf{x}^* &\in \Omega, \end{aligned}$$

for some matrices $\mathbf{M}_\mu \in \mathbb{R}^{c_= \times d}$ and $\mathbf{M}_\lambda \in \mathbb{R}^{c_\leq \times d}$.

(C3) The equilibrium set $\mathcal{E}$ is nonempty and bounded.

(C4) The EDM and PDM are δ-dissipative characterized by $(\sigma_{EDM}, \boldsymbol{\Pi}_{EDM})$ and $(\sigma_{PDM}, \boldsymbol{\Pi}_{PDM})$, respectively, and for every $(\mathbf{x}, \mathbf{q}) \in \Delta \times \mathbb{R}^d$ it holds that

$$\left\| \begin{matrix} \dot{\boldsymbol{\mathcal{H}}}(\mathbf{q}, \mathbf{x}) \\ \overline{\boldsymbol{\mathcal{V}}}(\mathbf{x}, \mathbf{q}) \end{matrix} \right\|^2_{\boldsymbol{\Pi}_{EDM}} + \left\| \begin{matrix} \overline{\boldsymbol{\mathcal{V}}}(\mathbf{x}, \mathbf{q}) \\ \dot{\boldsymbol{\mathcal{H}}}(\mathbf{q}, \mathbf{x}) \end{matrix} \right\|^2_{\boldsymbol{\Pi}_{PDM}} \leq 0,$$

where

J. Martinez-Piazuelo et al., *Generalized Nash Equilibrium Seeking in Population Games*, Communications and Control Engineering,
https://doi.org/10.1007/978-3-032-06081-5_5

$$\overline{\mathcal{V}}(\mathbf{x}, \mathbf{q}) = \mathcal{V}\left(\mathbf{x}, \mathcal{H}(\mathbf{q}, \mathbf{x})\right)$$
$$\dot{\mathcal{H}}(\mathbf{q}, \mathbf{x}) = \mathrm{D}_{\mathbf{q}}\mathcal{H}(\mathbf{q}, \mathbf{x})\mathcal{W}(\mathbf{q}, \mathbf{x}) + \mathrm{D}_{\mathbf{x}}\mathcal{H}(\mathbf{q}, \mathbf{x})\overline{\mathcal{V}}(\mathbf{x}, \mathbf{q}).$$

(C5) The equilibrium set $\mathcal{E}$ is the largest invariant set of the EDM–PDM pair contained in $\mathcal{R}$, with

$$\mathcal{R} = \left\{ (\mathbf{x}, \mathbf{q}) \in \Delta \times \mathbb{R}^d : \begin{array}{r} \sigma_{EDM}\left(\mathbf{x}, \mathcal{H}(\mathbf{q}, \mathbf{x}), \mathbf{0}_n\right) = 0 \\ \sigma_{PDM}\left(\mathbf{q}, \mathbf{x}, \mathbf{0}_n\right) = 0 \end{array} \right\}.$$

Alternatively, Theorem 4.2 stated a similar result to assert the asymptotic stability of GNE($\tilde{\mathbf{f}}$), with $\tilde{\mathbf{f}}(\mathbf{x}) = \mathbf{f}(\mathbf{x}) - \nabla_{\mathbf{x}}\varpi(\mathbf{x})$ representing the perturbed fitness function introduced in Definition 2.10. In this latter case, the EDM–PDM pair is assumed to satisfy Conditions (A) and (C2)–(C5), where Condition (A) is stated as:

(A) The EDM satisfies the ϖ-perturbed Nash stationarity property:

$$\mathcal{V}\left(\mathbf{x}^*, \mathbf{p}^*\right) = \mathbf{0}_n \Leftrightarrow \mathbf{x}^* \in \arg\max_{\mathbf{x}\in \mathrm{int}_{\varpi}(\Delta)} \left(\mathbf{x}^\top \mathbf{p}^* - \varpi\left(\mathbf{x}\right)\right),$$

with $\varpi(\cdot)$ and $\mathrm{int}_{\varpi}(\Delta)$ defined as in Definition 2.10.

In addition, Corollaries 4.1 and 4.2 provide simplified sufficient conditions to ensure the asymptotic stability of GNE($\mathbf{f}$) and GNE($\tilde{\mathbf{f}}$), respectively, under some particular classes of EDM–PDM pairs and in the unconstrained game setup with $\Omega = \mathbb{R}^n$. More precisely, Corollary 4.1 states that the set GNE($\mathbf{f}$) is asymptotically stable under the EDM–PDM dynamics, granted that the EDM–PDM pair satisfies Conditions (D1)–(D2), while Corollary 4.2 guarantees the asymptotic stability of GNE($\tilde{\mathbf{f}}$) granted that the EDM–PDM pair satisfies Conditions (D2) and (D3). These conditions are summarized next for convenience:

(D1) The EDM is Nash stationary and δ-passive characterized by an informative σ_{EDM}.
(D2) The PDM is bounded, δ-antipassive characterized by an informative σ_{PDM}, satisfies Condition (A), and the set $\mathcal{E}_{\mathcal{W}}(\mathbf{x}^*) = \left\{\mathbf{q}^* \in \mathbb{R}^d : \mathcal{W}(\mathbf{q}^*, \mathbf{x}^*) = \mathbf{0}_d\right\}$ is nonempty for every $\mathbf{x}^* \in \mathrm{NE}(\mathbf{f})$.
(D3) The EDM satisfies Condition (A) and is δ-passive characterized by an informative σ_{EDM}.

Additionally, Condition (S) reads as follows:

(S) For every $(\mathbf{q}^*, \mathbf{x}^*) \in \mathbb{R}^d \times \Delta$, the PDM satisfies that $\mathcal{W}(\mathbf{q}^*, \mathbf{x}^*) = \mathbf{0}_d$ implies

$$\mathcal{H}\left(\mathbf{q}^*, \mathbf{x}^*\right) = \mathbf{f}\left(\mathbf{x}^*\right).$$

This chapter builds on the above theoretical results to design revision protocols and payoff mechanisms whose associated EDM–PDM dynamics are guaranteed to converge to a GNE of the underlying population game. These design methodologies enable large-scale multi-agent systems to converge, in expectation, to a GNE.

The rest of this chapter is organized as follows. Section 5.1 focuses on prescriptive control approaches aimed at designing revision protocols to enforce constraint satisfaction among the agents' decisions, while ensuring stability of the EDM–PDM dynamics. Section 5.2 addresses the design of centralized and distributed payoff mechanisms that steer the society to achieve constraint satisfaction without prescribing the agents' decision-making behavior. Each of the proposed design approaches is illustrated through numerical simulations based on the applications discussed in Sect. 1.2.

5.1 Prescriptive Control Approaches

This section focuses on scenarios in which the payoff mechanisms are fixed, and the system designer has influence only over the revision process governing the decision-making of agents. Specifically, their revision protocols.

The central objective is to design these protocols in a way that guarantees satisfaction of coupled constraints across the decisions of multiple agents. Naturally, this paradigm is relevant for engineered systems, where the system designer has the capacity to implement and enforce specific revision protocols to shape collective agent behavior.

5.1.1 Constraint Satisfaction via Protocol Design

Throughout this section, consider the scenario where the constraint set Ω is of the form

$$\Omega = \left\{\mathbf{x} \in \mathbb{R}^n : \mathbf{x} \preceq \overline{\mathbf{x}}\right\}. \tag{5.1}$$

Here, $\overline{\mathbf{x}} \in \mathbb{R}^n_{>0}$ upper-bounds the masses of agents that can simultaneously choose each strategy of the game. For the setup to remain meaningful, it is assumed that

$$\sum_{i \in \mathcal{S}^k} \overline{x}_i^k > m^k, \quad \forall k \in \mathcal{P}. \tag{5.2}$$

That is, the relative interior of $\mathcal{X} = \Delta \cap \Omega$ is nonempty.

Assuming that the payoff mechanism is given and cannot be designed, the satisfaction of the constraints in Ω must be enforced via the revision protocols of the decision-making agents. In other words, the agents must take into account the constraint set Ω when revising and updating their strategies. A possibility to achieve this goal is by considering revision protocols of the form

$$\rho_{ij}^k\left(\mathbf{p}^k,\mathbf{x}^k\right)=\left[\overline{x}_j^k-x_j^k\right]_+\phi_j^k\left(p_j^k-p_i^k\right),\quad \forall i,j\in\mathcal{S}^k, \tag{5.3}$$

where $[\cdot]_+=\max\{\cdot,0\}$ and $\phi_j^k:\mathbb{R}\to\mathbb{R}_{\geq 0}$ is a Lipschitz continuous function that satisfies

$$\begin{cases}\phi_j^k(\alpha)>0, & \text{if } \alpha>0\\ \phi_j^k(\alpha)=0, & \text{if } \alpha\leq 0.\end{cases}$$

Similar to SEPT and IPC protocols (see Sect. 2.3.2), for the revision protocol in (5.3) it is required that

$$\sum_{j\in\mathcal{S}^k}\left[\overline{x}_j^k-x_j^k\right]_+\phi_j^k\left(p_j^k-p_i^k\right)\leq 1,\qquad \forall\left(\mathbf{x}^k,\mathbf{p}^k\right)\in\Delta^k\times\mathbb{R}^{n^k}, \tag{5.4}$$

so that it yields valid switching probabilities. Note that if the payoff mechanism under consideration is characterized by a bounded PDM, then condition (5.4) is only required to hold for all $\left(\mathbf{x}^k,\mathbf{p}^k\right)\in\Delta^k\times\mathcal{B}^k$ (see Remark 2.4).

Revision protocols of the form (5.3) can be interpreted as *weighted* IPC protocols, where the switching probabilities are modulated by the remaining capacity of the target strategy to "receive" new agents. Specifically, if $x_j^k\geq\overline{x}_j^k$, then $\rho_{ij}^k\left(\mathbf{p}^k,\mathbf{x}^k\right)=0$ for all $i\in\mathcal{S}^k$, indicating that the probability of any agent switching to strategy $j\in\mathcal{S}^k$ is zero. Thus, these protocols capture scenarios where agents engage in pairwise comparisons between their current strategy and alternatives, as in classical IPC protocols, while adjusting their switching behavior to respect the constraints encoded in Ω. Canonical examples of such capacity-weighted IPC protocols are characterized by $\phi_j^k(\cdot)$ functions of the form

$$\phi_j^k(\alpha)=\frac{1}{n^k\gamma^k}\min\left\{\max\{\alpha,\ 0\},\ \gamma^k\right\},$$

for some arbitrary $\gamma^k\in\mathbb{R}_{>0}$; or

$$\phi_j^k(\alpha)=\frac{1}{\zeta^k}\max\{\alpha,\ 0\},$$

for some sufficiently large $\zeta^k\in\mathbb{R}_{>0}$, and subject to the requirement that the underlying PDM is bounded. Furthermore, similar to both IPC and SEPT protocols, revision protocols of the form (5.3) are in general not exact, and so their diagonal entries $\rho_{ii}^k(\cdot,\cdot)$ are mostly taken as placeholders.

Replacing (5.3) into the Poisson EDM of Definition 2.6 yields the *capacity-constrained* IPC EDM, as defined next.

Definition 5.1 (*Capacity-Constrained IPC EDM*) Replacing the revision protocol (5.3) into (2.15) yields the capacity-constrained IPC EDM given by

$$\dot{x}_i^k(t) = R^k \sum_{j \in \mathcal{S}^k} \Big(x_j^k(t) \left[\overline{x}_i^k - x_i^k(t)\right]_+ \phi_i^k \left(p_i^k(t) - p_j^k(t)\right) - x_i^k(t) \left[\overline{x}_j^k - x_j^k(t)\right]_+ \phi_j^k \left(p_j^k(t) - p_i^k(t)\right) \Big),$$

for all $i \in \mathcal{S}^k$ and all $k \in \mathcal{P}$. The canonical example of such an EDM is the capacity-constrained *Smith* EDM, which results from setting $\phi_j^k(\cdot)$ as in (2.12) with $\zeta^k = R^k$, and assuming that R^k is sufficiently large so that (2.14) holds. The resulting dynamics are given by

$$\dot{x}_i^k(t) = \sum_{j \in \mathcal{S}^k} \Big(x_j^k(t) \left[\overline{x}_i^k - x_i^k(t)\right]_+ \left[p_i^k(t) - p_j^k(t)\right]_+ - x_i^k(t) \left[\overline{x}_j^k - x_j^k(t)\right]_+ \left[p_j^k(t) - p_i^k(t)\right]_+ \Big),$$

for all $i \in \mathcal{S}^k$ and all $k \in \mathcal{P}$. ◇

It turns out that the considered set $\mathcal{X}$ is positively invariant under every capacity-constrained IPC EDM. That is, given a initial condition $\mathbf{x}(0) \in \mathcal{X}$, it holds that $\mathbf{x}(t) \in \mathcal{X}$ for all $t \geq 0$, with $\mathbf{x}(t)$ evolving according to the EDM of Definition 5.1. This result is formalized next.

Lemma 5.1 *Let* $\mathbf{x}(t)$ *evolve according to the EDM in Definition 5.1. If* $\mathbf{x}(0) \in \mathcal{X}$, *then* $\mathbf{x}(t) \in \mathcal{X}$ *for all* $t \geq 0$. ◇

Proof First, note that Δ is positively invariant under the considered dynamics. To see this fact, observe that

$$\begin{aligned}
\sum_{i \in \mathcal{S}^k} \dot{x}_i^k(t) &= R^k \sum_{i \in \mathcal{S}^k} \sum_{j \in \mathcal{S}^k} \left(x_j^k(t)\rho_{ji}^k(t) - x_i^k(t)\rho_{ij}^k(t)\right) \\
&= R^k \left(\sum_{i \in \mathcal{S}^k} \sum_{j \in \mathcal{S}^k} x_j^k(t)\rho_{ji}^k(t) - \sum_{i \in \mathcal{S}^k} \sum_{j \in \mathcal{S}^k} x_i^k(t)\rho_{ij}^k(t) \right) \\
&= R^k \left(\sum_{i \in \mathcal{S}^k} \sum_{j \in \mathcal{S}^k} x_j^k(t)\rho_{ji}^k(t) - \sum_{i \in \mathcal{S}^k} \sum_{j \in \mathcal{S}^k} x_j^k(t)\rho_{ji}^k(t) \right) \\
&= 0,
\end{aligned}$$

for all $k \in \mathcal{P}$. Here, $\rho_{ij}^k(t) := \rho_{ij}^k\left(\mathbf{p}^k(t), \mathbf{x}^k(t)\right)$. Therefore,

$$\sum_{i \in \mathcal{S}^k} x_i^k(0) = m^k \quad \Rightarrow \quad \sum_{i \in \mathcal{S}^k} x_i^k(t) = m^k$$

for all $t \geq 0$ and all $k \in \mathcal{P}$. Additionally, notice that $x_i^k(t) = 0$ implies that $\dot{x}_i^k(t) \geq 0$, for all $i \in \mathcal{S}^k$ and all $k \in \mathcal{P}$. Thus, $\mathbf{x}(0) \in \mathbb{R}^n_{\geq 0} \Rightarrow \mathbf{x}(t) \in \mathbb{R}^n_{\geq 0}$, for all $t \geq 0$.

Second, from (5.3) it follows that

$$\left[\overline{x}_i^k - x_i^k(t)\right]_+ = 0 \quad \Rightarrow \quad \rho_{ji}^k(t) = 0, \ \forall j \in \mathcal{S}^k \quad \Rightarrow \quad \dot{x}_i^k(t) \leq 0.$$

Hence,

$$\overline{\mathbf{x}} - \mathbf{x}(0) \succeq \mathbf{0}_n \quad \Rightarrow \quad \overline{\mathbf{x}} - \mathbf{x}(t) \succeq \mathbf{0}_n, \ \forall t \geq 0.$$

Marshaling both of these facts yields the desired result. □

On the other hand, from Lemma 5.1 one may anticipate that the capacity-constrained IPC EDM of Definition 5.1 does not satisfy the Nash stationarity property:

$$\boldsymbol{\mathcal{V}}\left(\mathbf{x}^*, \mathbf{p}^*\right) = \mathbf{0}_n \Leftrightarrow \mathbf{x}^* \in \arg\max_{\mathbf{x} \in \Delta} \mathbf{x}^\top \mathbf{p}^*.$$

Instead, it satisfies a stronger constrained Nash stationarity property as stated next in Lemma 5.2.

Lemma 5.2 *The capacity-constrained IPC EDM of Definition 5.1 satisfies the constrained Nash stationarity property:*

$$\boldsymbol{\mathcal{V}}\left(\mathbf{x}^*, \mathbf{p}^*\right) = \mathbf{0}_n \Leftrightarrow \mathbf{x}^* \in \arg\max_{\mathbf{x} \in \mathcal{X}} \mathbf{x}^\top \mathbf{p}^*.$$

That is, $(\mathbf{x}^*, \mathbf{p}^*) \in \Delta \times \mathbb{R}^n$ *is an equilibrium point of the capacity-constrained IPC EDM if and only if* $\mathbf{x}^* \in \arg\max_{\mathbf{x} \in \mathcal{X}} \mathbf{x}^\top \mathbf{p}^*$. ◊

Proof Throughout this proof, let

$$\overline{\mathcal{S}}^k(\mathbf{x}^k) = \left\{i \in \mathcal{S}^k : x_i^k < \overline{x}_i^k\right\}$$

denote the subset of strategies of population $k \in \mathcal{P}$ that have non-zero remaining capacity at the strategic distribution $\mathbf{x}^k$. Using such a notation and with Ω given in (5.1), it follows that

$$\arg\max_{\mathbf{x} \in \mathcal{X}} \mathbf{x}^\top \mathbf{p}^* = \left\{\mathbf{x} \in \mathcal{X} : \begin{array}{l} x_i^k > 0 \Rightarrow p_i^{k*} \geq p_j^{k*}, \\ \forall i \in \mathcal{S}^k, \ j \in \overline{\mathcal{S}}^k(\mathbf{x}^k), \ k \in \mathcal{P} \end{array}\right\}.$$

Based on this notation, the considered claim can be proved as follows.

(Sufficiency) Let $\mathbf{x}^* \in \arg\max_{\mathbf{x} \in \mathcal{X}} \mathbf{x}^\top \mathbf{p}^*$. If $x_i^{k*} > 0$, then $p_i^{k*} \geq p_j^{k*}$, for all $j \in \overline{\mathcal{S}}^k(\mathbf{x}^{k*})$. Moreover, if $p_i^{k*} > p_j^{k*}$, then $x_j^{k*} = 0$ and

$$\begin{aligned} x_j^{k*}\left[\overline{x}_i^k - x_i^{k*}\right]_+ \phi_i^k\left(p_i^{k*} - p_j^{k*}\right) &= 0 \\ x_i^{k*}\left[\overline{x}_j^k - x_j^{k*}\right]_+ \phi_j^k\left(p_j^{k*} - p_i^{k*}\right) &= 0. \end{aligned}$$

Thus, $\boldsymbol{\mathcal{V}}(\mathbf{x}^*, \mathbf{p}^*) = \mathbf{0}_n$.

(Necessity) Let $\boldsymbol{\mathcal{V}}(\mathbf{x}^*, \mathbf{p}^*) = \mathbf{0}_n$, but suppose that $\mathbf{x}^* \in \mathcal{X} \setminus \arg\max_{\mathbf{x}\in\mathcal{X}} \mathbf{x}^\top \mathbf{p}^*$. Hence, there is some $x_i^* > 0$ such that $p_i^{k*} < p_j^{k*}$ for some $j \in \overline{\mathcal{S}}(\mathbf{x}^{k*})$. Therefore,

$$x_i^{k*}\left[\overline{x}_j^k - x_j^{k*}\right]_+ \phi_j^k\left(p_j^{k*} - p_i^{k*}\right) > 0,$$

leading to a contradiction on the assumption that $\boldsymbol{\mathcal{V}}(\mathbf{x}^*, \mathbf{p}^*) = \mathbf{0}_n$. □

Finally, it can also be shown that the capacity-constrained IPC EDM of Definition 5.1 is δ-passive characterized by an informative σ. Such a result is formalized next in Lemma 5.3.

Lemma 5.3 *The capacity-constrained IPC EDM of Definition 5.1 is δ-passive characterized by the informative σ_{CC} function given by*

$$\sigma_{CC}(\mathbf{x}, \mathbf{p}) = -\sum_{k\in\mathcal{P}} R^k \sum_{i\in\mathcal{S}^k} \mathcal{V}_i^k\left(\mathbf{x}^k, \mathbf{p}^k\right) \sum_{\ell\in\mathcal{S}^k}\left(\left(\overline{x}_\ell^k - x_\ell^k\right)\int_0^{p_\ell^k - p_i^k} \phi_\ell^k(\zeta)\, d\zeta - x_\ell^k \int_0^{p_i^k - p_\ell^k} \phi_i^k(\zeta)\, d\zeta\right), \tag{5.5}$$

with

$$\mathcal{V}_i^k\left(\mathbf{x}^k, \mathbf{p}^k\right) = R^k \sum_{j\in\mathcal{S}^k}\left(x_j^k\left[\overline{x}_i^k - x_i^k\right]_+ \phi_i^k\left(p_i^k - p_j^k\right) - x_i^k[\overline{x}_j^k - x_j^k]_+ \phi_j^k\left(p_j^k - p_i^k\right)\right),$$

for all $i \in \mathcal{S}^k$ and all $k \in \mathcal{P}$. ◇

Proof Given the forward time invariance of $\mathcal{X}$ provided by Lemma 5.1, throughout this proof the analysis is confined to the domain $\mathbf{x} \in \mathcal{X}$ without any loss of generality.

To show the δ-passivity of the capacity-constrained IPC EDM (Definition 5.1), consider the δ-storage function defined as

$$S_{CC}(\mathbf{x}, \mathbf{p}) = \sum_{k\in\mathcal{P}} R^k \sum_{i\in\mathcal{S}^k} x_i^k \sum_{\ell\in\mathcal{S}^k}\left(\overline{x}_\ell^k - x_\ell^k\right)\int_0^{p_\ell^k - p_i^k} \phi_\ell^k(\zeta)\, d\zeta,$$

and the σ_{CC} function in (5.5). It is not too difficult to show that[1] $S_{CC}(\cdot,\cdot)$ and $\sigma_{CC}(\cdot,\cdot)$ are always non-negative, and that

$$S_{CC}(\mathbf{x}, \mathbf{p}) = 0 \quad\Leftrightarrow\quad \boldsymbol{\mathcal{V}}(\mathbf{x}, \mathbf{p}) = \mathbf{0}_n$$
$$\sigma_{CC}(\mathbf{x}, \mathbf{p}) = 0 \quad\Leftrightarrow\quad \boldsymbol{\mathcal{V}}(\mathbf{x}, \mathbf{p}) = \mathbf{0}_n.$$

[1] These claims can be proven following a similar argument as in [8, Appendix A.4].

These facts imply that $S_{CC}(\cdot,\cdot)$ can be taken as a valid δ-storage function candidate, and that σ_{CC} is informative in the sense of Definition 3.1.

Consequently, to show that the capacity-constrained IPC EDM is δ-passive with informative σ_{CC} it suffices to prove that

$$\nabla_{\mathbf{x}} S_{CC}(\mathbf{x},\mathbf{p})^\top \boldsymbol{\mathcal{V}}(\mathbf{x},\mathbf{p}) + \nabla_{\mathbf{p}} S_{CC}(\mathbf{x},\mathbf{p})^\top \mathbf{v} \leq -\sigma_{CC}(\mathbf{x},\mathbf{p}) + \mathbf{v}^\top \boldsymbol{\mathcal{V}}(\mathbf{x},\mathbf{p}), \quad (5.6)$$

for every $\mathbf{v} \in \mathbb{R}^n$.

Accordingly, note that[2]

$$\frac{\partial S_{CC}(\mathbf{x},\mathbf{p})}{\partial x^\kappa_\iota} = R^\kappa \sum_{\ell\in\mathcal{S}^\kappa} \left(\overline{x}^\kappa_\ell - x^\kappa_\ell\right) \int_0^{p^\kappa_\ell - p^\kappa_\iota} \phi^\kappa_\ell(\zeta)\, d\zeta$$
$$- R^\kappa \sum_{i\in\mathcal{S}^\kappa} x^\kappa_i \int_0^{p^\kappa_\iota - p^\kappa_i} \phi^\kappa_\iota(\zeta)\, d\zeta,$$

and

$$\frac{\partial S_{CC}(\mathbf{x},\mathbf{p})}{\partial p^\kappa_\iota} = R^\kappa \sum_{i\in\mathcal{S}^\kappa} x^\kappa_i \sum_{\ell\in\mathcal{S}^\kappa} \left(\overline{x}^\kappa_\ell - x^\kappa_\ell\right) \phi^\kappa_\ell\left(p^\kappa_\ell - p^\kappa_i\right) \left(\frac{\partial p^\kappa_\ell}{\partial p^\kappa_\iota} - \frac{\partial p^\kappa_i}{\partial p^\kappa_\iota}\right)$$
$$= R^\kappa \sum_{i\in\mathcal{S}^\kappa}\sum_{\ell\in\mathcal{S}^\kappa} \left(x^\kappa_i \left(\overline{x}^\kappa_\ell - x^\kappa_\ell\right)\phi^\kappa_\ell\left(p^\kappa_\ell - p^\kappa_i\right) - x^\kappa_\ell\left(\overline{x}^\kappa_i - x^\kappa_i\right)\phi^\kappa_i\left(p^\kappa_i - p^\kappa_\ell\right)\right) \frac{\partial p^\kappa_\ell}{\partial p^\kappa_\iota}$$
$$= \mathcal{V}^\kappa_\iota\left(\mathbf{x}^\kappa,\mathbf{p}^\kappa\right),$$

which implies that $\nabla_{\mathbf{p}} S_{CC}(\mathbf{x},\mathbf{p})^\top \mathbf{v} = \mathbf{v}^\top \boldsymbol{\mathcal{V}}(\mathbf{x},\mathbf{p})$ and

$$\nabla_{\mathbf{x}} S_{CC}(\mathbf{x},\mathbf{p})^\top \boldsymbol{\mathcal{V}}(\mathbf{x},\mathbf{p})$$
$$= \sum_{k\in\mathcal{P}} R^k \sum_{i\in\mathcal{S}^k} \mathcal{V}^k_i\left(\mathbf{x}^k,\mathbf{p}^k\right) \sum_{\ell\in\mathcal{S}^k} \left(\left(\overline{x}^k_\ell - x^k_\ell\right) \int_0^{p^k_\ell - p^k_i} \phi^k_\ell(\zeta)\, d\zeta \right.$$
$$\left. - x^k_\ell \int_0^{p^k_i - p^k_\ell} \phi^k_i(\zeta)\, d\zeta \right)$$
$$= -\sigma_{CC}(\mathbf{x},\mathbf{p}).$$

Therefore, (5.6) holds with equality and the capacity-constrained IPC EDM is indeed δ-passive with informative σ_{CC}. □

Lemmas 5.2 and 5.3 show that the capacity-constrained ICP EDM of Definition 5.1 satisfies a condition similar to (D1) but tailored to the constrained Nash stationarity property of Lemma 5.2. In consequence, a result similar to Corollary 4.1 can be readily obtained for such an EDM. This result is formally stated next in Corollary 5.1.

[2] Observe that the indices in the partial derivatives are κ and ι rather than k and i.

Corollary 5.1 *Consider a population game* $G = (\mathcal{P}, \mathcal{X}, \mathbf{f}(\cdot))$ *with* $\mathcal{X} = \Delta \cap \Omega$ *and* Ω *given in (5.1), and consider an EDM–PDM pair where the EDM is a capacity-constrained IPC EDM as in Definition 5.1, and the PDM satisfies Condition (D2). Then, the set* $\mathrm{GNE}(\mathbf{f})$ *is asymptotically stable.* ◇

Proof The proof is virtually identical to that of Corollary 4.1 and is supported by Lemmas 5.1–5.3. □

Note that, in contrast to Corollary 4.1, which concerns the asymptotic stability of the (unconstrained) set $\mathrm{NE}(\mathbf{f})$, Corollary 5.1 establishes the asymptotic stability of $\mathrm{GNE}(\mathbf{f})$ in the (constrained) case where Ω is defined in (5.1). Furthermore, Lemma 5.1 guarantees that these constraints are satisfied at all times, provided the initial condition is feasible. These properties are particularly relevant in large-scale control engineering applications where constraint satisfaction is required throughout the whole system's trajectory, not just at equilibrium. The next section presents an illustrative example to further support and motivate these results.

5.1.2 Simulation Example

To illustrate the application of the developed capacity-constrained revision protocol and its corresponding EDM in a practical setup, recall the *multi-robot garbage collection problem* of Sect. 1.2.3.

Namely, a fleet of autonomous robots represents a single population ($P = 1$) of decision-making agents. Each robot autonomously selects a zone to perform some garbage collection task. Thus, the set of strategies $\mathcal{S}$ represents the available zones that the robots can choose from.

The payoff signals associated with these zones are ruled by dynamics of the form

$$\dot{q}_i(t) = \psi_i\left(q_i(t), \gamma_i - x_i(t)\right) \tag{5.7a}$$

$$p_i(t) = f_i\left(x_i(t)\right) + q_i(t), \tag{5.7b}$$

for all $i \in \mathcal{S}$. Here, the fitness function

$$f_i\left(x_i(t)\right) = -x_i(t)$$

encourages a baseline incentive to keep the allocation across zones balanced; the constant $\gamma_i \in \mathbb{R}_{>0}$ is the nominal *garbage accumulation rate* at zone i; the state $q_i(t) \in \mathbb{R}_{\geq 0}$ represents the amount of garbage in zone i at time t; and $\psi_i : \mathbb{R} \times \mathbb{R} \to \mathbb{R}$ is locally Lipschitz continuous and satisfies

$$\begin{cases} \alpha \le 0 \Rightarrow \psi_i(\alpha, \beta) \ge 0, \ \forall \beta \in \mathbb{R} \\ \beta > 0 \Rightarrow \psi_i(\alpha, \beta) > 0, \ \forall \alpha \in \mathbb{R}_{\ge 0} \\ \beta < 0 \Rightarrow \psi_i\,(\alpha, \beta) < 0, \ \forall \alpha \in \mathbb{R}_{>0} \\ \beta = 0 \Rightarrow \psi_i\,(\alpha, \beta) = 0, \ \forall \alpha \in \mathbb{R}_{>0}. \end{cases}$$

Notably, the properties of $\psi_i(\cdot, \cdot)$ imply that

- $q_i(t)$ is always non-negative (provided that $q_i(0) \ge 0$).
- $q_i(t)$ increases when $x_i(t)$ is less than the nominal accumulation rate γ_i.
- $q_i(t)$ decreases (bounded within $\mathbb{R}_{\ge 0}$) when $x_i(t)$ exceeds γ_i.
- $q_i(t)$ remains constant when $x_i(t) = \gamma_i$.

Under this model, the payoff $p_i(t)$ captures both the accumulated garbage in zone i, represented by $q_i(t)$, and the baseline incentive to maintain a balanced allocation across zones, given by $f_i\,(x_i(t))$. Consequently, a robot's incentive to operate in a specific zone dynamically depends on both the current level of accumulated garbage and the distribution of robots across zones, making it more likely for robots to move toward zones that are more polluted and less populated.

A simple canonical example of a function $\psi_i(\cdot)$ satisfying all these properties is given by

$$\psi_i(q_i, \gamma_i - x_i) = \left[\overline{q}_i - q_i\right]_+ \left[\gamma_i - x_i\right]_+ - q_i \left[x_i - \gamma_i\right]_+, \tag{5.8}$$

where $\overline{q}_i \in \mathbb{R}_{>0}$ is the *garbage accumulation capacity* of zone i. The interpretation of the resulting PDM (5.7) and (5.8) is as follows:

- When $\gamma_i > x_i$, the state q_i increases with rate $\left[\overline{q}_i - q_i\right]_+ (\gamma_i - x_i)$. Thus, the fewer the robots working in zone i, the faster the garbage accumulates, as expected from the imbalance between the nominal garbage generation rate γ_i and the cleaning capacity provided by the robots. Moreover, the closer q_i is to the upper bound $\overline{q}_i$, the slower the rate of accumulation. This reflects the fact that zones have finite spatial capacity: once a zone is nearly full, additional garbage tends to accumulate closer to existing piles, making it easier to collect together. In other words, a large pile of garbage concentrated in a single location can be collected more efficiently than smaller pieces scattered across the zone.
- When $\gamma_i < x_i$, the state q_i decreases with rate $q_i\,(x_i - \gamma_i)$. In this case, there are more robots cleaning the zone than the nominal garbage accumulation rate, resulting in a net reduction of garbage. The decrease is proportional to the current amount of garbage q_i: if there is a large backlog, robots remove it more rapidly, whereas if the zone is already nearly clean ($q_i \approx 0$), the cleaning rate naturally slows down. This captures the realistic effect that cleaning efforts are more effective when there is more to clean and diminish as the zone approaches a clean state.
- When $\gamma_i = x_i$, the state q_i remains constant, meaning that the number of robots exactly matches the nominal garbage accumulation rate. In this equilibrium situation, the garbage level in the zone is maintained without further growth or reduction.

Finally, it is straightforward to verify that if $q_i(0) \leq \overline{q}_i$, then $q_i(t)$ never exceeds $\overline{q}_i$ for any $t \geq 0$, thereby guaranteeing boundedness of the garbage state. In fact, the dynamics (5.7) and (5.8) constitute a specific instance of the bounded dual-Lagrangian dynamics introduced in Examples 2.5 and 3.4, and discussed in detail in Sect. 3.4. Accordingly, based on the results in Sect. 3.4, it follows that this PDM readily satisfies Condition (D2). In consequence, by Corollary 5.1, the asymptotic stability of GNE($\mathbf{f}$) is guaranteed under the capacity-constrained IPC EDM. This fact implies that the decision-making process of the multiple robots will asymptotically converge toward a GNE.

As an illustration, consider a fleet of 50 autonomous robots tasked with garbage collection across $n = 4$ zones. The maximum allowable mass of robots in each zone is given by $\bar{\mathbf{x}} = [1,\ 0.2,\ 0.2,\ 0.2]^\top$, meaning that zone 1 can host the entire fleet, while the remaining zones are capacity-constrained up to 8 robots. Each zone $i \in \mathcal{S}$ has a nominal garbage accumulation rate $\gamma_i \in [0, 1]$ and an upper bound $\bar{q}_i \in [0, 2]$ for the garbage level, both randomly (uniformly) sampled to introduce heterogeneity. Robots revise their decisions asynchronously following a Poisson revision process with rate $R = 1$ (in arbitrary time units) and using a capacity-constrained Smith revision protocol, ensuring no zone exceeds its capacity $\bar{\mathbf{x}}$. Regarding the initial condition, it is assumed that all robots start in zone 1, i.e., $\mathbf{x}(0) = [\,1,\ 0,\ 0,\ 0\,]^\top$, and that the zones are initially clean, i.e., $\mathbf{q}(0) = \mathbf{0}_n$.

Figures 5.1, 5.2, 5.3 and 5.4 show the simulation results, illustrating how the population reallocates over time toward a balanced configuration where zones with higher garbage accumulation receive more attention while respecting capacity limits. Specifically, Fig. 5.1 depicts how the number of robots in each zone evolves over time, showing that zones eventually reach, but never exceed, their maximum capacities. The corresponding payoff signals and garbage levels, shown in Figs. 5.2 and 5.3, confirm that zones with higher garbage accumulation are associated with higher payoffs, while garbage levels converge to a steady state. Finally, Fig. 5.4 illustrates

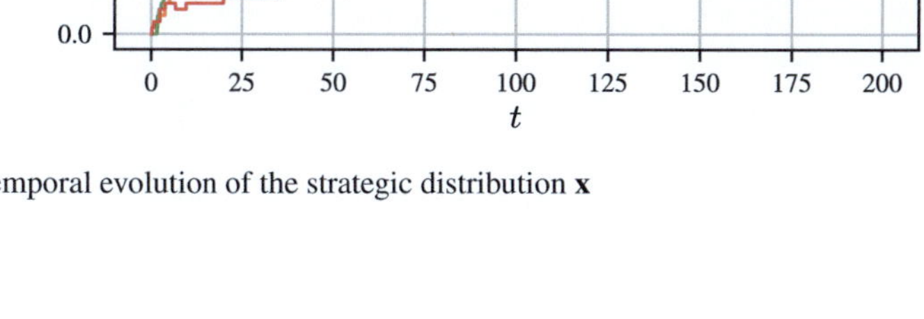

Fig. 5.1 Temporal evolution of the strategic distribution $\mathbf{x}$

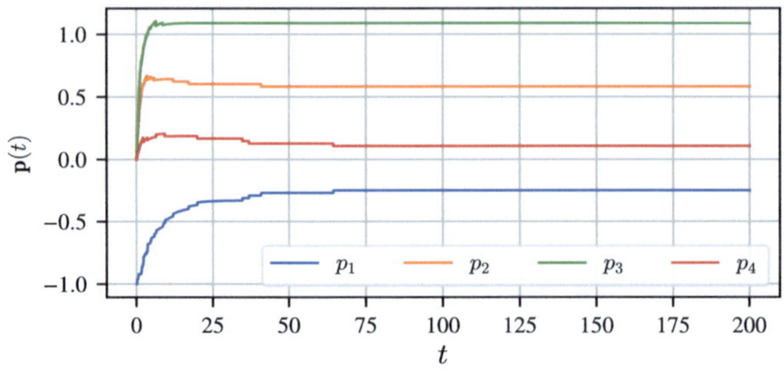

Fig. 5.2 Temporal evolution of the payoff vector $\mathbf{p}$

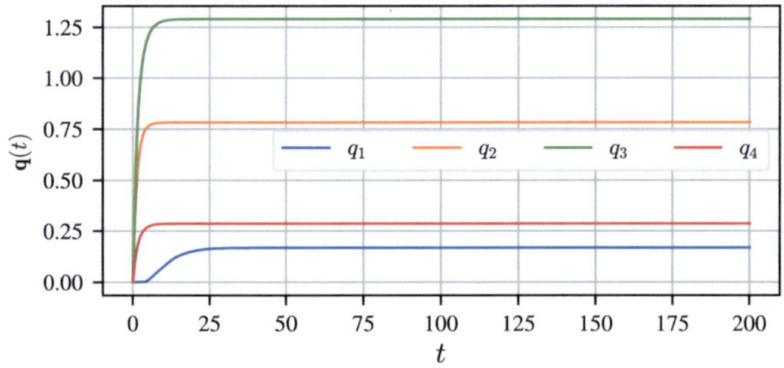

Fig. 5.3 Temporal evolution of the garbage accumulation levels $\mathbf{q}$

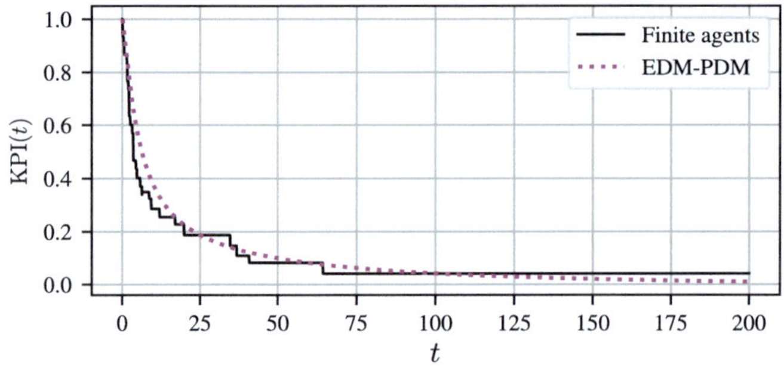

Fig. 5.4 Temporal evolution of $\text{KPI}(t) = \|\mathbf{x}(t) - \mathbf{x}^*\|_2 \,/\, \|\mathbf{x}(0) - \mathbf{x}^*\|_2$, with respect to $\mathbf{x}^* = [0.4,\ 0.2,\ 0.2,\ 0.2]^\top$, for both the finite-agent simulation and the EDM–PDM approximation

the evolution of the selected performance index, validating the convergence of the population toward a GNE.

The *Python* code to replicate these simulations is provided in Listing 5.1 at the end of this chapter.

5.2 Non-prescriptive Control Approaches

Section 5.1 examined scenarios where the payoff mechanism is fixed, and convergence to a GNE is achieved by appropriately designing the agents' revision protocols to enforce constraint satisfaction. While this approach ensures constraint satisfaction over time via invariance arguments (see Lemma 5.1), the specific constraint set Ω defined in (5.1) is limited in scope and may not capture the broader range of constraints encountered in several practical applications. Moreover, it relies on the assumption that the revision protocols of agents can be designed or imposed, which is typically feasible only in engineered systems.

In contrast, this section considers the complementary setting in which the agents' revision processes are fixed and cannot be directly influenced. Instead, the system designer shapes the behavior of the system by modifying the payoff mechanism. Specifically, it is assumed that the payoffs consist of an exogenous component, which cannot be altered, and a designable component, which is under the control of the system designer. The objective is to leverage this designable portion of the payoff mechanism to guide the system toward convergence to a GNE, while accommodating a more general form of the constraint set Ω.

5.2.1 A Structured Class of EDM–PDM Pairs

This section introduces a class of structured EDM–PDM pairs that inherently satisfy either Conditions (C1)–(C5) or Conditions (A), (C2)–C5). As a result, their asymptotic stability properties can be directly certified via Theorem 4.1 or Theorem 4.2, respectively.

Although the EDM is fixed and not subject to design, it is assumed to satisfy either Condition (D1) or Condition (D3). This requirement is formalized in Assumption 5.1.

Assumption 5.1 The EDM satisfies either Condition (D1) or Condition (D3). ◊

On the other hand, the PDM is assumed to be of the form

$$\mathcal{W}(\mathbf{q}, \mathbf{x}) = \begin{bmatrix} \mathcal{W}_a(\mathbf{q}_a, \mathbf{x}) \\ \mathcal{W}_b(\mathbf{q}_b, \mathbf{x}) \end{bmatrix} \tag{5.9a}$$

$$\mathcal{H}(\mathbf{q}, \mathbf{x}) = \mathcal{H}_a(\mathbf{q}_a, \mathbf{x}) + \mathcal{H}_b(\mathbf{q}_b, \mathbf{x}), \tag{5.9b}$$

where $\mathbf{q}_s \in \mathbb{R}^{d_s}$ and $\mathcal{W}_s : \mathbb{R}^{d_s} \times \Delta \to \mathbb{R}^{d_s}$, for some $d_s \in \mathbb{Z}_{\geq 0}$, for all $s \in \{a, b\}$, and $\mathbf{q} = \mathrm{col}(\mathbf{q}_a, \mathbf{q}_b)$. Namely, as depicted in Fig. 5.5, the PDM is the parallel interconnection of two subsystems, Σ_a and Σ_b, with dynamics of the form

$$\Sigma_s : \begin{cases} \dot{\mathbf{q}}_s(t) = \mathcal{W}_s(\mathbf{q}_s(t), \mathbf{x}(t)) \\ \mathbf{p}_s(t) = \mathcal{H}_s(\mathbf{q}_s(t), \mathbf{x}(t)), \end{cases} \qquad \forall s \in \{a, b\}.$$

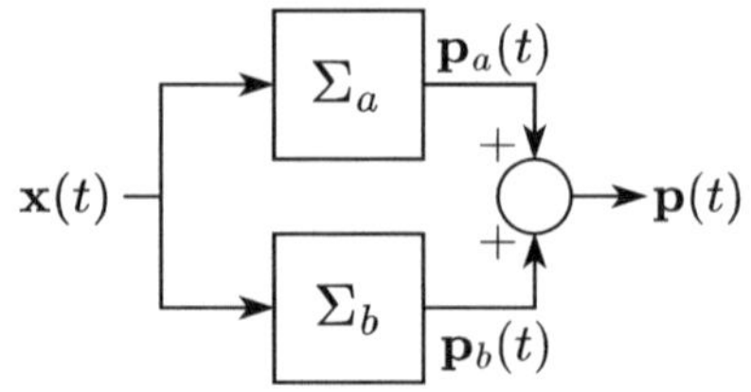

Fig. 5.5 The PDM (5.9) viewed as the parallel interconnection of two subsystems: Σ_a and Σ_b

The fact that the PDM (5.9) is the parallel interconnection of two subsystems has the following implication.

Lemma 5.4 *Let Σ_s be δ-antipassive characterized by σ_s, for all $s \in \{a, b\}$. Then, the PDM (5.9) is δ-antipassive characterized by $(\sigma_a + \sigma_b)$.* ◊

Proof This result follows immediately from Lemma 3.8. □

Now, subsystem Σ_a is assumed to satisfy Assumption 5.2, as presented below.

Assumption 5.2 The next conditions hold:

(i) Σ_a is δ-antipassive characterized by some σ_a.
(ii) For every $\left(\mathbf{x}^*, \mathbf{q}_a^*\right) \in \Delta \times \mathbb{R}^{d_a}$, $\boldsymbol{\mathcal{W}}_a\left(\mathbf{q}_a^*, \mathbf{x}^*\right) = \mathbf{0}_{d_a}$ implies that

$$\boldsymbol{\mathcal{H}}_a\left(\mathbf{q}_a^*, \mathbf{x}^*\right) = \mathbf{f}\left(\mathbf{x}^*\right).$$

(iii) If $d_a \geq 1$, then the set $\mathcal{E}_a\left(\mathbf{x}^*\right) = \left\{\mathbf{q}_a^* \in \mathbb{R}^{d_a} : \boldsymbol{\mathcal{W}}_a\left(\mathbf{q}_a^*, \mathbf{x}^*\right) = \mathbf{0}_{d_a}\right\}$ is nonempty, bounded, and asymptotically stable under Σ_a, for every $\mathbf{x}^* \in \mathrm{GNE}(\mathbf{f})$.

◊

Observe that, in contrast to Condition (S), Assumption 5.2 does not require σ_a to be informative. Moreover, according to Assumption 5.2-(ii), the role of subsystem Σ_a is to relate the steady-state output of the PDM (5.9) to the overall fitness function $\mathbf{f}(\cdot)$, ensuring that the payoff signal perceived by the decision-making agents reflects the strategic environment of the underlying population game G.

On the other hand, subsystem Σ_b is responsible for incorporating the constraints defined by Ω into the payoff signal at the output of the PDM (5.9). This is approached from a dual optimization perspective, in which Σ_b computes the Lagrange multipliers associated with the constraints in Ω and outputs the corresponding dual-Lagrangian terms. The remainder of this section presents two specific instances of subsystem Σ_b, addressing both centralized and distributed implementations.

5.2.2 Centralized Multipliers Dynamics Model

In centralized implementations, the payoff mechanism is assumed to follow the structure shown in Fig. 5.6. This scheme relies on a single centralized entity that has

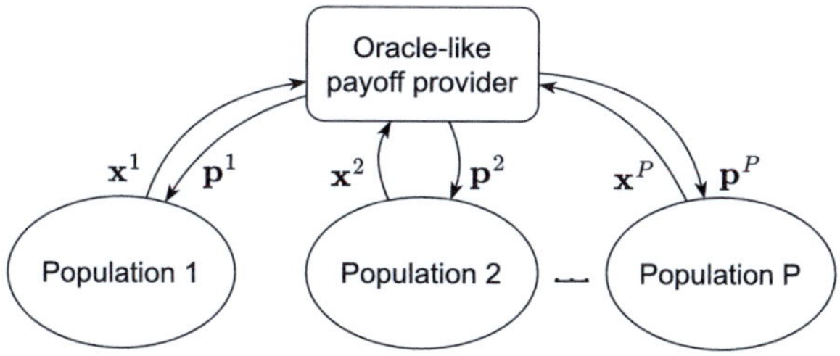

Fig. 5.6 Centralized full-decision information setup: a single oracle-like payoff provider has full information regarding the strategic distribution of all populations and broadcasts the payoffs to all society agents

full information regarding the strategic distribution of the society and is responsible for generating and providing the corresponding payoff signals to the decision-making agents of all populations. It is noted, however, that while this entity serves as a useful conceptual model, in practice the centralized payoff mechanism may not correspond to a tangible component, but rather to an abstract dynamic process that governs the evolution and flow of payoff signals.

Throughout this section, it is assumed that Ω has the general form

$$\Omega = \left\{ \mathbf{x} \in \mathbb{R}^n : \begin{array}{r} \mathbf{A}\mathbf{x} - \mathbf{b} = \mathbf{0}_{c_=} \\ \mathbf{g}(\mathbf{x}) \preceq \mathbf{0}_{c_\leq} \end{array} \right\}$$

and complies with Assumption 2.1. Furthermore, it is assumed that the information required to evaluate the constraints defined by Ω is available centrally for the payoff mechanism. Accordingly, subsystem Σ_b is specified as

$$\boldsymbol{\mathcal{W}}_b\left(\mathbf{q}_b, \mathbf{x}\right) = \tau_b \begin{bmatrix} \mathbf{A}\mathbf{x} - \mathbf{b} \\ \left[\mathbf{g}\left(\mathbf{x}\right)\right]_+ - \mathbf{q}_{b2} \odot \left[-\mathbf{g}\left(\mathbf{x}\right)\right]_+ \end{bmatrix} \tag{5.10a}$$

$$\boldsymbol{\mathcal{H}}_b\left(\mathbf{q}_b, \mathbf{x}\right) = -\mathbf{A}^\top \mathbf{q}_{b1} - \mathrm{D}\,\mathbf{g}(\mathbf{x})^\top \mathbf{q}_{b2}, \tag{5.10b}$$

where $\mathbf{q}_{b1} \in \mathbb{R}^{c_=}$, $\mathbf{q}_{b2} \in \mathbb{R}^{c_\leq}$, $\mathbf{q}_b = \mathrm{col}\left(\mathbf{q}_{b1}, \mathbf{q}_{b2}\right)$, $\tau_b \in \mathbb{R}_{>0}$ is an arbitrary positive constant, and $[\cdot]_+ = \max\{\cdot, 0\}$ is applied element-wise to the input vector. Throughout, the next requirement on the initial condition of the dynamics (5.10a) is imposed.

Assumption 5.3 It holds that $\mathbf{q}_{b2}(0) \succeq \mathbf{0}_{c_\leq}$. ◊

Under the considered setup the following properties are satisfied.

Lemma 5.5 *Let Σ_b be characterized by (5.10) subject to Assumption 5.3. It holds that:*

(i) *For every $\left(\mathbf{x}^*, \mathbf{q}_b^*\right) \in \Delta \times \mathbb{R}^{d_b}$, $\boldsymbol{\mathcal{W}}_b\left(\mathbf{q}_b^*, \mathbf{x}^*\right) = \mathbf{0}_{d_b}$ implies that $\mathbf{x}^* \in \Omega$ and $\mathbf{q}_{b2}^* \odot \mathbf{g}\left(\mathbf{x}^*\right) = \mathbf{0}_{c_\leq}$.*

(ii) $\mathbf{q}_{b2}(t) \succeq \mathbf{0}_{c_\leq}$, *for all* $t \geq 0$.
(iii) Σ_b *is* δ*-antipassive characterized by some* σ_b. ◊

Proof Each claim is now proven.

Proof of Claim (i): From (5.10a) it readily follows that $\mathcal{W}\left(\mathbf{q}_b^*, \mathbf{x}^*\right) = \mathbf{0}_{d_b}$ implies that $\mathbf{A}\mathbf{x}^* - \mathbf{b} = \mathbf{0}_{c_=}$ and $\mathbf{g}\left(\mathbf{x}^*\right) \preceq \mathbf{0}_{c_\leq}$. Thus, $\mathbf{x}^* \in \Omega$. Moreover, if $\mathbf{g}\left(\mathbf{x}^*\right) \preceq \mathbf{0}_{c_\leq}$, then $\mathbf{q}_{b2}^* \odot \left[-\mathbf{g}\left(\mathbf{x}^*\right)\right]_+ = \mathbf{0}_{c_\leq}$ implies $\mathbf{q}_{b2}^* \odot \mathbf{g}\left(\mathbf{x}^*\right) = \mathbf{0}_{c_\leq}$.

Proof of Claim (ii): Denote $C_\leq = \{1, 2, \ldots, c_\leq\}$, let $h \in C_\leq$ be arbitrary, and let $q_{b2h} \in \mathbb{R}$ be the hth element of $\mathbf{q}_{2b} = \operatorname{col}\left(q_{b2h}\right)_{h\in C_\leq}$. From (5.10a), note that $q_{b2h}(t) = 0$ implies that $\dot{q}_{b2h}(t) \geq 0$. Hence, $q_{b2h}(0) \geq 0$ implies that $q_{b2h}(t) \geq 0$, for all $t \geq 0$. Therefore, since h is arbitrary, Assumption 5.3 implies that $\mathbf{q}_{b2}(t) \succeq 0$, for all $t \geq 0$.

As a consequence of Claim (ii) and Assumption 5.3, without loss of generality, for the remainder of this proof the analysis can be restricted to the domain where $q_{b2h} \geq 0$, for all $h \in C_\leq$.

Proof of Claim (iii): Denote

$$\begin{aligned}\mathcal{W}_{b1}(\mathbf{x}) &= \tau_b\left(\mathbf{A}\mathbf{x} - \mathbf{b}\right)\\ \mathcal{W}_{b2}\left(\mathbf{q}_{b2}, \mathbf{x}\right) &= \tau_b\left(\left[\mathbf{g}\left(\mathbf{x}\right)\right]_+ - \mathbf{q}_{b2} \odot \left[-\mathbf{g}\left(\mathbf{x}\right)\right]_+\right),\end{aligned}$$

and note that $\mathrm{D}\,\mathbf{g}(\mathbf{x})^\top \mathbf{q}_{b2} = \sum_{h\in C_\leq} q_{b2h} \nabla g_h\left(\mathbf{x}\right)$. Then, it follows that

$$\mathbf{w}_b(\mathbf{q}_b, \mathbf{x}, \mathbf{v}_b) = -\mathbf{A}^\top \mathcal{W}_{b1}(\mathbf{x}) - \mathrm{D}\,\mathbf{g}(\mathbf{x})^\top \mathcal{W}_{b2}\left(\mathbf{q}_{b2}, \mathbf{x}\right) - \sum_{h\in C_\leq} q_{b2h} \nabla^2 g_h(\mathbf{x})\mathbf{v}_b.$$

Here, $\mathbf{v}_b$ and $\mathbf{w}_b$ play the roles of $\mathbf{v}$ and $\mathbf{w}$ in Definition 3.1, respectively. Now, consider the storage function

$$S\left(\mathbf{q}_b, \mathbf{x}\right) = \frac{1}{2\tau_b}\left\|\mathcal{W}_{b1}(\mathbf{x})\right\|^2 + \frac{\tau_b}{2}\sum_{h\in C_\leq} \tilde{S}_h\left(q_{b2h}, \mathbf{x}\right),$$

where $\tilde{S}_h(q_{b2h}, \mathbf{x}) = [g_h(\mathbf{x})]_+^2 + q_{b2h}\left[-g_h(\mathbf{x})\right]_+^2$, for all $h \in C_\leq$. It follows that

$$\begin{aligned}\frac{\partial \tilde{S}_h(q_{b2h}, \mathbf{x})}{\partial q_{b2h}} &= \left[-g_h(\mathbf{x})\right]_+^2,\\ \frac{\partial \tilde{S}_h(q_{b2h}, \mathbf{x})}{\partial x_i^k} &= 2\left(\left[g_h(\mathbf{x})\right]_+ - q_{b2h}\left[-g_h(\mathbf{x})\right]_+\right)\frac{\partial g_h(\mathbf{x})}{\partial x_i^k},\end{aligned}$$

for all $h \in C_\leq$, all $i \in \mathcal{S}^k$, and all $k \in \mathcal{P}$. Hence,

$$\nabla_{\mathbf{q}_b} S(\mathbf{q}_b, \mathbf{x}) = \left[\mathbf{0}_{C_=}^\top,\ \frac{\tau_b}{2}\left[-\mathbf{g}(\mathbf{x})\right]_+^2\right]^\top$$
$$\nabla_{\mathbf{x}} S(\mathbf{q}_b, \mathbf{x}) = \mathbf{A}^\top \boldsymbol{\mathcal{W}}_{b1}(\mathbf{x}) + \mathrm{D}\,\mathbf{g}(\mathbf{x})^\top \boldsymbol{\mathcal{W}}_{b2}(\mathbf{q}_{b2}, \mathbf{x}),$$

where $[\cdot]_+^2$ is applied element-wise. Then,

$$\begin{aligned}\dot{S}(\mathbf{q}_b, \mathbf{x}) &= \frac{\tau_b}{2}\left(\left[-\mathbf{g}(\mathbf{x})\right]_+^2\right)^\top \boldsymbol{\mathcal{W}}_{b2}(\mathbf{q}_{b2}, \mathbf{x}) \\ &\quad + \mathbf{v}_b^\top \mathbf{A}^\top \boldsymbol{\mathcal{W}}_{b1}(\mathbf{x}) + \mathbf{v}_b^\top \mathrm{D}\,\mathbf{g}(\mathbf{x})^\top \boldsymbol{\mathcal{W}}_{b2}(\mathbf{q}_{b2}, \mathbf{x}) \\ &= -\sigma_b(\mathbf{q}_b, \mathbf{x}, \mathbf{v}_b) - \mathbf{v}_b^\top \mathbf{w}_b(\mathbf{q}_b, \mathbf{x}, \mathbf{v}_b),\end{aligned}$$

with

$$\begin{aligned}\sigma_b(\mathbf{q}_b, \mathbf{x}, \mathbf{v}_b) &= \frac{\tau_b^2}{2}\left(\mathbf{q}_{b2} \odot \left[-\mathbf{g}(\mathbf{x})\right]_+\right)^\top \left[-\mathbf{g}(\mathbf{x})\right]_+^2 \\ &\quad + \sum_{h \in C_\leq} q_{b2h} \mathbf{v}_b^\top \nabla^2 g_h(\mathbf{x}) \mathbf{v}_b.\end{aligned}$$

Notice that $\sigma_b(\cdot,\cdot,\cdot)$ is indeed non-negative. To see this fact, recall that $\mathbf{q}_{b2}$ is a vector with non-negative elements, and from Assumption 2.1, $\nabla^2 g_h(\mathbf{x}) \succeq 0$, for all $h \in C_\leq$. Marshaling all these facts yields that Claim (iii) holds. □

Exploiting Lemma 5.5 it is possible to obtain a result analogous to Theorem 4.1 but tailored to the considered setup. This fact is formalized next in Theorem 5.1.

Theorem 5.1 *Consider an EDM–PDM pair where the EDM satisfies Assumption 5.1 and the PDM is characterized by (5.9) and (5.10), and satisfies Assumptions 5.2 and 5.3. If Assumption 5.1 holds by means of (D1), then Conditions (C1)–(C5) hold and the set* GNE(**f**) *is asymptotically stable (by Theorem 4.1). Alternatively, If Assumption 5.1 holds by means of (D3), then Conditions (A) and (C2)–(C5) hold and the set* GNE($\tilde{\mathbf{f}}$) *is asymptotically stable (by Theorem 4.2).* ◇

Proof Consider first the scenario in which Assumption 5.1 holds due to Condition (D1). The objective is to show that the set GNE(**f**) is asymptotically stable. This is achieved by verifying that Conditions (C1)–(C5) are satisfied and applying Theorem 4.1.

Specifically, Condition (C1) follows from Assumption 5.1. Condition (C2) follows by construction from (5.9b) and (5.10b) (by taking $\mathbf{q}_{b1}^* = \mathbf{M}_\mu \mathbf{q}^*$ and $\mathbf{q}_{b2}^* = \mathbf{M}_\lambda \mathbf{q}^*$), in conjunction with Assumption 5.2-(ii) and Claims (i) and (ii) of Lemma 5.5. To show that (C3) holds, on the other hand, let $\mathbf{x}^* \in$ GNE(**f**) (recall that GNE(**f**) is nonempty and compact) and note that from Assumption 5.2-(iii) there exists a bounded $\mathbf{q}_a^*$ such that $(\mathbf{q}_a^*, \mathbf{x}^*)$ is an equilibrium point of Σ_a. Moreover, note that every $(\mathbf{x}^*, \mathbf{q}_{b1}^*, \mathbf{q}_{b2}^*)$ that satisfies

$$\begin{aligned}
&\mathbf{x}^* \in \underset{\mathbf{x}\in\Delta}{\operatorname{argmax}} \left(\mathbf{x}^\top \mathbf{f}\left(\mathbf{x}^*\right) - \mathbf{x}^\top \mathbf{A}^\top \mathbf{q}_{b1}^* - \mathbf{g}\left(\mathbf{x}\right)^\top \mathbf{q}_{b2}^*\right) \\
&\mathbf{x}^* \in \Omega \\
&\mathbf{q}_{b2}^* \odot \mathbf{g}\left(\mathbf{x}^*\right) = \mathbf{0}_{c_\leq} \\
&\mathbf{q}_{b2}^* \succeq \mathbf{0}_{c_\leq}
\end{aligned}$$

is an equilibrium point of Σ_b and, in consequence, $\left(\mathbf{x}^*, \operatorname{col}\left(\mathbf{q}_a^*, \mathbf{q}_{b1}^*, \mathbf{q}_{b2}^*\right)\right)$ is an equilibrium point of the EDM–PDM pair. In fact, Proposition 4.1 and Assumption 5.2 yield that $\left(\mathbf{x}^*, \operatorname{col}\left(\mathbf{q}_a^*, \mathbf{q}_{b1}^*, \mathbf{q}_{b2}^*\right)\right) \in \mathcal{E}$ if and only if $\left(\mathbf{x}^*, \mathbf{q}_{b1}^*, \mathbf{q}_{b2}^*\right)$ is a KKT point associated with GNE($\mathbf{f}$) and $\mathbf{q}_a^* \in \mathcal{E}_a(\mathbf{x}^*)$. Therefore, $\mathcal{E}$ is nonempty and bounded [7] and Condition (C3) holds. Now, from Lemmas 3.8 and 5.5-(iii), together with Assumption 5.2-(i), it follows that the PDM (5.9) is δ-antipassive characterized by $(\sigma_a + \sigma_b)$. Hence, given that the EDM satisfies Assumption 5.1, one can conclude that the resulting EDM–PDM pair satisfies (C4). Finally, to show that Condition (C5) holds, let $\mathcal{I} \subseteq \mathcal{R}$ be the largest invariant set within $\mathcal{R}$. From Proposition 4.3 it holds that $\mathcal{I}$ is asymptotically stable. Thus, the goal is to show that $\mathcal{I} = \mathcal{E}$. Let $\mathcal{T} = \mathcal{I} \setminus \mathcal{E}$, suppose that $\mathcal{T}$ is nonempty, and let $\left(\mathbf{x}(t'), \mathbf{q}(t')\right) \in \mathcal{T}$ at some time t' (which implies that $(\mathbf{x}(t), \mathbf{q}(t)) \in \mathcal{I} \subseteq \mathcal{R}$, for all $t \geq t'$). From the definition of $\mathcal{R}$ and the informativity property of σ_{EDM} in Assumption 5.1, it follows that $\mathbf{x}(t) = \mathbf{x}(t')$ is fixed for all $t \geq t'$, i.e., the EDM is in steady state and so $\mathbf{x}$ is fixed. Since $\mathbf{x}(t)$ is fixed for all $t \geq t'$, by Assumption 5.2-(iii) it follows that $\mathbf{q}_a(t) \to \mathcal{E}_a\left(\mathbf{x}(t')\right)$ as $t \to \infty$. On the other hand, note that if $\mathbf{x}(t') \notin \Omega$, then by (5.10a) it holds that $\|\mathbf{q}_b(t)\| \to \infty$ as $t \to \infty$, which contradicts the Lyapunov stability of $\mathcal{E}$ guaranteed by Proposition 4.3. Thus, it must hold that $\mathbf{x}(t') \in \Omega$. Also, by Lemma 5.5-(ii) it follows that $\mathbf{q}_{b2}(t) \succeq \mathbf{0}_{c_\leq}$, for all $t \geq 0$. In consequence, $\mathbf{q}_{b2}(t) \to \left\{\mathbf{q}_{b2} : \mathbf{q}_{b2} \odot \mathbf{g}\left(\mathbf{x}(t') = \mathbf{0}_{c_\leq}\right)\right\}$ as $t \to \infty$. Marshaling all these facts yields that if $\left(\mathbf{x}(t'), \mathbf{q}(t')\right) \in \mathcal{T}$ at some time $t' \leq t$, then $(\mathbf{x}(t), \mathbf{q}(t)) \to \mathcal{E}$ as $t \to \infty$. Hence, $\mathcal{E}$ is the largest invariant set within $\mathcal{R}$.

Consider now the second scenario, in which Assumption 5.1 holds due to Condition (D3). In this case, the objective is to show that the set GNE($\tilde{\mathbf{f}}$) is asymptotically stable. As before, this is achieved by verifying that Conditions (D3) and (C2)–(C5) are satisfied, and by applying Theorem 4.2.

In particular, Condition (D3) is guaranteed by Assumption 5.1. The verification of Conditions (C2)–(C5) follows the same reasoning as in the first scenario and is therefore omitted. □

5.2.3 *Distributed Multipliers Dynamics Model*

In distributed implementations, on the other hand, the payoff mechanism follows a conceptual structure in which each population is associated with a dedicated payoff provider. As illustrated in Fig. 5.7, each provider has access only to the strategic distribution of its corresponding population and generates the associated payoff signals accordingly. These providers may exchange information over a communication

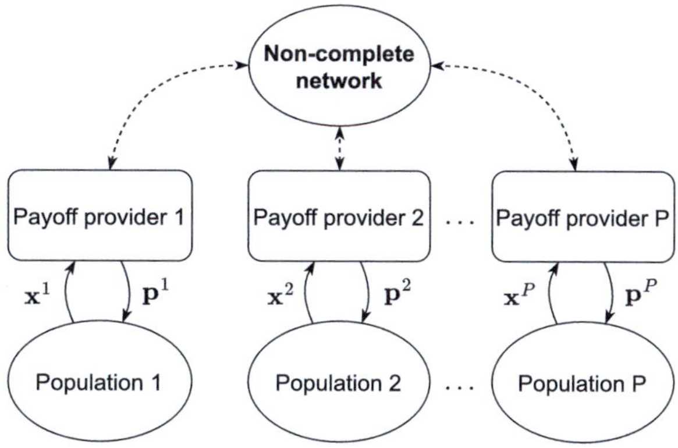

Fig. 5.7 Distributed partial-decision information setup: each population is associated with its own payoff provider, which only observes the local strategic distribution and communicates with other payoff providers over a possibly non-complete network

network, which need not be fully connected. Importantly, similar to the centralized case, this distributed scheme serves as a conceptual model to describe the flow and computation of payoff signals in a decentralized context. It does not necessarily correspond to a physical or explicitly deployed infrastructure, but rather captures the dynamics ruling payoff signals and the information flow.

To model the considered distributed scheme, let each population $k \in \mathcal{P}$ be represented as a node in a network characterized by the directed graph (digraph) $\mathcal{G} = (\mathcal{P}, \mathcal{L}, \mathbf{W})$. Here, $\mathcal{P}$ is the set of nodes corresponding to the populations (i.e., there is exactly one node associated to each population), $\mathcal{L} \subseteq \mathcal{P} \times \mathcal{P}$ is the set of links of possible communication between the nodes, and $\mathbf{W} \in \mathbb{R}_{\geq 0}^{P \times P}$ is the weighted adjacency matrix that captures the topology of the digraph. A link $(\ell, k) \in \mathcal{L}$ if and only if node k can receive information from node ℓ and, as convention, it is assumed that $(k, k) \notin \mathcal{L}$. Besides, $w_{k\ell} > 0$ for all $(\ell, k) \in \mathcal{L}$, and $w_{k\ell} = 0$ otherwise. Here, $w_{k\ell}$ is the (k, ℓ)th element of $\mathbf{W}$. Moreover, let $\mathbf{L} = \operatorname{diag}(\mathbf{W}\mathbf{1}_P) - \mathbf{W}$ be the Laplacian matrix associated to $\mathcal{G}$.

Assumption 5.4 The digraph $\mathcal{G}$ is strongly connected and weight-balanced.[3] ◊

Throughout this section, it is assumed that Ω has the form[4]

$$\Omega = \left\{ \mathbf{x} \in \mathbb{R}^n : \sum_{k \in \mathcal{P}} \left(\mathbf{A}^k \mathbf{x}^k - \mathbf{b}^k \right) = \mathbf{0}_{c_=} \right\},$$

where $\mathbf{A} = \left[\mathbf{A}^1, \mathbf{A}^2, \cdots, \mathbf{A}^P \right]$ and $\mathbf{b} = \sum_{k \in \mathcal{P}} \mathbf{b}^k$. Under the considered setup, it is assumed that Ω complies with Assumption 2.1, and that the information regarding the block $\mathbf{A}^k$ and $\mathbf{b}^k$ is available only at the corresponding node k. Consequently, centralized dynamics as the ones in (5.10) are not suitable to cope with such information-related constraints. As an alternative, for each node $k \in \mathcal{P}$ the following dynamics are introduced:

---

[3] Then, $\operatorname{rank}(\mathbf{L}) = P - 1$, $\mathbf{1}_P^\top \mathbf{L} = \mathbf{0}_P^\top$, $\lambda_{\min}\left(\mathbf{L} + \mathbf{L}^\top\right) = 0$, and $\mathbf{L}\boldsymbol{\zeta} = \mathbf{0}_P \Leftrightarrow \boldsymbol{\zeta} \in \operatorname{span}(\mathbf{1}_P)$ [3].

[4] Note that there are no inequality constraints in Ω and therefore $c_{\leq} = 0$.

$$\frac{1}{\tau_b}\dot{\mathbf{q}}_{b1}^k(t) = -\sum_{\ell\in\mathcal{P}} w_{k\ell}\left(\mathbf{q}_{b1}^k(t) - \mathbf{q}_{b1}^\ell(t)\right)$$
$$-\sum_{\ell\in\mathcal{P}} w_{\ell k}\left(\mathbf{q}_{b2}^k(t) - \mathbf{q}_{b2}^\ell(t)\right) + \mathbf{A}^k\mathbf{x}^k(t) - \mathbf{b}^k \tag{5.12a}$$
$$\frac{1}{\tau_b}\dot{\mathbf{q}}_{b2}^k(t) = \sum_{\ell\in\mathcal{P}} w_{k\ell}\left(\mathbf{q}_{b1}^k(t) - \mathbf{q}_{b1}^\ell(t)\right) \tag{5.12b}$$
$$\mathbf{p}_b^k(t) = -\mathbf{A}^{k\top}\mathbf{q}_{b1}^k(t), \tag{5.12c}$$

where $\mathbf{q}_{b1}^k, \mathbf{q}_{b2}^k \in \mathbb{R}^{c_=}$, and $\tau_b \in \mathbb{R}_{>0}$ is an arbitrary positive constant. Writing the dynamics in (5.12) compactly for all nodes yields the subsystem Σ_b given by

$$\mathcal{W}_b\left(\mathbf{q}_b, \mathbf{x}\right) = \tau_b \begin{bmatrix} -\overline{\mathbf{L}}\mathbf{q}_{b1} - \overline{\mathbf{L}}^\top\mathbf{q}_{b2} + \overline{\mathbf{A}}\mathbf{x} - \overline{\mathbf{b}} \\ \overline{\mathbf{L}}\mathbf{q}_{b1} \end{bmatrix} \tag{5.13a}$$
$$\mathcal{H}_b\left(\mathbf{q}_b, \mathbf{x}\right) = -\overline{\mathbf{A}}^\top\mathbf{q}_{b1}, \tag{5.13b}$$

where $\mathbf{q}_b = \operatorname{col}(\mathbf{q}_{b1}, \mathbf{q}_{b2})$, $\mathbf{q}_{bz} = \operatorname{col}\left(\mathbf{q}_{bz}^k\right)_{k\in\mathcal{P}}$, for all $z \in \{1, 2\}$, $\overline{\mathbf{L}} = \mathbf{L}\otimes\mathbf{I}_{c_=}$, $\overline{\mathbf{A}} = \operatorname{diag}\left(\mathbf{A}^k\right)_{k\in\mathcal{P}}$, and $\overline{\mathbf{b}} = \operatorname{col}\left(\mathbf{b}^k\right)_{k\in\mathcal{P}}$. Notice that (5.13) can be computed in a distributed fashion under the information-related constraints imposed by the digraph $\mathcal{G}$. Besides, the properties outlined in Lemma 5.6 are satisfied.

Lemma 5.6 *Let Σ_b be characterized by (5.13). It holds that:*

(i) For every $\left(\mathbf{x}^*, \mathbf{q}_b^*\right) \in \Delta\times\mathbb{R}^{d_b}$, $\mathcal{W}_b\left(\mathbf{q}_b^*, \mathbf{x}^*\right) = \mathbf{0}_{d_b}$ implies that $\mathbf{x}^* \in \Omega$.
(ii) Σ_b is δ-antipassive characterized by some σ_b.

◊

Proof Each claim is proven sequentially.

Proof of Claim (i): If $\mathcal{W}_b\left(\mathbf{q}_b^*, \mathbf{x}^*\right) = \mathbf{0}_{d_b}$, then, due to Assumption 5.4, summing (5.12a) over all $k\in\mathcal{P}$ yields that $\mathbf{0}_{c_=} = \sum_{k\in\mathcal{P}}\left(\mathbf{A}^k\mathbf{x}^{k*} - \mathbf{b}^k\right)$, and so $\mathbf{x}^* \in \Omega$.

Proof of Claim (ii): Throughout, consider an equivalent reduced-order formulation of the PDM in (5.13), which follows from the change of variables

$$\mathbf{q}_{b2} = \left(\mathbf{U}\otimes\mathbf{I}_{c_=}\right)\left[\tilde{\mathbf{q}}_{b2}^\top, \hat{\mathbf{q}}_{b2}^\top\right]^\top,$$

with $\tilde{\mathbf{q}}_{b2} \in \mathbb{R}^{(P-1)c_=}$ and $\hat{\mathbf{q}}_{b2} \in \mathbb{R}^{c_=}$, where $\mathbf{U} = \left[\tilde{\mathbf{U}}, \hat{\mathbf{u}}\right] \in \mathbb{R}^{P\times P}$ be an orthonormal matrix, with $\hat{\mathbf{u}} = (1/\sqrt{P})\mathbf{1}_P$. It can be shown[5] that under matching initial conditions,

[5] Refer for instance to [10, Remark 5]

the PDM (5.13) has the same input–output trajectories as the reduced-order PDM given by

$$\mathcal{W}_b(\mathbf{q}_b, \mathbf{x}) = \tau_b \begin{bmatrix} -\overline{\mathbf{L}}\mathbf{q}_{b1} - \overline{\mathbf{L}}_{\tilde{\mathrm{U}}}^\top \tilde{\mathbf{q}}_{b2} + \overline{\mathbf{A}}\mathbf{x} - \overline{\mathbf{b}} \\ \overline{\mathbf{L}}_{\tilde{\mathrm{U}}}\mathbf{q}_{b1} \end{bmatrix} \tag{5.14a}$$

$$\mathcal{H}_b(\mathbf{q}_b, \mathbf{x}) = -\overline{\mathbf{A}}^\top \mathbf{q}_{b1}, \tag{5.14b}$$

where $\overline{\mathbf{L}}_{\tilde{\mathrm{U}}} = \left(\tilde{\mathrm{U}}^\top \mathbf{L}\right) \otimes \mathbf{I}_{c_=}$. Here, $\hat{\mathbf{q}}_{b2}$ has been eliminated as it corresponds to an uncontrollable state variable that does not affect the remaining state variables or the output. As such, to analyze (5.13), it suffices to consider (5.14).

Let $\mathcal{W}_b(\mathbf{q}_b, \mathbf{x})$ and $\mathcal{H}_b(\mathbf{q}_b, \mathbf{x})$ in (5.14) be rewritten as

$$\mathcal{W}_b(\tilde{\mathbf{q}}_b, \mathbf{x}) = \mathbf{F}_b\tilde{\mathbf{q}}_b + \mathbf{E}_b\left(\overline{\mathbf{A}}\mathbf{x} - \overline{\mathbf{b}}\right)$$

and $\mathcal{H}(\tilde{\mathbf{q}}_b, \mathbf{x}) = -\overline{\mathbf{A}}^\top \mathbf{E}_b^\top \tilde{\mathbf{q}}_b$, respectively, where $\tilde{\mathbf{q}}_b = \mathrm{col}(\mathbf{q}_{b1}, \tilde{\mathbf{q}}_{b2})$ and

$$\mathbf{F}_b = \tau_b \begin{bmatrix} -\overline{\mathbf{L}} & -\overline{\mathbf{L}}_{\tilde{\mathrm{U}}}^\top \\ \overline{\mathbf{L}}_{\tilde{\mathrm{U}}} & \mathbf{0}_{(P-1)c_= \times (P-1)c_=} \end{bmatrix}$$

$$\mathbf{E}_b = \tau_b\left[\mathbf{I}_{Pc_=}, \mathbf{0}_{Pc_= \times (P-1)c_=}\right]^\top.$$

Besides, consider the functions given by

$$S_b(\tilde{\mathbf{q}}_b, \mathbf{x}) = \frac{1}{2\tau_b}\left\|\mathcal{W}_b(\tilde{\mathbf{q}}_b, \mathbf{x})\right\|^2$$

$$\sigma_b(\tilde{\mathbf{q}}_b, \mathbf{x}) = \left\|\mathcal{W}_b(\tilde{\mathbf{q}}_b, \mathbf{x})\right\|_{\mathbf{R}_b}^2,$$

where $\mathbf{R}_b = -\frac{1}{2}\left(\mathbf{F}_b + \mathbf{F}_b^\top\right)$, i.e.,

$$\mathbf{R}_b = \mathrm{diag}\left(\frac{\tau_b}{2}\left(\overline{\mathbf{L}} + \overline{\mathbf{L}}^\top\right), \mathbf{0}_{(P-1)c_= \times (P-1)c_=}\right).$$

Here, note that $\mathbf{R}_b \succeq 0$. It follows that

$$\nabla_{\tilde{\mathbf{q}}_b} S_b(\tilde{\mathbf{q}}_b, \mathbf{x}) = \mathbf{F}_b^\top \mathcal{W}_b(\tilde{\mathbf{q}}_b, \mathbf{x})$$

$$\nabla_{\mathbf{x}} S_b(\tilde{\mathbf{q}}_b, \mathbf{x}) = \overline{\mathbf{A}}^\top \mathbf{E}_b^\top \mathcal{W}_b(\tilde{\mathbf{q}}_b, \mathbf{x}).$$

Therefore,

$$\dot{S}_b := -\left\|\mathcal{W}_b(\tilde{\mathbf{q}}_b, \mathbf{x})\right\|_{\mathbf{R}_b}^2 + \mathcal{W}_b(\tilde{\mathbf{q}}_b, \mathbf{x})^\top \mathbf{E}_b\overline{\mathbf{A}}\mathbf{v}_b.$$

On the other hand, $\dot{\mathcal{H}} := -\overline{\mathbf{A}}^{\top}\mathbf{E}_b^{\top}\mathcal{W}_b(\tilde{\mathbf{q}}_b, \mathbf{x})$, and in consequence $-\dot{\mathcal{H}}^{\top}\mathbf{v}_b = \mathcal{W}_b(\tilde{\mathbf{q}}_b, \mathbf{x})^{\top}\mathbf{E}_b\overline{\mathbf{A}}\mathbf{v}_b$. Combining these facts yields $\dot{S}_b = -\sigma_b(\tilde{\mathbf{q}}_b, \mathbf{x}) - \dot{\mathcal{H}}^{\top}\mathbf{v}_b$, which implies that (5.14) is δ-passive characterized by σ_b, and so Claim (ii) holds. □

An analog of Theorem 4.1, adapted to the distributed setup under consideration, can be established using Lemma 5.6. This result is formalized in Theorem 5.2.

Theorem 5.2 *Consider an EDM–PDM pair where the EDM satisfies Assumption 5.1 and the PDM is characterized by (5.9) and (5.13), and satisfies Assumption 5.2. If Assumption 5.1 holds by means of (D1), then Conditions (C1)–(C5) hold and the set* GNE($\mathbf{f}$) *is asymptotically stable (by Theorem 4.1). Alternatively, If Assumption 5.1 holds by means of (D3), then Conditions (A) and (C2)–(C5) hold and the set* GNE($\tilde{\mathbf{f}}$) *is asymptotically stable (by Theorem 4.2).* ◊

Proof As in the proof of Theorem 5.1, consider first the scenario where Assumption 5.1 holds as a consequence of Condition (D1). The aim is then to establish the asymptotic stability of the set GNE($\mathbf{f}$), and this is done by verifying that Conditions (C1)–(C5) are satisfied and by invoking Theorem 4.1.

In fact, Condition (C1) follows from Assumption 5.1. Condition (C2) follows by construction from (5.13) (by taking $\mathbf{q}_{b1}^* = \mathbf{M}_{\mu}\mathbf{q}^*$), in conjunction with Assumption 5.2-(ii) and Claim (i) of Lemma 5.6. To show that (C3) holds, without loss of generality, let Σ_b be given by the reduced-order PDM (5.14) (recall that, under matching initial conditions, the input–output trajectories of (5.13) and (5.14) are the same [10, Remark 5]). Furthermore, let $\mathbf{x}^* \in$ GNE($\mathbf{f}$) (recall that GNE($\mathbf{f}$) is nonempty and compact) and note that from Assumption 5.2-(iii) there exists a bounded $\mathbf{q}_a^*$ such that $\left(\mathbf{q}_a^*, \mathbf{x}^*\right)$ is an equilibrium point of Σ_a. Now, consider the optimization problem given by

$$\max_{\mathbf{q}_{b1} \in \mathbb{R}^{Pc_=}} \quad -\frac{1}{2}\mathbf{q}_{b1}^{\top}\overline{\mathbf{L}}\mathbf{q}_{b1} + \mathbf{x}^{*\top}\overline{\mathbf{A}}^{\top}\mathbf{q}_{b1} - \overline{\mathbf{b}}^{\top}\mathbf{q}_{b1} \tag{5.15a}$$

$$\text{s.t.} \quad \overline{\mathbf{L}}_{\tilde{\mathbb{U}}}\mathbf{q}_{b1} = \mathbf{0}_{(P-1)c_=}. \tag{5.15b}$$

Let $\tilde{\mathbf{q}}_{b2} \in \mathbb{R}^{(P-1)c_=}$ represent the Lagrange multiplier associated to the constraint (5.15b). It follows that $\mathbf{q}_{b1}^*$ solves (5.15) if and only if $\left(\mathbf{q}_{b1}^*, \tilde{\mathbf{q}}_{b2}^*\right)$ satisfies the corresponding KKT conditions:

$$-\overline{\mathbf{L}}\mathbf{q}_{b1}^* + \overline{\mathbf{A}}\mathbf{x}^* - \overline{\mathbf{b}} - \overline{\mathbf{L}}_{\tilde{\mathbb{U}}}^{\top}\tilde{\mathbf{q}}_{b2}^* = \mathbf{0}_{Pc_=} \tag{5.16a}$$

$$\overline{\mathbf{L}}_{\tilde{\mathbb{U}}}\mathbf{q}_{b1}^* = \mathbf{0}_{(P-1)c_=}. \tag{5.16b}$$

Given that $\overline{\mathbf{L}}_{\tilde{\mathbb{U}}}$ is full-row rank, it follows that the set of KKT points associated to (5.16) is nonempty and bounded for any $\mathbf{x}^* \in$ GNE($\mathbf{f}$) [7]. Besides, by construction, $(\mathbf{x}^*, \mathbf{q}_{b1}^*, \tilde{\mathbf{q}}_{b2}^*)$ satisfies (5.16) if and only if $(\mathbf{q}_{b1}^*, \tilde{\mathbf{q}}_{b2}^*)$ is an equilibrium point of (5.14) under the corresponding input $\mathbf{x}^*$. Therefore, the set $\mathcal{E}_b(\mathbf{x}^*) = \left\{\tilde{\mathbf{q}}_b^* \in \mathbb{R}^{(2P-1)c_=} : \mathcal{W}_b\left(\tilde{\mathbf{q}}_b^*, \mathbf{x}^*\right) = \mathbf{0}_{(2P-1)c_=}\right\}$ is nonempty and bounded, for every $\mathbf{x}^* \in \mathrm{GNE}(\mathbf{f})$. Combining these facts together with Assumption 5.1 yields that $\mathcal{E}$ is nonempty and bounded and so (C3) holds. Next, from Lemmas 3.8 and 5.6-(ii), in conjunction with Assumption 5.2-(i), it follows that the PDM (5.9) is δ-antipassive characterized by $(\sigma_a + \sigma_b)$. Hence, given that the EDM satisfies Assumption 5.1, one may conclude that the resulting EDM–PDM pair satisfies (C4). Finally, it remains to show that Condition (C5) holds. For such, consider again the reduced-order PDM (5.14) for Σ_b. Now, let $\mathcal{I} \subseteq \mathcal{R}$ be the largest invariant set within $\mathcal{R}$. From Proposition 4.3 it holds that $\mathcal{I}$ is asymptotically stable. It is intended to show that $\mathcal{I} = \mathcal{E}$. Let $\mathcal{T} = \mathcal{I} \setminus \mathcal{E}$, suppose that $\mathcal{T}$ is nonempty, and let $\left(\mathbf{x}(t'), \mathbf{q}(t')\right) \in \mathcal{T}$ at some time t' (which implies that $(\mathbf{x}(t), \mathbf{q}(t)) \in \mathcal{I} \subseteq \mathcal{R}$, for all $t \geq t'$). From the definition of $\mathcal{R}$ and the informativity property of $\sigma_{EDM}(\cdot,\cdot,\cdot)$ in Assumption 5.1, it follows that $\mathbf{x}(t) = \mathbf{x}(t')$ is fixed for all $t \geq t'$, i.e., the EDM is in steady state and so $\mathbf{x}$ is fixed. Since $\mathbf{x}(t)$ is fixed for all $t \geq t'$, by Assumption 5.2-(iii) it follows that $\mathbf{q}_a(t) \to \mathcal{E}_a\left(\mathbf{x}(t')\right)$ as $t \to \infty$. Furthermore, recall from the proof of Lemma 5.6 that Σ_b is δ-antipassive characterized by σ_b with $\sigma_b(\tilde{\mathbf{q}}_b, \mathbf{x}) = \left\|\mathcal{W}_b(\tilde{\mathbf{q}}_b, \mathbf{x})\right\|^2_{\mathbf{R}_b}$. It then follows that $(\mathbf{x}^*, \mathbf{q}^*) \in \mathcal{R}$ implies that (5.16a) holds. Therefore, $\left(\mathbf{x}(t'), \mathbf{q}(t')\right) \in \mathcal{T}$ implies that $\mathbf{q}_{b1}(t) = \mathbf{q}_{b1}(t')$ is fixed, for all $t \geq t'$. Since $\mathcal{T} \cap \mathcal{E} = \emptyset$, it must hold that $\overline{\mathbf{L}}_{\tilde{\mathbf{U}}}\mathbf{q}_{b1}(t') \neq \mathbf{0}_{(P-1)c_=}$, which implies that $\|\tilde{\mathbf{q}}_{b2}(t)\| \to \infty$ as $t \to \infty$, contradicting the Lyapunov stability of $\mathcal{E}$ guaranteed by Proposition 4.3. Consequently, $\mathcal{T}$ must be empty and $\mathcal{I} = \mathcal{E}$. In conclusion, (C5) holds.

Consider now the second scenario, where Assumption 5.1 holds as a result of Condition (D3). The goal is to establish the asymptotic stability of the set $\mathrm{GNE}(\tilde{\mathbf{f}})$ in the distributed setting. As in the previous case, this is accomplished by verifying that Conditions (D3) and (C2)–(C5) are satisfied and by applying Theorem 4.2.

Specifically, Condition (D3) is ensured by Assumption 5.1, while the verification of Conditions (C2)–(C5) proceeds analogously to the first scenario and is therefore omitted. □

5.2.4 *Simulation Example*

To illustrate the application of the non-prescriptive control approach developed in this chapter, a case study on *distributed energy management in microgrids* is considered, as outlined in Sect. 1.2.6.

Specifically, a set of 5 populations, $\mathcal{P} = \{1, 2, 3, 4, 5\}$, is considered. Populations $\{1, 2\}$ represent energy producers, whereas populations $\{3, 4, 5\}$ correspond to energy consumers. Furthermore, these populations are divided into two microgrids, as depicted in Fig. 5.8, and are interconnected through the information-exchange graph $\mathcal{G}$ with adjacency matrix

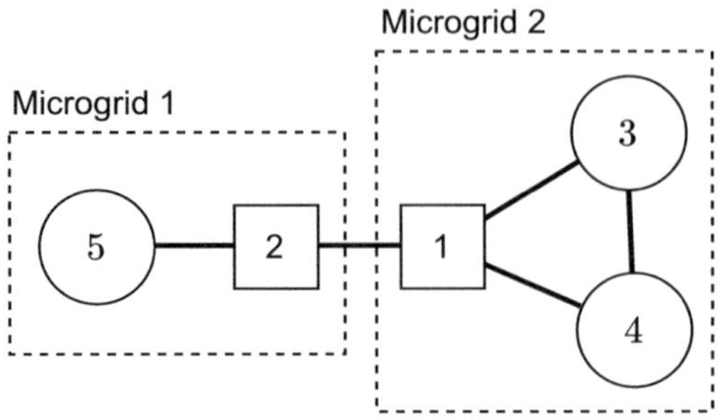

Fig. 5.8 Topology of the considered system. Each node represents a population in the set $\mathcal{P} = \{1, 2, 3, 4, 5\}$. Solid squares denote producer-type populations, circles represent consumer-type populations, and dashed rectangles indicate the two microgrids forming the system. Solid lines between nodes illustrate the possible communication links for information exchange

$$\mathbf{W} = \begin{bmatrix} 0 & 1 & 1 & 1 & 0 \\ 1 & 0 & 0 & 0 & 1 \\ 1 & 0 & 0 & 1 & 0 \\ 1 & 0 & 1 & 0 & 0 \\ 0 & 1 & 0 & 0 & 0 \end{bmatrix}.$$

Namely, nodes within the same microgrid are able to communicate with each other.

Each agent within a population selects its energy management policy from a predefined set of available options. For example:

- An agent in a producer-type population could correspond to a renewable energy source with battery storage, selecting a pair (*time slot*, *power*) from a finite set of combinations that specify the fixed power supplied during the chosen time period.
- An agent in a consumer-type population could represent a plug-in electric vehicle, selecting a pair (*time slot*, *charging rate*) from a finite set of combinations that specify the constant charging rate during the chosen time window.

Thus, the strategy set of each population corresponds to its available energy management policies. For simplicity, but without loss of generality, it is assumed that all populations have the same number of available options over identical time slots, and that the generation powers in producer-type populations match the demand rates in consumer-type populations.

The overall system must satisfy the supply-demand balance constraint

$$\mathbf{x}^1 + \mathbf{x}^2 = \mathbf{x}^3 + \mathbf{x}^4 + \mathbf{x}^5.$$

This condition ensures that the net energy produced by populations 1 and 2 equals the net energy consumed by populations 3, 4, and 5. The associated set of constraints Ω can therefore be written as

$$\Omega = \left\{ \mathbf{x} \in \mathbb{R}^n : \sum_{k \in \mathcal{P}} \left(\mathbf{A}^k \mathbf{x}^k - \mathbf{b}^k \right) = \mathbf{0}_{c_=} \right\},$$

where $\mathbf{A}^\ell = \mathbf{I}_{n^\ell}$ for all $\ell \in \{1, 2\}$, $\mathbf{A}^z = -\mathbf{I}_{n^z}$ for all $z \in \{3, 4, 5\}$, and $\mathbf{b}^k = \mathbf{0}_{n^k}$ for all $k \in \mathcal{P}$. Moreover, to ensure the nonemptiness of the set $\mathcal{X} = \Delta \cap \Omega$, populations $\{1, 3, 4, 5\}$ are modeled with a unit mass $m^k = 1$, whereas population 2 is assigned double the mass, $m^2 = 2$. Hence, coordination between the two microgrids becomes necessary to guarantee feasibility.

To encourage compliance with these constraints, a dynamic pricing scheme is implemented within the microgrid network. In particular, dynamic energy prices serve as payoff signals perceived by the agents, and the corresponding payoff mechanism is represented by the instance of the structured PDM (5.9b) given by

$$\mathcal{W}(\mathbf{q}, \mathbf{x}) = \begin{bmatrix} \mathbf{0}_n \\ \mathcal{W}_b\left(\mathbf{q}_b, \mathbf{x}\right) \end{bmatrix}, \quad \mathcal{H}\left(\mathbf{q}, \mathbf{x}\right) = \mathbf{f}(\mathbf{x}) + \mathcal{H}_b\left(\mathbf{q}_b, \mathbf{x}\right),$$

where the fitness function $\mathbf{f}\left(\mathbf{x}\right) = -\boldsymbol{\alpha}$, with $\boldsymbol{\alpha} \in \mathbb{R}^n_{\geq 0}$, captures the nominal (fixed) *time-of-use* prices. Meanwhile, $\mathcal{W}_b(\cdot, \cdot)$ and $\mathcal{H}_b(\cdot, \cdot)$, given by the distributed multipliers model (5.13) associated with the communication graph $\mathcal{G}$, govern the dynamic constraint-related prices (paid *by* consumers *to* producers) that incentivize supply-demand balance throughout the system.

As an illustrative case, yet without loss of generality, a scenario with 3 strategies per population is considered. The *time-of-use* prices are set as $\alpha^k = \begin{bmatrix} 1, & 1, & 3 \end{bmatrix}$ for all $k \in \{1, 2\}$, and $\alpha^\ell = \begin{bmatrix} 3, & 1, & 1 \end{bmatrix}$ for all $\ell \in \{3, 4, 5\}$. This implies that, for producer-type populations, the third strategy is the least profitable, whereas for consumer-type populations, the first strategy is the least profitable.

Moreover, populations $\{1, 3, 4, 5\}$ are assumed to consist of 100 agents each, while population 2 comprises 200 agents (reflecting its doubled mass). Agents in the producer-type populations $\{1, 2\}$ are assumed to follow IPC-type revision protocols, whereas agents in the consumer-type populations $\{3, 4, 5\}$ adopt IPC-, SEPT-, and PBR-type revision protocols, respectively. In all cases, agents operate under Poisson alarm clocks with a unit rate (in arbitrary time units).

Figure 5.9 illustrates that the society collectively converges asymptotically toward GNE($\tilde{\mathbf{f}}$) of the considered game, with $\tilde{\mathbf{f}}(\cdot)$ defined as in Corollary 2.2. This equilibrium corresponds to the pure state in which all populations select strategy 2, as this configuration is equally profitable for both producer- and consumer-type populations. In contrast, Figs. 5.10, 5.11, 5.12, 5.13, and 5.14 depict the temporal evolution of the strategic distributions of all populations. These results show that, although different revision protocols lead to distinct transient trajectories, the mixed society ultimately converges near a GNE as predicted by Corollary 4.2.

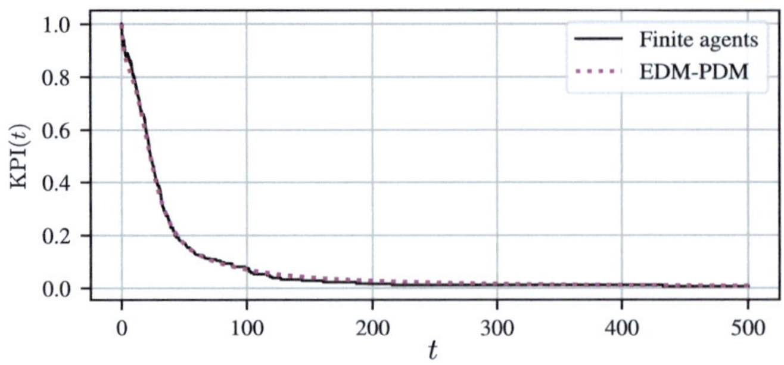

Fig. 5.9 Temporal evolution of $\text{KPI}(t) = \|\mathbf{x}(t) - \mathbf{x}^*\|_2 \,/\, \|\mathbf{x}(0) - \mathbf{x}^*\|_2$, with respect to $\mathbf{x}^* = \left[0,\ 1,\ 0,\ 0,\ 2,\ 0,\ 0,\ 1,\ 0,\ 0,\ 1,\ 0,\ 0,\ 1,\ 0\right]^\top$

Fig. 5.10 Evolution of population 1, under the Smith revision protocol

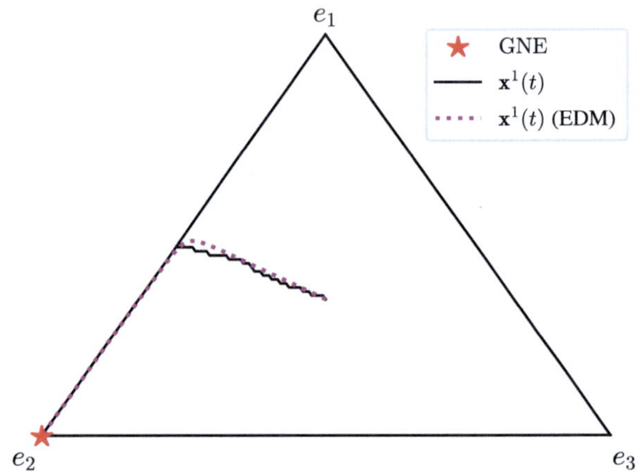

Fig. 5.11 Evolution of population 2, under the Smith revision protocol

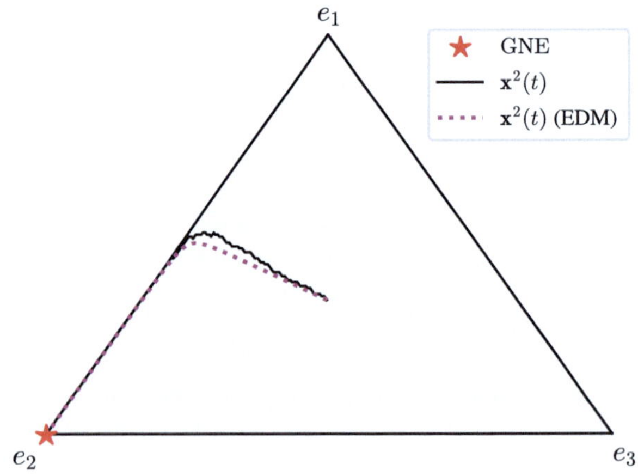

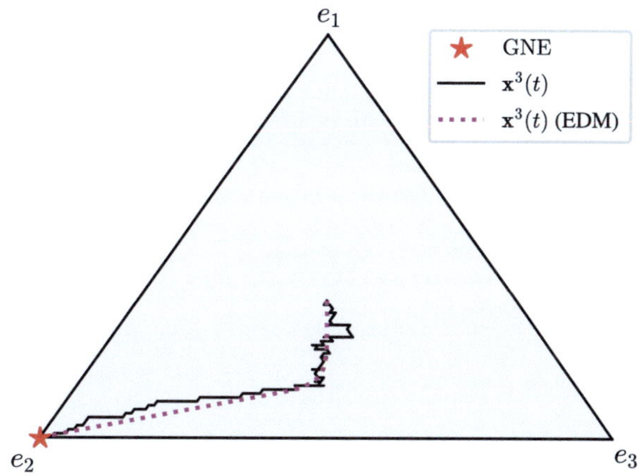

Fig. 5.12 Evolution of population 3, under the BNN revision protocol

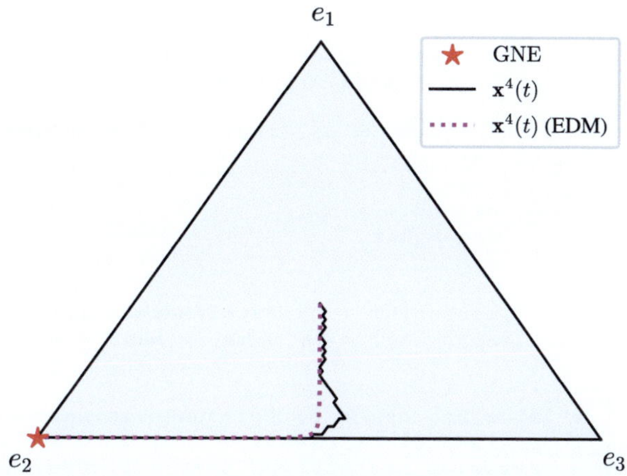

Fig. 5.13 Evolution of population 4, under the Smith revision protocol

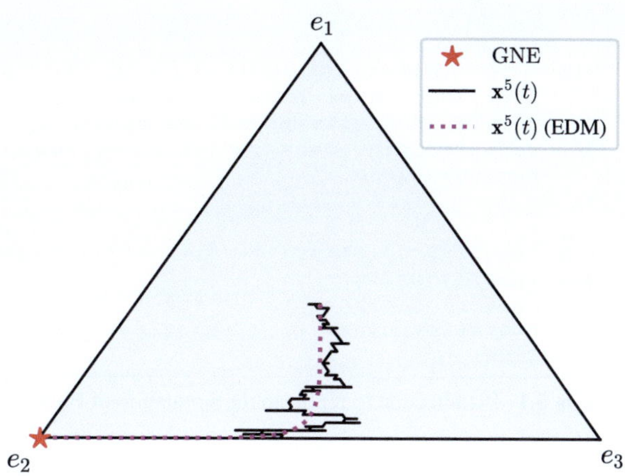

Fig. 5.14 Evolution of population 5, under the *logit-choice* revision protocol

Python Scripts

```
import numpy as np
import popgames as pg

np.random.seed(698) # For reproducibility

n = 4 # Number of zones
x_bar = np.array([1, 0.2, 0.2, 0.2]).reshape(n, 1)
q_bar = np.random.uniform(0, 2, (n, 1))
gamma = np.random.uniform(0, 1, (n, 1))

max0 = lambda x: np.maximum(x, 0) # max(., 0)

def f(x): return -x

def h_map(q, x): return f(x) + q

def w_map(q, x):
    return max0(q_bar - q) * max0(gamma - x) \
        - q * max0(x - gamma)

population_game = pg.SinglePopulationGame(
    num_strategies=n,
    fitness_function=f,
    A_ineq=np.eye(n),
    b_ineq=x_bar,
)

payoff_mechanism = pg.PayoffMechanism(
    h_map=h_map, n=n, w_map=w_map, d=n,
)

revision_process = pg.PoissonRevisionProcess(
    Poisson_clock_rate=1,
    revision_protocol=pg.protocol.CCSmith(scale=0.1,x_bar=
        x_bar)
)

sim = pg.Simulator(
    population_game=population_game,
    payoff_mechanism=payoff_mechanism,
    revision_processes=revision_process,
    num_agents=50,
)

sim.reset(x0=np.array([1, 0, 0, 0]).reshape(n, 1))
sim.run(T_sim=200)
sim.plot(plot_type='univariate')
sim.plot(plot_type='kpi', plot_deterministic_approximation=
    True)
```

Listing 5.1 Python code to replicate the simulation of Figs. 5.1, 5.2, 5.3 and 5.4

```python
import numpy as np
from scipy.linalg import block_diag
import popgames as pg
W = np.array([[0, 1, 1, 1, 0], [1, 0, 0, 0, 1], [1, 0, 0, 1,
    0],
              [1, 0, 1, 0, 0], [0, 1, 0, 0, 0]])
In = np.eye(3)
L_bar = np.kron(np.diag(W.sum(axis=1)) - W, In)
A_bar = block_diag(In, In, -In, -In, -In)
b_bar = np.zeros((15, 1))
tou1, tou2 = np.array([[1], [1], [3]]), np.array([[3], [1],
    [1]])
alpha, taub = np.vstack([tou1, tou1, tou2, tou2, tou2]), 0.1
def f(x): return -alpha
def w_map(q, x):
    qa, qb1, qb2 = q[:15], q[15:30], q[30:]
    dqa = np.zeros((15, 1))
    dqb1 = -np.dot(L_bar, qb1) - np.dot(L_bar.T, qb2) \
           + np.dot(A_bar, x) - b_bar
    dqb2 = np.dot(L_bar, qb1)
    return np.vstack([dqa, taub*dqb1, taub*dqb2])
def h_map(q, x):
    return f(x) - np.dot(A_bar.T, q[15:30])
population_game = pg.PopulationGame(
    num_populations=5, num_strategies=[3, 3, 3, 3, 3],
    fitness_function=f, masses=[1, 2, 1, 1, 1],
    A_eq=np.hstack([In, In, -In, -In, -In]),
    b_eq=np.zeros((3, 1)),
)
payoff_mechanism = pg.PayoffMechanism(
    h_map=h_map, n=15, w_map=w_map, d=45
)
revision_processes = [
    pg.PoissonRevisionProcess(1.0, pg.protocol.Smith(0.1)),
    pg.PoissonRevisionProcess(1.0, pg.protocol.Smith(0.1)),
    pg.PoissonRevisionProcess(1.0, pg.protocol.BNN(0.1)),
    pg.PoissonRevisionProcess(1.0, pg.protocol.Smith(0.1)),
    pg.PoissonRevisionProcess(1.0, pg.protocol.Softmax(0.01)),
]
sim = pg.Simulator(
    population_game=population_game,
    payoff_mechanism=payoff_mechanism,
    revision_processes=revision_processes,
    num_agents=[100, 200, 100, 100, 100]
)
aux = (1/3)*np.ones((3, 1))
sim.reset(x0=np.vstack([aux, 2*aux, aux, aux, aux]),
          q0=np.zeros((45, 1)))
sim.run(T_sim=500)
sim.plot(plot_type='kpi', plot_deterministic_approximation=
    True)
sim.plot(plot_type='ternary',
         plot_deterministic_approximation=True)
```

Listing 5.2 Python code to replicate the simulation of Figs. 5.9, 5.10, 5.11, 5.12, 5.13 and 5.14

Table 5.1 Comparison of related literature with respect to key features. A checkmark (✓) indicates that the corresponding reference considers coupling constraints (CC), payoff dynamics models (PDM), or δ-dissipativity tools to handle a broad class of dynamics (δD)

Reference	CC	PDM	δD
[17]	–	–	–
[2]	✓	–	–
[12]	✓	–	–
[13]	✓	✓	–
[14–16]	–	✓	✓
[1]	–	✓	✓
[9]	–	✓	✓
[10]	–	✓	✓
[11]	✓	✓	✓
This book	✓	✓	✓

Notes

The structured EDM–PDM framework developed in this chapter recovers, extends, and unifies several results from the recent literature on population games. Table 5.1 summarizes this comparison by contrasting key references according to their treatment of coupling constraints, payoff dynamics models, and the use of δ-dissipativity tools for analyzing broad classes of dynamics.

The seminal work of [17] establishes the foundational framework of population games and provides convergence guarantees for various revision protocols but considers only memoryless payoff mechanisms and standard Nash equilibria, without coupling constraints. In contrast, [2, 12] address GNE seeking under box-type coupling constraints via prescriptive control approaches, as discussed in Sect. 5.1. Building on this, [13] introduces a payoff dynamics model for handling general affine equality and convex inequality constraints, representing a non-prescriptive control approach aligned with Sect. 5.2.

The smoothing-anticipatory learning dynamics introduced in [16] and revisited in Sect. 4.3.2 for NE seeking can be embedded within the structured PDM framework to guarantee convergence toward a GNE. This is achieved by employing them as subsystem Σ_a together with either centralized or distributed multiplier dynamics for Σ_b. Such an approach generalizes the setups considered in [1, 9, 14, 15] from NE seeking to GNE seeking. Furthermore, the distributed consensus-based PDM proposed in [10] for NE seeking in strongly contractive games can be extended to distributed GNE seeking by coupling it with the distributed multiplier model in (5.13), thereby preserving partial-decision information constraints.

It is worth emphasizing that the results obtained in this chapter are not restricted to population games or evolutionary dynamics. More generally, the structured PDM (5.9) can be interconnected in a closed loop with alternative dynamical systems governing the evolution of primal optimization variables, such as continuous-time gradient ascent schemes. If such dynamics satisfy suitable δ-dissipativity properties, asymptotic stability of the resulting equilibria can be established using the same analysis tools. This observation opens new perspectives for the design of continuous-time primal-dual algorithms for GNE seeking in classical multi-player games, complementing existing passivity-based approaches for NE and GNE seeking [4–6]. This idea is explored in Chap. 6.

References

1. Arcak M, Martins NC (2021) Dissipativity tools for convergence to Nash equilibria in population games. IEEE Trans Control Netw Syst 8(1):39–50. https://doi.org/10.1109/TCNS.2020.3029990
2. Barreiro-Gomez J, Tembine H (2018) Constrained evolutionary games by using a mixture of imitation dynamics. Automatica 97:254–262. https://doi.org/10.1016/j.automatica.2018.08.014
3. Bullo F (2022) Lectures on Network Systems, 1.6 edn. Kindle Direct Publishing
4. De Persis C, Grammatico S (2019) Distributed averaging integral Nash equilibrium seeking on networks. Automatica 110:108548. https://doi.org/10.1016/j.automatica.2019.108548
5. Fu Z, Cenedese C, Cucuzzella M, Yu W, Scherpen JMA (2024) Distributed control of islanded DC microgrids: a passivity-based game theoretical approach. IEEE Trans Control Syst Technol 32(6):2207–2222. https://doi.org/10.1109/TCST.2024.3405711
6. Gadjov D, Pavel L (2019) A passivity-based approach to Nash equilibrium seeking over networks. IEEE Trans Autom Control 64(3):1077–1092. https://doi.org/10.1109/TAC.2018.2833140
7. Gauvin J (1977) A necessary and sufficient regularity condition to have bounded multipliers in nonconvex programming. Math Program 12:136–138
8. Hofbauer J, Sandholm WH (2009) Stable games and their dynamics. J Econ Theory 144(4):1665–1693
9. Kara S, Martins NC (2023) Pairwise comparison evolutionary dynamics with strategy-dependent revision rates: stability and δ-passivity. IEEE Trans Control Netw Syst 10(4):1656–1668. https://doi.org/10.1109/TCNS.2023.3237485
10. Martinez-Piazuelo J, Ocampo-Martinez C, Quijano N (2024) Distributed Nash equilibrium seeking in strongly contractive aggregative population games. IEEE Trans Autom Control 69(7): 4427–4442. https://doi.org/10.1109/TAC.2023.3321208
11. Martinez-Piazuelo J, Ocampo-Martinez C, Quijano N (2024) Generalized Nash equilibrium seeking in a class of contractive population games over networks. In: Proceedings of the 63rd IEEE conference on decision and control (CDC), Milan, Italy, 2319–2324. https://doi.org/10.1109/CDC56724.2024.10886517
12. Martinez-Piazuelo J, Quijano N, Ocampo-Martinez C (2022) Nash equilibrium seeking in full-potential population games under capacity and migration constraints. Automatica 141:110285. https://doi.org/10.1016/j.automatica.2022.110285
13. Martinez-Piazuelo J, Quijano N, Ocampo-Martinez C (2022) A payoff dynamics model for generalized Nash equilibrium seeking in population games. Automatica 140:110227. https://doi.org/10.1016/j.automatica.2022.110227

14. Park S, Barreiro-Gomez J (2023) Payoff mechanism design for coordination in multi-agent task allocation games. In: Proceedings of the 62nd IEEE conference on decision and control (CDC), pp 8116–8121. https://doi.org/10.1109/CDC49753.2023.10383344
15. Park S, Certorio J, Martins NC, La RJ (2025) Epidemic population games and perturbed best response dynamics. IEEE Trans Autom Control 1–13. https://doi.org/10.1109/TAC.2025.3554601
16. Park S, Martins NC, Shamma JS (2019) From population games to payoff dynamics models: a passivity-based approach. In: Proceedings of the 58th IEEE conference on decision and control (CDC), pp 6584–6601 (2019). https://doi.org/10.1109/CDC40024.2019.9029756
17. Sandholm WH (2010) Population games and evolutionary dynamics. MIT Press

Chapter 6
Analogy-Based Design: Connections with N-Player Games

In this chapter, the role of population games and evolutionary dynamics in the control of *large-scale complex systems* (LSCSs) is discussed. Namely, it is shown that LSCSs can be viewed as non-cooperative N-player games, and that the underlying optimization-based control task can be cast as a GNE-seeking problem, possibly subject to partial-decision information. Moreover, it is shown that a population game is itself an LSCS, and so it can be interpreted as an N-player game where the populations play the role of (macro) players. Consequently, a connection between the concepts of Chap. 2 and the optimization-based control of LSCSs is provided.

This chapter is organized as follows. Section 6.1 explains the parallelism between optimization-based control of LSCSs and game theory, covering both non-cooperative N-player games (Sect. 6.1.1) and population games (Sect. 6.1.2). Section 6.2 provides a brief discussion of recent methods and applications for NE and GNE seeking in classical non-cooperative N-player settings. Section 6.3 then introduces an analogy-based perspective, showing how population games and evolutionary dynamics can be applied to LSCSs even when they do not represent a genuine population of autonomous decision-makers. Finally, Sect. 6.4 presents an illustrative example.

6.1 On the Control of LSCSs via Game Theory

Throughout this chapter all discussions will turn around the setup depicted in Fig. 6.1. Namely, consider an LSCS comprised of $N \in \mathbb{Z}_{\geq 1}$ controllable subsystems, which is to be controlled by a set of controllers that might exchange information over a (possibly non-complete) communication network. Besides, let $\mathbf{u}^k \in \mathbb{R}^{d_{u^k}}$ and $\mathbf{y}^k \in \mathbb{R}^{d_{y^k}}$ be the input and output vectors of subsystem $k \in \{1, 2, \ldots, N\}$, respectively.

J. Martinez-Piazuelo et al., *Generalized Nash Equilibrium Seeking in Population Games*, Communications and Control Engineering,
https://doi.org/10.1007/978-3-032-06081-5_6

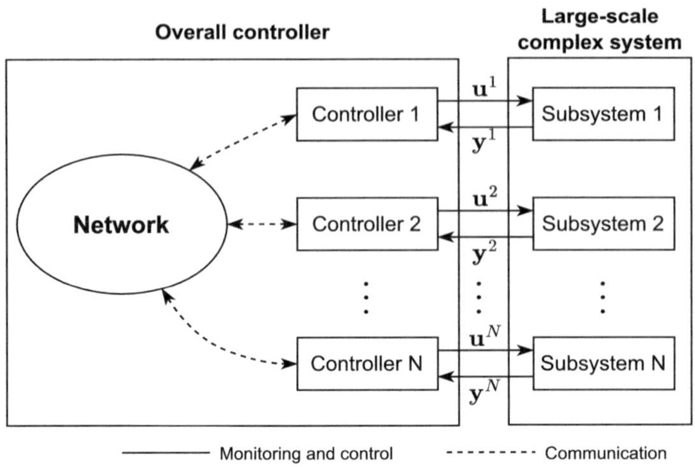

Fig. 6.1 Considered (non-centralized) control setup for the LSCS

Note that each controller is able to measure/observe the output vector ($\mathbf{y}^k$) of its corresponding subsystem and is able to establish its subsystems' control input vector ($\mathbf{u}^k$). In general, subsystem k is characterized by dynamics of the form[1]

$$\dot{\mathbf{z}}^k(t) = \boldsymbol{\mathcal{F}}^k\left(\mathbf{z}(t), \mathbf{u}^k(t), \mathbf{d}(t)\right), \tag{6.1a}$$

$$\mathbf{y}^k(t) = \boldsymbol{\mathcal{H}}^k\left(\mathbf{z}(t), \mathbf{u}^k(t), \mathbf{d}(t)\right), \tag{6.1b}$$

for all $k \in \{1, 2, \ldots, N\}$. Here, $\mathbf{z}^k \in \mathbb{R}^{d_{z^k}}$ denotes the state vector of the kth subsystem; $\mathbf{z} = \text{col}\left(\mathbf{z}^1, \mathbf{z}^2, \ldots, \mathbf{z}^N\right)$ is the state vector of the entire LSCS; $\mathbf{d} \in \mathbb{R}^{d_d}$ represents uncertainties or disturbances of the overall LSCS; and $\boldsymbol{\mathcal{F}}^k(\cdot,\cdot,\cdot)$ and $\boldsymbol{\mathcal{H}}^k(\cdot,\cdot,\cdot)$ correspond to (possibly nonlinear) vector-valued maps. Note that if $d_{z^k} = 0$ and the internal dynamics (6.1a) are neglected, then (6.1) would correspond to a memoryless map for subsystem k. That is, the subsystems might be either dynamic or static systems depending on the application. On the other hand, let $\mathbf{u}^{-k} = \text{col}\left(\mathbf{u}^1, \ldots, \mathbf{u}^{k-1}, \mathbf{u}^{k+1}, \ldots, \mathbf{u}^N\right) \in \mathbb{R}^{\sum_{l=1, l\neq k}^{N} u_{dl}}$ and $\mathbf{y}^{-k} = \text{col}\left(\mathbf{y}^1, \ldots, \mathbf{y}^{k-1}, \mathbf{y}^{k+1}, \ldots, \mathbf{y}^N\right) \in \mathbb{R}^{\sum_{l=1, l\neq k}^{N} y_{dl}}$ denote the stacked vectors of all inputs and outputs except the kth one, respectively. Finally, let $\psi^k : \mathbb{R}^{d_{y^k}} \times \mathbb{R}^{\sum_{l=1, l\neq k}^{N} y_{dl}} \to \mathbb{R}$ be a scalar-valued map that represents the utility function of the kth controller.

Remark 6.1 Alternatively, $\psi^k(\cdot,\cdot)$ could represent a cost function associated to the kth controller. Nonetheless, $\psi^k(\cdot,\cdot)$ is usually interpreted as a utility function to better cope with the framework of population games, where the decision-making agents seek to maximize their perceived payoffs (see Chap. 2). ◊

[1] Discrete-time dynamics may also be considered, although discussions are mainly focused on continuous-time dynamics to ease the exposition.

Here, it is assumed that each controller seeks to solve the optimization-based control problem given by

$$\max_{\mathbf{u}^k} \ \psi^k\left(\mathbf{y}^k, \mathbf{y}^{-k}\right) \quad \text{s.t.} \begin{cases} (6.1) \\ \left(\mathbf{u}^k, \mathbf{u}^{-k}\right) \in \mathcal{U} \\ \left(\mathbf{y}^k, \mathbf{y}^{-k}\right) \in \mathcal{Y}. \end{cases}$$

Besides, $\mathcal{U} \subseteq \mathbb{R}^{\sum_{k=1}^{N} d_{uk}}$ and $\mathcal{Y} \subseteq \mathbb{R}^{\sum_{k=1}^{N} d_{yk}}$ are the constraints set for the input and output vectors, respectively. Observe that, in general, $\mathcal{U}$ might couple the inputs of multiple subsystems, while $\mathcal{Y}$ might couple the outputs of multiple subsystems. Moreover, implicit coupling constraints might also exist through the utility functions or the dynamics of the subsystems. Therefore, the overall control task for the LSCS is comprised of multiple (possibly coupled) optimization-based control problems. In addition, the following observations are remarked:

- The utility functions of the different controllers are not necessarily aligned with a global utility to be maximized, i.e., the controllers might in general be self-interested (non-cooperative).
- The communication network of the controllers might be non-complete, meaning that the controllers could be subject to partial information regarding the inputs and outputs of other subsystems different than their corresponding ones.

Overall, the considered non-centralized control setup is particularly relevant to systems with a large number of controlled variables, where a centralized control scheme might suffer from the dimensionality of the problem, or to spatially distributed systems, where centralized monitoring and control are not viable. Some practical examples of such systems include distributed power generators, water distribution networks, fleets of plug-in electric vehicles, mobile sensor networks, among others.

6.1.1 LSCSs Viewed as Non-cooperative N-Player Games

The considered control setup of Fig. 6.1 can be viewed as a non-cooperative N-player game as depicted in Fig. 6.2. To do so, each controller-subsystem pair is interpreted as a strategic player, and $\mathbf{u}^k$ and $\mathbf{y}^k$ are defined as the action and output of player $k \in \{1, 2, \ldots, N\}$, respectively. Under such a perspective, the concept of a GNE can be taken as a suitable solution for the overall control task. Namely, in terms of the overall output $\mathbf{y} = \text{col}\left(\mathbf{y}^1, \mathbf{y}^2, \ldots, \mathbf{y}^N\right)$, a GNE is defined as a vector $\mathbf{y}^* = \text{col}\left(\mathbf{y}^{*1}, \mathbf{y}^{*2}, \ldots, \mathbf{y}^{*N}\right) \in \mathcal{Y}$ such that

$$\psi^k\left(\mathbf{y}^{*k}, \mathbf{y}^{*-k}\right) \geq \psi^k\left(\mathbf{y}^k, \mathbf{y}^{*-k}\right), \quad \forall \mathbf{y}^k \in \mathcal{Y}^k\left(\mathbf{y}^{*-k}\right), \quad \forall k \in \{1, 2, \ldots, N\}.$$

Here, $\mathcal{Y}^k\left(\mathbf{y}^{-k}\right)$ denotes the set of feasible outputs for player k as a function of the outputs of other players. Thus, at a GNE the coupled constraints are satisfied, and no

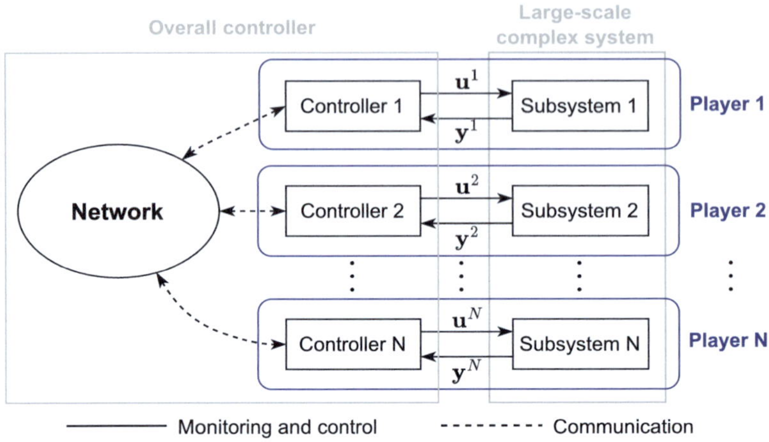

Fig. 6.2 Control setup of Fig. 6.1 viewed as an N-player game

player can increase their utility by unilaterally deviating from the GNE. As such, a GNE is a self-enforceable solution concept for a non-cooperative game with explicit coupled constraints over the players [11]. Furthermore, if there are no explicit coupled constraints over the players outputs, i.e., $\mathcal{Y}^k\left(\mathbf{y}^{-k}\right) = \mathcal{Y}^k \subseteq \mathbb{R}^{d_{yk}}$ is independent of $\mathbf{y}^{-k}$, then the notion of a GNE reduces to the one of an NE [36].

Under the considered formulation, the goal is then for each player k to set their corresponding action $\mathbf{u}^k \in \mathcal{U}^k\left(\mathbf{u}^{-k}\right)$ (here, $\mathcal{U}^k\left(\mathbf{u}^{-k}\right)$ denotes the set of feasible actions for player k as a function of the actions of other players), so that the output $\mathbf{y}^k$ of their corresponding subsystem is driven to an NE/GNE with respect to the other players (assuming of course that an NE/GNE exists). Moreover, given that each player might have only partial information of the entire system, the players should exchange information over the network in order to respond strategically to the other players' actions. Therefore, the problem of reaching an NE/GNE in such a distributed fashion is in general a challenging task and it has received significant research attention in the recent years (see the *Notes* section at the end of this chapter).

Remark 6.2 It is highlighted that the considered non-cooperative framework is fairly general as it allows to model self-interested players, which arise in several large-scale systems (e.g., in fleets of plug-in electric vehicles). Moreover, note that if $\psi^k(\cdot,\cdot) = \psi^\ell(\cdot,\cdot)$, for all $k, \ell \in \{1, 2, \ldots, N\}$, then under the considered framework it is also possible to model global optimization objectives that coordinate the multiple players of the game. However, as in general $\psi^k(\cdot,\cdot) \neq \psi^\ell(\cdot,\cdot)$ for all $k \neq \ell$, the framework of non-cooperative game theory is (in some sense) more general than the one of global optimization. $\diamond$

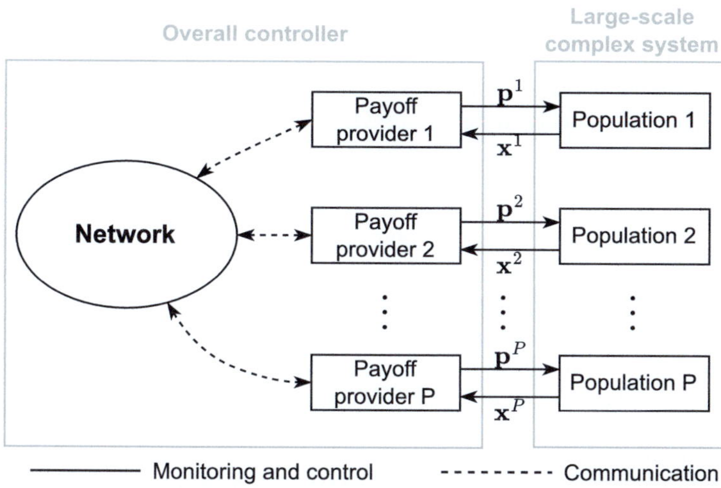

Fig. 6.3 Population game viewed as an LSCSs as in Fig. 6.1

6.1.2 Population Games Viewed as LSCSs

A population game can be viewed as an LSCS as depicted in Fig. 6.3. Namely, think of each population of decision-making agents as a subsystem of an LSCS (thus, $N = P$). In fact, following the ideas of Chap. 2, each population $k \in \mathcal{P}$ is a system with input $\mathbf{u}^k = \mathbf{p}^k$ and output $\mathbf{y}^k = \mathbf{x}^k$, where $\mathbf{p}^k$ and $\mathbf{x}^k$ are the payoff vector and strategic distribution of population k, respectively. The controllers associated to such an LSCS are the so-called payoff providers of the multiple populations (see Remark 6.3), which are high-level entities that yield the payoff vector to their associated population. That is, the payoff provider of population k is a system with input $\mathbf{x}^k$ and output $\mathbf{p}^k$, and the multiple payoff providers communicate through the network.

Remark 6.3 (*Payoff Providers*) In the study of population games and evolutionary dynamics the actual process of how payoff signals are generated is often modeled implicitly rather than explicitly. The reason being that in many applications, particularly in the context of sociology, the payoff signals depend on quite complex dynamics related to how the agents perceive and learn the outcomes of their strategic behaviors. As such, high-level models are often used to implicitly capture the agents' learning processes by accounting for effects such as anticipation, delay, and inertia [38, 39]. For instance, the smoothing-anticipatory PDM of Example 2.4 is an illustration of such high-level modeling of the underlying learning dynamics. On the other hand, in many practical engineering applications the payoffs are indeed given by a tangible entity which can be easily modeled. For example, in the demand response application studied in [46], the payoff signals are explicitly computed and broadcast by the electric power utility, and they are given by the gradients associated to a monetary-incentive minimization problem (i.e., the maximization of monetary profit).

To maintain a flexible framework, the notion of payoff providers has been established in Chap. 1. Namely, a payoff provider could be either a tangible physical entity explicitly computing the payoff signals, or an intangible entity modeling an underlying learning process. ♢

Under the considered analogy, the dynamics of the subsystems (populations) are thus given by the underlying EDM that models the decision-making process of the society agents, while the dynamics of the controllers (payoff providers) are captured by the PDM ruling the temporal evolution of the payoff signals (see Sects. 2.2 and 2.4). Besides, the solution concept for the overall control task of the LSCSs is the GNE as described in Sect. 2.5 (see Definition 2.12).

Remark 6.4 (*Population Games as Non-cooperative N-Player Games*) Notice that since population games can be viewed as LSCSs and LSCSs can be viewed as non-cooperative N-player games (see Sect. 6.1.1), then it follows that population games can be viewed as non-cooperative N-player games as well (with $N = P$). Under such a perspective, each population and payoff provider pair can be seen as a virtual macro-player engaged in a game with the other populations. Yet, the actual decision-makers of such a game are the agents that comprise each population. ♢

The Full-Decision and Partial-Decision Information Setups

When population games are viewed as LSCSs as in Fig. 6.3, two possible setups might arise depending on the properties of the information-exchange network. Namely, when the network is complete, meaning that all payoff providers are able to communicate with each other, it can be conceptualized the set of payoff providers as a single oracle-like payoff provider that has full information regarding the LSCSs and the control signals. Such a setup is depicted in Fig. 5.6. On the other hand, when the network is non-complete, meaning that there is not all-to-all communication between the payoff providers, multiple separate payoff providers can be considered, each having only partial information regarding the LSCSs and the control signals. This second setup is illustrated in Fig. 5.7.

6.2 GNE Seeking in N-Player Games

This section synthesizes foundational and emerging ideas in NE/GNE seeking for N-player games. The discussion is organized to emphasize conceptual frameworks, prevailing methodologies, and their role in LSCSs, leaving the reference of specific literature sources or citations for the *Notes* section at the end of the chapter.

Distributed equilibrium-seeking challenges arise in settings where multiple strategic agents interact, each agent making decisions based on local objectives and information, and where global coordination is either impractical or not permitted. The literature distinguishes between scenarios where agents, or players, are modeled as static systems (i.e., each player's decision directly yields their system's output) and dynamic systems (where agents' behaviors evolve according to prescribed dynamics over time).

In NE seeking problems, players act independently to optimize individual objectives. In the static case, reported approaches often blend gradient-play methods, where each player steers their action according to the gradient of their own utility function, with consensus protocols that provide estimates of other players' actions. These formulations assume players have only partial access to others' decisions and cannot directly observe the global action profile. Key developments have focused on enabling such algorithms to operate robustly even when the network topology is undirected, directed, time-varying, or subject to intermittent communication. Initially, many approaches required players to know global network or game parameters for convergence, but more recent work has aimed at fully distributed algorithms where such assumptions are eliminated. A notable challenge that arises in large-scale games is scalability: when each player must track the actions of many others, maintaining and exchanging high-dimensional information becomes impractical. In response, the study of aggregative games has emerged, in which each player's cost is influenced not by all individual actions but by an aggregate function, such as the sum or average, of all actions. This structural property allows significant reduction in communication and estimation requirements, as players only need to estimate and share a low-dimensional aggregate, leading to more scalable solutions.

Beyond gradient play, alternative approaches have been explored in the static-player setting. Extremum seeking methods dispense with the need for explicit knowledge of the cost function's mathematical form, relying instead on sampled evaluations to drive equilibrium seeking. Another prominent direction considers coalition-based games, where groups of agents coordinate as virtual macro-players, enabling hierarchical or multi-level equilibrium computation, and offering new avenues for dealing with learning and model uncertainty.

Discrete-time distributed NE seeking methods generally mirror the continuous-time developments, adopting stepwise updates of actions and estimates. Strategies have been extended to accommodate issues such as privacy: using randomized or perturbed consensus protocols to mask sensitive data. These trends reflect growing interest in digital, scalable implementation and practical demands in networked control and resource allocation systems.

The scenario becomes more complex when considering dynamic players, whose internal states and outputs evolve according to physical or algorithmic dynamics, often modeled as differential or difference equations. Here, NE seeking methods typically build upon the basic design of static-player algorithms, but now must also address stabilization and performance of the underlying dynamical systems. This often requires observer design to estimate unknown or external influences and, in some formulations, the incorporation of passivity-based control principles to ensure robustness and convergence. Privacy considerations can also be integrated, for example, by having players communicate only transformed outputs rather than states. Most advances in this area have focused on continuous-time dynamics, while discrete-time adaptations are less common, often limited to sampled-data versions of continuous-time schemes with additional considerations for uncertainty or partial feedback.

In contrast with NE seeking, GNE-seeking problems introduce explicit coupling constraints among players' feasible actions, leading to additional computational and coordination challenges. The dominant approach for addressing these problems, particularly for static players, is the primal-dual optimization framework. Here, each player not only updates their own strategy but also maintains auxiliary dual variables reflecting the extent to which currently chosen actions violate shared constraints. Consensus protocols are again key, enabling alignment and agreement of dual variables across the network. Primal-dual methods can be sensitive to the observability of constraint violations, leading to variants that leverage penalty functions as alternatives. Penalty-based approaches modify player costs to include terms penalizing constraint violations, trading the communication burden of dual variables for the careful selection and adaptation of penalty coefficients to guarantee constraint satisfaction. Aggregative game structures continue to play an important role, offering reduced complexity and communication through a focus on aggregate variables alone.

Discrete-time GNE-seeking methods follow closely, but frequently employ operator splitting or monotone inclusion techniques to formalize the solution process. These approaches are adapted to both full and partial information scenarios regarding players' actions and support asynchronous as well as synchronous communication patterns. There are also extensions for handling time-varying cost functions, stochastic objectives, regret-driven online learning problems, and aggregative structures, broadening the applicability of distributed GNE computation beyond simple static games.

When turning to dynamic players in the GNE context, available reported treatments are primarily in the continuous-time domain. Here, primal-dual algorithms are embedded within dynamic control architectures, extending equilibrium seeking to systems whose states and constraint satisfaction must be maintained continuously over time. These schemes often strike a balance between fully distributed update rules and hybrid arrangements, where some degree of centralized coordination (most commonly for dual variables) may be retained for technical or practical reasons. Despite promising methodologies and applications, many questions regarding closed-loop stability, robustness to model uncertainty, and truly distributed constraint enforcement remain open research avenues.

6.3 Population Games and LSCSs: The Analogy Approach

In contrast to classical N-player games, population games offer a complementary and influential framework for modeling and designing distributed decision-making processes, particularly in systems composed of a very large number of agents. In this setting, the analysis is focused on aggregate population-level outcomes rather than on the behavior of individual agents. Nevertheless, through suitable analogies, the established theory of population games and their evolutionary dynamics can be leveraged to design control mechanisms for the LSCSs introduced in Sect. 6.1.

Within LSCS applications, two principal paradigms are employed for population game formulations: the *natural* approach and the *analogy* approach. The natural approach is suitable when the system at hand genuinely consists of a large number of agents making discrete and sporadic choices among a finite set of strategies, under bounded rationality and with potentially limited computational capabilities. Models of this type arise in high-level decision-making for domains such as those presented in Sect. 1.2 and have been extensively studied throughout this book.

The analogy approach, in contrast, is more abstract and aims to exploit the structural properties of population games and their associated evolutionary dynamics to reformulate and solve general optimization-based control problems, even when the decision-making agents do not constitute populations in the literal sense.

To elaborate on this, consider an LSCS as the one depicted in Fig. 6.1, where the control variables $\mathbf{u}^k$ of each subsystem k are subject to a simplex-type constraint of the form

$$\mathbf{u}^k \succeq \mathbf{0}, \tag{6.2a}$$

$$\mathbf{1}^\top \mathbf{u}^k = c^k, \tag{6.2b}$$

where $c^k \in \mathbb{R}_{>0}$. The overall constraint set of these variables may also include additional conditions, all captured by the set $\mathcal{U}$. Therefore, $\mathcal{U}$ can be expressed as the intersection of a simplex and an additional set of constraints, closely resembling the constraint structure considered in population games, i.e.,

$$\mathcal{X} = \Delta \cap \Omega.$$

Here, $\mathcal{X}$ plays the role of $\mathcal{U}$, the set $\Delta = \prod_{k \in \mathcal{P}} \Delta^k$ corresponds to the simplex-type constraint (6.2) of the multiple subsystems, and Ω represents any remaining constraints.

Therefore, for simplex-constrained LSCSs, $\mathbf{u}^k$ can always be interpreted as the strategic distribution $\mathbf{x}^k$ of a population with mass $m^k = c^k$, while an appropriate set Ω can be defined to capture any additional constraints contained in $\mathcal{U}$. The analogy then becomes clear: the control variables are treated as societal states, and the admissible set $\mathcal{U}$ inherits the same structure as the feasible set in a population game. In other words, the decision variables of the LSCS, for example the quantities of a resource to be allocated, are interpreted as proportions of agents within a society.

As such, the associated EDMs then act as continuous-time optimization algorithms for the primal variables $\mathbf{u}$, while the PDMs govern the continuous-time evolution of the dual variables linked to the constraints in Ω, or any additional variables describing the dynamics of the subsystems.

A key feature of this analogy approach is that evolutionary dynamics exhibit invariance properties that inherently preserve simplex-related constraints, such as non-negativity and conservation of total mass. As a result, the solution trajectories remain within the physical constraint set at all times by construction, ensuring feasibility without requiring explicit projection steps. This property not only simplifies the implementation of distributed optimization-based control schemes but also enhances their robustness and scalability when applied to LSCSs.

To further illustrate the analogy approach, the next section presents an application example in the context of charging coordination for a fleet of plug-in electric vehicles.

6.4 An Illustrative Example

Future smart cities are expected to integrate novel technologies to reduce the environmental impact of modern industries. Among these, plug-in electric vehicles (PEVs) are envisioned not only to provide low-emission transportation but also to offer flexibility for supporting renewable energy integration. However, uncoordinated charging of large PEV fleets can cause power losses and overloads in distribution networks. Therefore, effective mechanisms are needed to coordinate charging, ensuring both grid reliability and economic efficiency.

Decentralized approaches have been widely explored for coordinating PEV fleet charging, often relying on iterative price-based schemes between vehicles and an aggregator. While these methods can achieve optimal schedules, they are sensitive to uncertainties such as random arrival and departure times and typically neglect coupled constraints like feeder capacity limits. Moreover, many algorithms guarantee feasibility only after convergence, limiting their applicability in real-time settings. To address these challenges, this section proposes a continuous-time primal-dual control strategy that enables real-time coordination under strict network constraints while leveraging stability and invariance properties to ensure safe and feasible operation.

6.4.1 Problem Statement

Consider a set of $n_{\mathcal{V}} \in \mathbb{Z}_{\geq 1}$ PEVs, $\mathcal{V} = \{1, 2, \ldots, n_{\mathcal{V}}\}$, which seeks to coordinate their charging profiles over a multi-time period $\mathcal{T}$. Let $\mathcal{T}^k = \{1, 2, \ldots, n_{\mathcal{T}^k}\}$, with $n_{\mathcal{T}^k} \in \mathbb{Z}_{\geq 1}$, be the charging period of the kth PEV, so that the multi-time period $\mathcal{T}$ is constructed as $\mathcal{T} = \cup_{k \in \mathcal{V}} \mathcal{T}^k = \{1, 2, \ldots, \max_{k \in \mathcal{V}} n_{\mathcal{T}^k}\}$. Here, 1 denotes the present time slot, 2 the next one, and so on. Throughout, it is assumed that all the time slots within $\mathcal{T}$ have the same duration $\tau \in \mathbb{R}_{>0}$ (with units of hours). Moreover, let $\mathcal{V}_i \subseteq \mathcal{V}$ be the set of PEVs that have the ith time slot within their charging period, i.e., $\mathcal{V}_i = \{k \in \mathcal{V} : i \in \mathcal{T}^k\}$, for all $i \in \mathcal{T}$. Finally, to ease the forthcoming discussions, let $\mathcal{T}_e^k = \mathcal{T}^k \cup \{0\}$ be the extended set of charging time slots of the kth PEV, which includes a fictitious charging time slot denoted by 0 that is used to allocate the excess of power.

Under the considered framework, for each PEV $k \in \mathcal{V}$ the charging power that is scheduled to be injected at the time slot $i \in \mathcal{T}_e^k$ is denoted by $x_i^k \in \mathbb{R}_{\geq 0}$ (with units of kW). Therefore, the scheduled charging profile of the kth PEV is given by the vector $\mathbf{x}^k = \left[x_0^k, x_1^k, x_2^k, \ldots, x_{n_{\mathcal{T}^k}}^k\right]^\top \in \mathbb{R}_{\geq 0}^{(n_{\mathcal{T}^k}+1)}$, where x_0^k is a fictitious charging power for all $k \in \mathcal{V}$, and the collection of the scheduled charging profiles of all the PEVs is given by $\mathbf{x} = \left[(\mathbf{x}^1)^\top, (\mathbf{x}^2)^\top, \cdots, (\mathbf{x}^{n_\mathcal{V}})^\top\right]^\top \in \mathbb{R}_{\geq 0}^n$, where $n = \sum_{k \in \mathcal{V}}(n_{\mathcal{T}^k} + 1)$. Throughout, it is considered that a collection of scheduled charging profiles $\mathbf{x}$ is feasible if and only if it satisfies the constraints

$$x_i^k \geq 0, \quad \forall i \in \mathcal{T}_e^k, \quad \forall k \in \mathcal{V}, \tag{6.3a}$$

$$x_i^k \leq d^k, \quad \forall i \in \mathcal{T}^k, \quad \forall k \in \mathcal{V}, \tag{6.3b}$$

$$x_0^k + \sum_{i \in \mathcal{T}^k} x_i^k = \frac{e^k(0)}{\eta^k \tau}, \quad \forall k \in \mathcal{V}, \tag{6.3c}$$

$$\sum_{i \in \mathcal{V}_i} \iota^{z,k} x_i^k \leq c_i^z, \quad \forall i \in \mathcal{T}, \quad \forall z \in \mathcal{F}. \tag{6.3d}$$

Here, the constraints in (6.3a) require for the scheduled charging powers to be non-negative; the constraints in (6.3b) require for the scheduled charging powers to be not greater than the maximum charging rate $d^k \in \mathbb{R}_{>0}$ (with units of kW) of the kth PEV; the constraints in (6.3c) require for the scheduled charging powers to avoid overcharging the battery of the kth PEV, where $e^k(0) \in \mathbb{R}_{\geq 0}$ is the remaining energy capacity (with units of kWh) of the battery of the kth PEV at the beginning of the present time slot 1, and $\eta^k \in (0, 1]$ is the charging efficiency of the kth PEV; and, finally, the constraints in (6.3d) require for the fleet of PEVs to satisfy the capacity constraints of the feeder lines of the distribution network. Regarding the constraints in (6.3d), $\mathcal{F} = \{1, 2, \ldots, n_\mathcal{F}\}$ denotes the set of feeder lines of the distribution network (with $n_\mathcal{F} \in \mathbb{Z}_{\geq 1}$); $c_i^z \in \mathbb{R}_{>0}$ is the maximum capacity of the feeder line $z \in \mathcal{F}$ at the time slot $i \in \mathcal{T}$ (with units of kW); and $\iota^{z,k} = 1$ if the PEV $k \in \mathcal{V}_i$ is fed by the feeder line $z \in \mathcal{F}$, and $\iota^{z,k} = 0$ otherwise. Throughout, it is assumed that the kth PEVs know their feeder dependency value $\iota^{z,k}$ for all $z \in \mathcal{F}$. Notice that while (6.3a), (6.3b), and (6.3c) regard individual PEV-level decoupled constraints, (6.3d) considers distribution network-level coupled constraints. In particular, the parameters d^k, $e^k(0)$, and η^k might depend on the characteristics of the charging port, the vehicle type, and the driving style of each PEV, while the parameters $\iota^{z,k}$ and c_i^z might depend on the topology and operational constraints of the distribution network. Moreover, observe that (6.3a) and (6.3c) imply that $\eta^k \tau \sum_{i \in \mathcal{T}^k} x_i^k \leq e^k(0)$, for all $k \in \mathcal{V}$. Hence, the (non-fictitious) scheduled charging energy to the kth PEV must not exceed its corresponding remaining energy capacity $e^k(0)$. It is straightforward to verify that under the considered framework, and due to the consideration of the fictitious charging powers x_0^k in (6.3c), the set of feasible collections of scheduled charging profiles is nonempty.

Remark 6.5 For every $d^k \in \mathbb{R}_{>0}$, $e^k \in \mathbb{R}_{\geq 0}$, $\eta^k \in (0, 1]$, $\tau \in \mathbb{R}_{>0}$, and $c_i^z \in \mathbb{R}_{>0}$, the set of feasible collections of scheduled charging profiles, $\mathcal{X} = \{\mathbf{x} \in \mathbb{R}^n : \mathbf{x}$ satisfies (6.3)$\}$, is nonempty. In fact, the relative interior of $\mathcal{X}$ is nonempty as well. That is, there exists some $\hat{\mathbf{x}} \in \mathcal{X}$ such that the inequality constraints in (6.3a), (6.3b), and (6.3d), do not bind at $\hat{\mathbf{x}}$. ◊

Considering the aforementioned constraints, the charging coordination task can be stated as the constrained optimization problem given by

$$\min_{\mathbf{x} \in \mathbb{R}^n} J(\mathbf{x}), \quad \text{s.t.} \quad \mathbf{x} \in \mathcal{X}, \tag{6.4}$$

where $J : \mathbb{R}^n \to \mathbb{R}$ is a cost function that captures the charging coordination objective of the PEVs and/or the electricity provider, and $\mathcal{X} = \{\mathbf{x} \in \mathbb{R}^n : \mathbf{x}$ satisfies (6.3)$\}$ is a compact convex set with nonempty relative interior (see Remark 6.5). For instance, the cost function $J(\cdot)$ might consider both the total energy generation cost as well as some local costs related to the battery degradation of each PEV [20, 33, 34, 50]. Throughout, it is assumed that $J(\cdot)$ satisfies the following assumptions.

Assumption 6.1 The cost function $J(\cdot)$ is convex and twice continuously differentiable. ◊

Assumption 6.2 For all $k \in \mathcal{V}$ and all $i \in \mathcal{T}_e^k$, the partial derivative $\partial J(\cdot)/\partial x_i^k$ can be evaluated using the local information available at the kth PEV, i.e., each PEV $k \in \mathcal{V}$ knows the functional form of $\partial J(\cdot)/\partial x_i^k$ and has enough local information to evaluate $\partial J(\mathbf{x})/\partial x_i^k$, for all $i \in \mathcal{T}_e^k$. ◊

Granted that the goal is to develop a semi-decentralized charging coordination algorithm, it is assumed that each PEV $k \in \mathcal{V}$ computes its own scheduled charging profile $\mathbf{x}^k$ using only local information available at the kth PEV. Such local information comprises both individual information of the corresponding PEV (e.g., battery level, charging period, battery degradation parameters, maximum charging rate, charging efficiency, and feeder dependency values $\iota^{z,k}$), as well as global aggregated variables sent to the fleet of PEVs by an aggregator node (e.g., aggregated scheduled charging profiles and distribution network-level variables). More formally, the following assumption is considered.

Assumption 6.3 There is an aggregator node that communicates with all the PEVs. The aggregator node is able to receive the scheduled charging profile of each PEV, and is able to send the aggregated scheduled charging profile (i.e., $\sum_{k \in \mathcal{V}_i} x_i^k$, for all $i \in \mathcal{T}$) and distribution network-level variables to the fleet of PEVs. Moreover, the aggregator has full information regarding the feeder dependency values $\iota^{z,k}$, for all $k \in \mathcal{V}$ and all $z \in \mathcal{F}$. ◊

To ease the exposition of the presented framework, let introduce some illustrative examples. In particular, Example 6.1 illustrates the constraints in (6.3d), and Example 6.2 provides some optimization objectives that satisfy Assumptions 6.1 and 6.2.

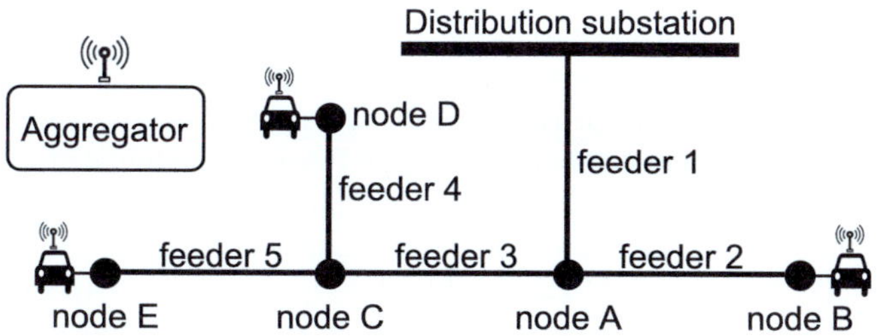

Fig. 6.4 Distribution network with 5 feeder lines and 3 charging nodes

Example 6.1 (*Distribution network constraints*) Consider the distribution network of Fig. 6.4. Without loss of generality, assume that there are only three PEVs to be charged (one at node B $\rightarrow$ 1, one at node D $\rightarrow$ 2, and one at node E $\rightarrow$ 3). Thus, $\mathcal{V} = \{1, 2, 3\}$ and $\mathcal{F} = \{1, 2, 3, 4, 5\}$. Moreover, suppose that $\mathcal{T}^1 = \{1\}$, $\mathcal{T}^2 = \{1, 2\}$, and $\mathcal{T}^3 = \{1, 2, 3\}$. Hence, $\mathcal{T} = \{1, 2, 3\}$, $\mathcal{V}_1 = \{1, 2, 3\}$, $\mathcal{V}_2 = \{2, 3\}$, and $\mathcal{V}_3 = \{3\}$. Under such a framework, the constraints in (6.3d) are as follows. For feeder 1: $x_1^1 + x_1^2 + x_1^3 \leq c_1^1$, $x_2^2 + x_2^3 \leq c_2^1$, and $x_3^3 \leq c_3^1$; for feeder 2: $x_1^1 \leq c_1^2$; for feeder 3: $x_1^2 + x_1^3 \leq c_1^3$, $x_2^2 + x_2^3 \leq c_2^3$, and $x_3^3 \leq c_3^3$; for feeder 4: $x_1^2 \leq c_1^4$, and $x_2^2 \leq c_2^4$; and, finally, for feeder 5: $x_1^3 \leq c_1^5$, $x_2^3 \leq c_2^5$, and $x_3^3 \leq c_3^5$. $\diamondsuit$

Example 6.2 (*Optimization objectives*) Perhaps the simplest optimization problem that might be considered under the proposed framework is

$$\min_{\mathbf{x} \in \mathcal{X}} \quad J_{chg}(\mathbf{x}) := \frac{1}{2} \sum_{k \in \mathcal{V}} \left(x_0^k\right)^2. \tag{6.5}$$

Solving such a problem means that the PEVs will charge as much as possible while satisfying the constraints in (6.3). Clearly, $J_{chg}(\cdot)$ satisfies Assumptions 6.1 and 6.2 (note that $\partial J_{chg}(\mathbf{x})/\partial x_0^k = x_0^k$ relies only on local information of the kth PEV). Moreover, although $J_{chg}(\cdot)$ is separable over the PEVs, the optimization problem in (6.5) is still coupled over the fleet of PEVs due to the constraints in (6.3d).

Another optimization problem that could be considered is

$$\min_{\mathbf{x} \in \mathcal{X}} \quad J(\mathbf{x}) := J_{chg}(\mathbf{x}) + J_{ref}(\mathbf{x}) + J_{deg}(\mathbf{x}), \tag{6.6}$$

where $J_{chg}(\cdot)$ is defined as in (6.5), and

$$J_{ref}(\mathbf{x}) = \frac{1}{2} \sum_{i \in \mathcal{T}} \left(r_i - \sum_{k \in \mathcal{V}_i} x_i^k \right)^2,$$

$$J_{deg}(\mathbf{x}) = \frac{1}{2} \sum_{k \in \mathcal{V}} \sum_{i \in \mathcal{T}^k} \left(\alpha^k \left(x_i^k\right)^2 + \beta^k x_i^k + \gamma^k \right).$$

Here, $J_{ref}(\cdot)$ is a cost associated to the deviation from a reference demand profile[2], and $r_i \in \mathbb{R}_{\geq 0}$ is the reference demand (in kW) to be consumed by the fleet of PEVs over the time interval $i \in \mathcal{T}$ (it is assumed that r_i, for all $i \in \mathcal{T}$, is broadcast by the aggregator to the PEVs). In contrast, $J_{deg}(\cdot)$ is a cost associated to the battery degradation[3] of the PEVs, and $\alpha^k, \beta^k, \gamma^k \in \mathbb{R}_{>0}$ are constants that depend on the voltage and energy capacity parameters of the battery of the kth PEV.

Clearly, the cost function $J(\cdot)$ defined in (6.6) is convex and twice continuously differentiable. Hence, Assumption 6.1 holds under such a cost. Moreover, note that

$$\frac{\partial J(\mathbf{x})}{\partial x_0^\ell} = x_0^\ell, \quad \forall \ell \in \mathcal{V},$$
$$\frac{\partial J(\mathbf{x})}{\partial x_j^\ell} = -r_j + \sum_{k \in \mathcal{V}_j} x_j^k + \alpha^\ell x_j^\ell + \frac{\beta^\ell}{2}, \quad \forall j \in \mathcal{T}^\ell, \quad \forall \ell \in \mathcal{V}.$$

Here, notice that besides the reference profile r_j and the aggregated term $\sum_{k \in \mathcal{V}_j} x_j^k$, all of the other terms in $\partial J(\mathbf{x}) / \partial x_j^\ell$ depend only on individual information of the ℓth PEV. Furthermore, since the aggregator is able to send distribution network-level variables and aggregated power profiles, the information regarding r_j and $\sum_{k \in \mathcal{V}_j} x_j^k$, for all $j \in \mathcal{T}$, is also available at every PEV. Therefore, Assumption 6.2 also holds under the cost defined in (6.6).

Finally, notice that in (6.6) one could replace $J_{ref}(\mathbf{x})$ with

$$J_{dem}(\mathbf{x}) = \frac{1}{2} \sum_{i \in \mathcal{T}} \left(D_i + \sum_{k \in \mathcal{V}_i} x_i^k \right)^2,$$

where D_i denotes the aggregated non-PEV demand at the time slot $i \in \mathcal{T}$ (broadcast by the aggregator to the PEVs), and thus consider the same optimization objectives as in [19, 20, 32–34, 50]. ◊

6.4.2 *Real-Time Decentralized Charging Control*

In this section, a continuous-time optimization-based control law to coordinate the charging process of a fleet of PEVs in real time is designed. The main motivation behind such a real-time approach is its capability to respond, in the sense of disturbance rejection, to unexpected arrivals or departures of the PEVs. Throughout, let

---

[2] Such a reference demand profile might be related, for instance, to some power bought by the electricity provider in a day-ahead market or to some forecast energy profile [25, 54].

[3] The considered cost $J_{deg}(\cdot)$ is consistent with battery degradation cost models of LiFePO4 battery cells [14], e.g., [33, Eq. (7)], [34, Eq. (5)], [50, Eq. (9)].

$$g_{d,i}^{k}(x_i^k) = x_i^k - d^k, \quad \forall i \in \mathcal{T}^k, \quad \forall k \in \mathcal{V},$$
$$g_{c,i}^{z}(\mathbf{x}) = \sum_{\ell \in \mathcal{V}_i} \iota^{z,\ell} x_i^{\ell} - c_i^z, \quad \forall i \in \mathcal{T}, \quad \forall z \in \mathcal{F}.$$

Using these functions, the constraints in (6.3b) can be written as $g_{d,i}^{k}(x_i^k) \leq 0$, and the constraints in (6.3d) as $g_{c,i}^{z}(\mathbf{x}) \leq 0$. Here, the sub-indices $_d$ and $_c$ are used to differentiate decoupled and coupled constraints, respectively. Finally, following such a notation, let $\lambda_{d,i}^{k} \in \mathbb{R}_{\geq 0}$ and $\lambda_{c,i}^{z} \in \mathbb{R}_{\geq 0}$ denote the Lagrange multipliers associated to the constraints $g_{d,i}^{k}(x_i^k) \leq 0$ and $g_{c,i}^{z}(\mathbf{x}) \leq 0$, respectively.

The approach here proposed is structured as follows. Let $t \in \mathbb{R}_{\geq 0}$ denote the continuous-time variable representing the current (continuous) time, and let the charging dynamics of the PEVs be characterized by

$$\dot{e}^k(t) = -\eta^k u^k(t) \quad \text{[with initial condition } e^k(0) \in \mathbb{R}_{\geq 0}],$$

for all $k \in \mathcal{V}$, where $u^k(t) \in \mathbb{R}_{\geq 0}$ denote the charging power (in kW) that is currently applied to the PEV $k \in \mathcal{V}$. Set

$$w^z(t) = \frac{c_1^z}{\max\left\{c_1^z,\ \sum_{\ell \in \mathcal{V}_1} \iota^{z,\ell} x_1^{\ell}(t)\right\}}, \quad \forall z \in \mathcal{F}, \tag{6.7a}$$

$$v^k(t) = \min_{\zeta \in \{z \in \mathcal{F} : \iota^{z,k} = 1\}} w^{\zeta}(t), \quad \forall k \in \mathcal{V}, \tag{6.7b}$$

$$u^k(t) = \min\left\{v^k(t) x_1^k(t),\ d^k\right\}, \quad \forall k \in \mathcal{V}, \tag{6.7c}$$

where $x_1^k(t)$ is the value of the optimization variable x_1^k at time t; $w^z(t)$ quantifies the capacity violation of the zth feeder at the time slot 1 under the collective scheduled charging profile $\mathbf{x}(t)$ (in particular, note that $0 < w^z(t) \leq 1$, for all $z \in \mathcal{F}$); and $v^k(t)$ is a scaling factor to scale down the value $x_1^k(t)$ for the kth PEV according to the relevant feeder capacity violations. Under such a mapping, it follows that $u^k(t) \geq 0$ whenever $x_1^k(t) \geq 0$, for all $k \in \mathcal{V}$; $\sum_{k \in \mathcal{V}_1} \iota^{z,k} u^k(t) \leq c_1^z$, for all $z \in \mathcal{F}$; $u^k(t) \leq d^k$, for all $k \in \mathcal{V}$; and $u^k(t) \leq x_1^k(t)$ with $u^k(t) = x_1^k(t)$ whenever $\mathbf{x}(t) \in \mathcal{X}$, for all $k \in \mathcal{V}$. Hence, under the control law in (6.7), the power capacity constraints of the system are satisfied for the present time slot $k = 1$ (c.f., (6.3b) and (6.3d)), and, whenever $\mathbf{x}(t)$ is feasible it holds that $u^k(t) = x_1^k(t)$, for all $k \in \mathcal{V}$. Moreover, note that $w^z(t)$ in (6.7a) is a distribution network-level variable, and thus can be sent to the PEVs by the aggregator (c.f., Assumption 6.3). Consequently, the scaling factor $v^k(t)$ in (6.7b) and the control law $u^k(t)$ in (6.7c) can be computed locally at the kth PEV.

Considering the aforementioned control law, it is needed to design some continuous-time optimization dynamics to update $x_i^k(t)$, for all $i \in \mathcal{T}_e^k$ and all $k \in \mathcal{V}$, that

(G1) satisfy the semi-decentralized structure of the system;
(G2) satisfy the constraints in (6.3a) and (6.3c) for all $t \geq 0$;
(G3) converge to the solution of (6.4).

Notice that Condition (G2) is important to exclude negative $u^k(t)$ values and to avoid overcharging the batteries of the PEVs. In what follows, it is proceeded to design some continuous-time optimization dynamics that satisfy all three conditions, namely (G1)–(G3).

Continuous-Time Optimization Dynamics

In this section, the design of a continuous-time dynamical system to update the optimization variables $x_i^k(t)$, for all $i \in \mathcal{T}_e^k$, all $k \in \mathcal{V}$, and all $t \geq 0$, is performed. To formulate such a dynamical system, the ideas on population games and evolutionary dynamics serve as supporting context.

Specifically, the EDM–PDM abstraction of Chap. 2 is relevant since the resulting ODEs have certain invariance properties that can be used to directly handle some of the constraints in (6.3). More precisely, the constraints in (6.3a) and (6.3c). On the other hand, motivated by the centralized multipliers model Sect. 5.2, a novel form of continuous-time primal-dual optimization dynamics that satisfy Conditions (G1), (G2), and (G3) is designed.

The proposed dynamics are stated as follows. On one hand, every PEV $k \in \mathcal{V}$ updates its variables x_i^k, for all $i \in \mathcal{T}_e^k$, as well as the variables $\lambda_{d,i}^k$, for all $i \in \mathcal{T}^k$, according to the expressions

$$\dot{x}_i^k(t) = \sum_{j \in \mathcal{T}_e^k} x_j^k(t)\rho_{ji}^k(t) - x_i^k(t) \sum_{j \in \mathcal{T}_e^k} \rho_{ij}^k(t), \tag{6.8a}$$

$$\dot{\lambda}_{d,i}^k(t) = \left[g_{d,i}^k\left(x_i^k(t)\right)\right]_+ - \lambda_{d,i}^k(t)\left[-g_{d,i}^k\left(x_i^k(t)\right)\right]_+, \tag{6.8b}$$

with $[\cdot]_+ \triangleq \max\{\cdot, 0\}$, and

$$\rho_{ij}^k(t) = \left[p_j^k(t) - p_i^k(t)\right]_+, \quad \forall k, j \in \mathcal{T}_e^k,$$

$$p_i^k(t) = -\frac{\partial J(\mathbf{x}(t))}{\partial x_i^k} - \lambda_{d,i}^k(t) - \sum_{z \in \mathcal{F}} \lambda_{c,i}^z(t)\iota^{z,k}, \quad \forall i \in \mathcal{T}^k,$$

$$p_0^k(t) = -\frac{\partial J(\mathbf{x}(t))}{\partial x_0^k}.$$

On the other hand, the aggregator updates the variables $\lambda_{c,i}^z$, for all $i \in \mathcal{T}$ and all $z \in \mathcal{F}$, according to the expressions

$$\dot{\lambda}_{c,i}^z(t) = \left[g_{c,i}^z(\mathbf{x}(t))\right]_+ - \lambda_{c,i}^z(t)\left[-g_{c,i}^z(\mathbf{x}(t))\right]_+. \tag{6.9}$$

Notice that the variables $\lambda_{c,i}^z$, for all $i \in \mathcal{T}$ and all $z \in \mathcal{F}$, are distribution network-level variables, and, therefore, can be broadcast by the aggregator to the fleet of PEVs. Hence, under Assumptions 6.2 and 6.3, the proposed dynamics can indeed be computed in the considered semi-decentralized fashion, and thus satisfy Condition (G1). Moreover, note that the number of optimization variables that the aggregator must update, i.e., $\lambda_{c,i}^z$, for all $i \in \mathcal{T}$ and all $z \in \mathcal{F}$, is independent of the number

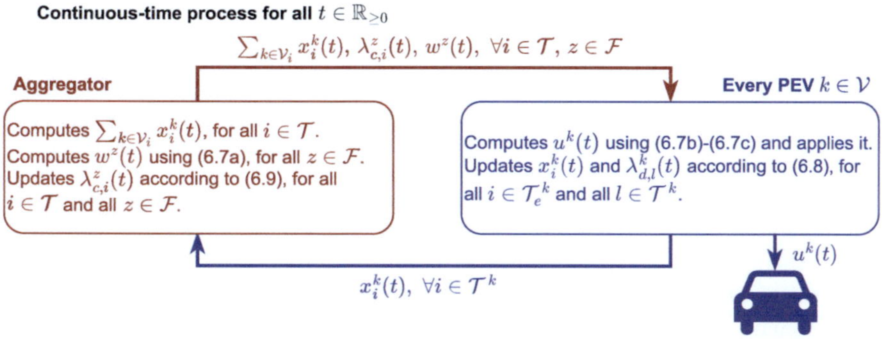

Fig. 6.5 Considered real-time optimization and control scheme. Notice that, depending on the objective $J(\mathbf{x})$, the aggregator may also send other distribution network-level variables to the PEVs (c.f., Example 6.2)

of PEVs. Hence, the local computational complexity (in terms of local computation resources) is independent of the number of PEVs. Consequently, the proposed framework is scalable to large PEV fleets. Furthermore, to exploit the invariance properties of the proposed dynamics, the following conditions on the initial state of the dynamical system are imposed.

Assumption 6.4 Let (i) $x_i^k(0) \geq 0$, for all $i \in \mathcal{T}_e^k$ and all $k \in \mathcal{V}$; (ii) $\sum_{i\in\mathcal{T}_e^k} x_i^k(0) = e^k(0)/(\eta^k \tau)$, for all $k \in \mathcal{V}$; (iii) $\lambda_{d,i}^k(0) \geq 0$, for all $i \in \mathcal{T}^k$ and all $k \in \mathcal{V}$; and (iv) $\lambda_{c,i}^z(0) \geq 0$, for all $i \in \mathcal{T}$ and all $z \in \mathcal{F}$. $\diamondsuit$

For the sake of clarity, the considered continuous-time optimization and control scheme is summarized in Fig. 6.5. Now, the proposed continuous-time optimization dynamics are analyzed.

6.4.3 Analysis of the Proposed Dynamics

In this section, the invariance properties and the equilibria set of the dynamics in (6.8) and (6.9) are characterized, and it is shown that such dynamics solve the optimization problem in (6.4). That is, the set of solutions of (6.4) is asymptotically stable under (6.8) and (6.9). Throughout, let

$$\begin{aligned}
\boldsymbol{\lambda}_d^k &= \left[\lambda_{d,1}^k,\ \lambda_{d,2}^k, \cdots, \lambda_{d,n_{\mathcal{T}^k}}^k\right]^\top \in \mathbb{R}^{n_{\mathcal{T}^k}},\\
\boldsymbol{\lambda}_c^z &= \left[\lambda_{c,1}^z,\ \lambda_{c,2}^z, \cdots, \lambda_{c,n_{\mathcal{T}}}^z\right]^\top \in \mathbb{R}^{n_{\mathcal{T}}},\\
\boldsymbol{\lambda} &= \left[\boldsymbol{\lambda}_d^1,\ \boldsymbol{\lambda}_d^2, \cdots, \boldsymbol{\lambda}_d^{n_{\mathcal{V}}},\ \boldsymbol{\lambda}_c^1,\ \boldsymbol{\lambda}_c^2, \cdots, \boldsymbol{\lambda}_c^{n_{\mathcal{F}}}\right]^\top \in \mathbb{R}^{n_Q},
\end{aligned}$$

where $n_{\mathcal{T}} = \max_{k\in\mathcal{V}} n_{\mathcal{T}^k}$ and $n_Q = \sum_{k\in\mathcal{V}} n_{\mathcal{T}^k} + n_{\mathcal{F}} n_{\mathcal{T}}$, and let

$$\Delta = \left\{ \mathbf{x} \in \mathbb{R}^{n}_{\geq 0} : \sum_{i \in \mathcal{T}^k_e} x^k_i = \frac{e^k(0)}{\eta^k \tau}, \quad \forall k \in \mathcal{V} \right\}.$$

Namely, Δ is the set of vectors $\mathbf{x}$ that satisfies the constraints in (6.3a) and (6.3c).

Remark 6.6 To deduce the proposed optimization dynamics (6.8) and (6.9), the next analogy is followed. Each PEV $k \in \mathcal{V}$ is interpreted as a population with mass $m^k = e^k(0)/(\eta^k \tau)$, and the extended set of charging time slots $\mathcal{T}^k_e$ are interpreted as the set of strategies of population k. Consequently, the proposed dynamics (6.8) and (6.9) have the same form of the EDM–PDM of Chap. 2, which is characterized by the IPC-EDM of Definition 2.9 and the centralized multipliers PDM of (5.10) (here, the Lagrange multipliers $\boldsymbol{\mu}$ associated to the equality constraints are ignored and Σ_a is set as a memoryless system fully characterized by the fitness function $\mathbf{f}(\mathbf{x}) = \nabla_{\mathbf{x}} J(\mathbf{x})$).

◊

Invariance Properties

Here, some invariance properties of the proposed dynamics are characterized, which allow the satisfaction of the constraints in (6.3a) and (6.3c), as well as the ones regarding the non-negativity of the Lagrange multipliers, for all $t \geq 0$.

The following facts hold:

(F1) Δ is positively invariant under the dynamics (6.8a).
(F2) $\mathbb{R}^{nQ}_{\geq 0}$ is positively invariant under the dynamics (6.8b) and (6.9).

Fact (F1) readily holds by from Definition 2.5, since the every EDM's flow map is constrained within $\mathrm{T}\,\Delta(\mathbf{x})$, for all $\mathbf{x}$. Fact (F2), on the other hand, follows from Lemma 5.5. Consequently, these facts together with Assumption 6.4 guarantee that $\mathbf{x}(t) \in \Delta$ and $\boldsymbol{\lambda}(t) \in \mathbb{R}^{nQ}_{\geq 0}$, for all $t \geq 0$. Hence, the proposed dynamics indeed satisfy Condition (G2).

Equilibria Set of the Proposed Dynamics

In this subsection, it is shown that the equilibria set of the dynamics (6.8) and (6.9) coincides with the set of solutions of the optimization problem in (6.4) and is asymptotically stable under the proposed dynamics.

Lemma 6.1 *Consider the dynamics in (6.8) and (6.9). The following facts hold:*

(i) $\dot{\mathbf{x}}^k(t) = \mathbf{0}$ *if and only if*

$$x^k_i(t) > 0 \Rightarrow p^k_i(t) = \max_{j \in \mathcal{T}^k_e} p^k_j(t), \quad \forall i \in \mathcal{T}^k_e. \tag{6.10}$$

(ii) $\dot{\boldsymbol{\lambda}}^k_d(t) = \mathbf{0}$ *if and only if*

$$x^k_i(t) \leq d^k \text{ and } \lambda^k_{d,i}(t) \left(x^k_i(t) - d^k \right) = 0, \quad \forall i \in \mathcal{T}^k. \tag{6.11}$$

(iii) $\dot{\boldsymbol{\lambda}}_c^z(t) = \mathbf{0}$ *if and only if*

$$\sum_{i \in \mathcal{V}_i} \iota^{z,k} x_i^k(t) \leq c_i^z \text{ and } \lambda_{c,i}^z(t) \left(\sum_{i \in \mathcal{V}_i} \iota^{z,k} x_i^k(t) - c_i^z \right) = 0, \quad \forall i \in \mathcal{T}. \quad (6.12)$$

Here, $\mathbf{0}$ *denotes the zero vector of appropriate dimension.* ◊

Proof The proof follows from Lemmas 2.3 and 5.5. □

Using Lemma 6.1, the set of equilibria of the dynamics (6.8) and (6.9) is now formally characterized.

Corollary 6.1 *Consider the dynamics in (6.8) and (6.9), and let*

$$\mathcal{E} = \left\{ (\mathbf{x}, \boldsymbol{\lambda}) \in \Delta \times \mathbb{R}_{\geq 0}^{n_Q} : \begin{array}{l} (6.10) \text{ and } (6.11) \text{ hold for all } k \in \mathcal{V}, \\ (6.12) \text{ holds for all } z \in \mathcal{F}. \end{array} \right\}. \quad (6.13)$$

Then, $(\mathbf{x}^*, \boldsymbol{\lambda}^*) \in \Delta \times \mathbb{R}_{\geq 0}^{n_Q}$ *is an equilibrium point of the dynamics in (6.8) and (6.9) if and only if* $(\mathbf{x}^*, \boldsymbol{\lambda}^*) \in \mathcal{E}$. ◊

Proof The proof follows from Lemma 6.1 and Proposition 4.1. □

Having characterized the equilibria set of the dynamics in (6.8) and (6.9), the main results of this subsection are now stated.

Corollary 6.2 *Consider the set* $\mathcal{E}$ *in (6.13) and the optimization problem in (6.4). It follows that the set* $\mathcal{E}$ *is nonempty, compact, and for every* $(\mathbf{x}^*, \boldsymbol{\lambda}^*) \in \mathcal{E}$ *it holds that* $\mathbf{x}^* \in \arg\min_{\mathbf{x} \in \mathcal{X}} J(\mathbf{x})$. ◊

Proof The result follows from Lemmas 2.5 and 2.7 by taking $-J(\cdot)$ as the potential function of the underlying full-potential game. □

Corollary 6.3 *Consider the dynamics (6.8) and (6.9) and the set of equilibria* $\mathcal{E}$ *in (6.13). The set* $\mathcal{E}$ *is asymptotically stable under the considered dynamics.* ◊

Proof The result follows from Lemma 2.5 together with Theorem 5.1. □

Observe that Corollaries 6.2 and 6.3 guarantee that the set of equilibria of the dynamics (6.8) and (6.9) is not only asymptotically stable, but also coincides with the set of solutions of the optimization problem in (6.4). Therefore, the proposed continuous-time optimization dynamics indeed satisfy Condition (G3).

Remark 6.7 Note that in contrast with other related works, e.g., [50], the optimality and asymptotic stability results provided by Corollaries 6.2 and 6.3 hold for arbitrary sizes of PEV fleets, and not only for the limiting case where the number of PEVs is assumed to be infinite (quite large). Hence, the proposed analogy-based framework allows the consideration of a broader scope of applications. ◊

6.4.4 Numerical Simulations

In this section, the proposed method is applied to a charging coordination scenario considering the distribution network of Fig. 6.4 and the optimization problem in (6.6). As an illustration, a comparison assessment between the dynamics in (6.8) and (6.9) against the continuous-time primal-dual gradient dynamics of [42] (with unitary time constants) is performed. Note that the continuous-time primal-dual gradient dynamics in [42] not only have exponential stability guarantees, but have been also considered in related network optimization and power-grid applications, e.g., [13, 27, 70]. Moreover, the dynamics in (6.8) and (6.9) can be computed using the same elementary linear algebra operations as the dynamics in [42]. For these reasons, such dynamics are regarded as a relevant benchmark to compare the proposed approach.

Consider a fleet with $n_{\mathcal{V}} = 40$ PEVs distributed as follows: 13 PEVs are connected at node B; 13 PEVs are connected at node D; and 14 PEVs are connected at node E. The charging coordination is considered over the time period from 19:00 to 7:00,$\tau = 1$, the charging horizons are sampled according to the (truncated) normal distribution $n_{\mathcal{T}^k} \sim \mathcal{N}(5{:}00, 4\tau^2)$, for all $k \in \mathcal{V}$. Namely, for the reported simulations, $n_{\mathcal{T}} = 12$. Besides, the PEVs are assumed to be heterogeneous, and, without loss of generality, the corresponding parameters are uniformly sampled for all $k \in \mathcal{V}$ as follows: $d^k \sim \mathcal{U}(2, 4)$; $e^k(0) \sim \mathcal{U}(5, 20)$; $\eta^k \sim \mathcal{U}(0.8, 0.9)$; $\alpha^k \sim \mathcal{U}(0.003, 0.005)$; $\beta^k \sim \mathcal{U}(0.06, 0.09)$; and $\gamma^k \sim \mathcal{U}(0.002, 0.004)$. In particular, notice that the battery degradation cost parameters are consistent with the values reported in the literature [33]. Furthermore, regarding the capacities of the feeder lines, $c_i^1 = 12$, $c_i^2 = c_i^4 = c_i^5 = 3.6$, and $c_i^3 = 6$, for all $i \in \{1, 2, 3\}$; while $c_i^1 = 40$, $c_i^2 = c_i^4 = c_i^5 = 12$, and $c_i^3 = 20$, for all $i \in \{4, 5, \cdots, 12\}$. Finally, for the reference demand profile, the one depicted in Fig. 6.6 is considered.

In Fig. 6.6, the optimal collective scheduled charging profile for the aforementioned scenario is provided, while in Figs. 6.7 and 6.8, the first 100 s of the continuous-time behavior of the dynamics in (6.8), (6.9) and [42] are presented when applied to the considered scenario. The constraint violation (CV) terms in Fig. 6.8 are computed as

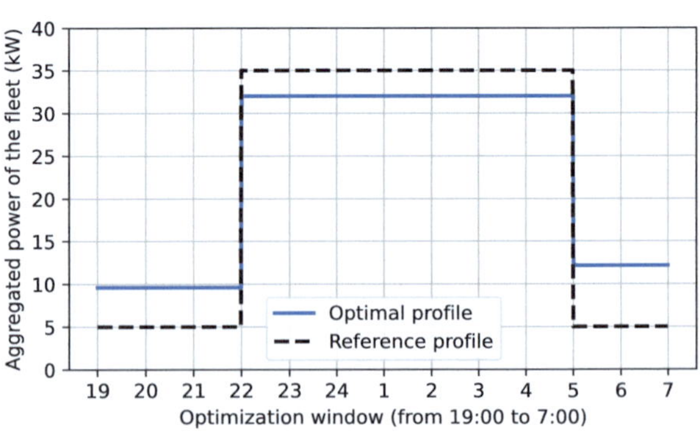

Fig. 6.6 Reference demand profile and optimal aggregated power profile of the fleet for the considered charging coordination problem

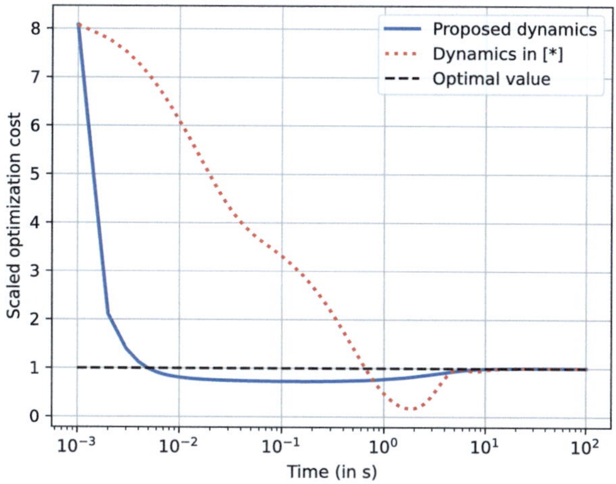

Fig. 6.7 Optimization cost $J(\mathbf{x}(t))$ over the first 100 s of the continuous-time optimization process. All values are scaled over the optimal. [*] corresponds to [42]

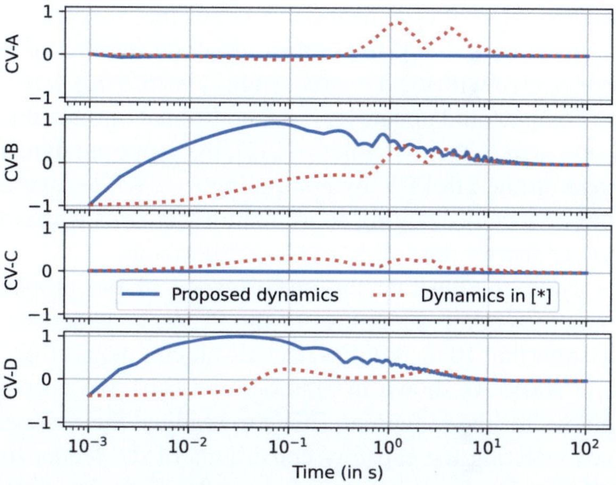

Fig. 6.8 Constraints violations over the first 100 s. Positive CV-terms' values denote constraints' violations. All values are scaled to $[-1, 1]$. [*] corresponds to [42]

- CV-A: $\max_{i \in \mathcal{T}_e^k, k \in \mathcal{V}} -x_i^k(t)$.
- CV-B: $\max_{i \in \mathcal{T}^k, k \in \mathcal{V}} g_{d,i}^k\left(x_i^k(t)\right)$.
- CV-C: $\max_{k \in \mathcal{V}} \left|\mathbf{1}^\top \mathbf{x}^k(t) - e^k(0)/\eta^k \tau\right|$.
- CV-D: $\max_{i \in \mathcal{T}, z \in \mathcal{F}} g_{c,i}^z(\mathbf{x}(t))$.

Clearly, both continuous-time primal-dual gradient dynamics converge to the solution of the considered optimization problem with comparable performance. However, notice that while the proposed dynamics (6.8) and (6.9) satisfy the constraints in (6.3a) and (6.3c) for all times, the dynamics in [42] do not satisfy such constraints over the trajectories of the optimization process. This property of the proposed dynamics has the following advantage. Notice that since the constraints in (6.3a) and (6.3c) are handled by the invariance properties of the dynamics in (6.8a), to solve the considered optimization problem using (6.8) and (6.9) it is not necessary to compute

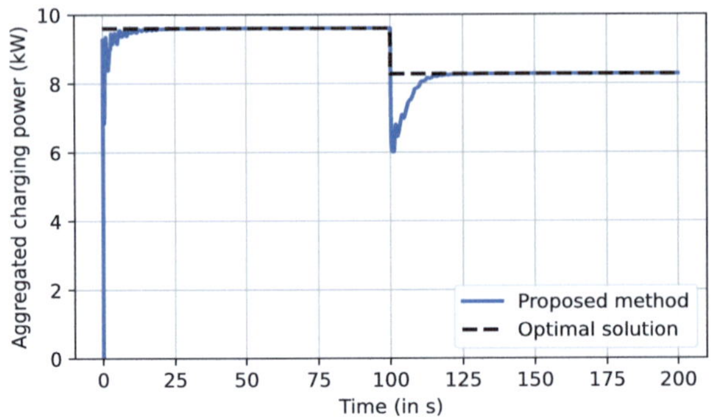

Fig. 6.9 Aggregated charging power given by $\sum_{k\in\mathcal{V}_1} u^k(t)$

or update any dual-Lagrange multipliers associated to such constraints. Hence, the total number of optimization variables that are considered with the in (6.8) and (6.9) is given by $n + n_Q$. In contrast, the total number of optimization variables that are considered with the dynamics of [42] is $2n + n_Q + n_{\mathcal{V}}$ (which requires for each PEV to compute and update $n_{\mathcal{T}^k} + 2$ additional optimization variables). Therefore, when compared to the dynamics of [42], the proposed dynamics reduce the computational load on the kth PEV by computing $n_{\mathcal{T}^k} + 2$ less variables. This fact is of especial importance in charging coordination scenarios with small τ and large $n_{\mathcal{T}}$ values, i.e., under shorter time slots and larger horizons.

Now, to illustrate the performance of the proposed real-time control method under uncertain departures of the PEVs, consider the aforementioned scenario, but assume that 10 randomly selected vehicles depart from their charging stations at time $t = 100$ s. As shown in Figs. 6.9 and 6.10, the proposed real-time control strategy allows for the remaining PEVs to readjust their charging profiles in real time without violating the capacity constraints of the feeder lines. Hence, the fleet of PEVs effectively responds in real time to the corresponding disturbance.

Notes

Tables 6.1 and 6.2 summarize the aforementioned approaches in the context of distributed NE and GNE seeking in N-player non-cooperative games, respectively. On the other hand, Table 6.3 points to some relevant literature related to practical applications in population games to LSCSs. The table organizes the works based on the type of approach: natural versus analogy.

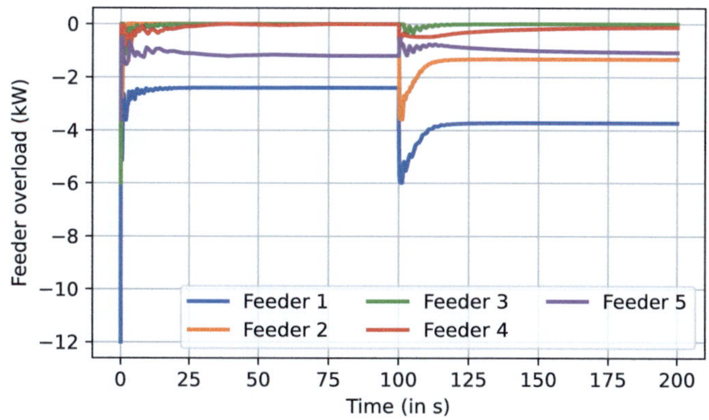

Fig. 6.10 The feeder overloads are given by $\sum_{k\in\mathcal{V}_1} \iota^{z,k} u^i(t) - c_1^z$, for all $z \in \mathcal{F}$. Hence, positive values coincide with feeder overloads

Table 6.1 Classification of distributed NE seeking methods based on the nature of the method (CT: continuous time, DT: discrete time) and the type of players

	Static players	Dynamic players
CT	[7, 17, 58–62, 64, 71]	[9, 21–23, 56, 57, 69]
DT	[5, 12, 37, 44, 45, 52, 63]	[47]

Table 6.2 Classification of distributed GNE-seeking methods based on the nature of the method (CT: continuous time, DT: discrete time) and the type of coupled constraints

	Affine equality	Affine inequality	Convex inequality
CT	[8, 10, 28, 29, 31, 48]	[6, 26, 49]	[4, 48, 68, 72]
DT	[55, 65, 67]	[3, 15, 18, 41, 65, 66]	[16, 30, 51]

Table 6.3 Examples of practical applications of population games by type of approach

References	Type of approach
[24, 35, 46, 53]	Natural approach
[1, 2, 40, 43]	Analogy approach

References

1. Barreiro-Gomez J, Mas I, Giribet J, Moreno P, Ocampo-Martinez C, Sánchez-Peña R, Quijano N (2021) Distributed data-driven uav formation control via evolutionary games: experimental results. J Franklin Inst 358(10):5334–5352. https://doi.org/10.1016/j.jfranklin.2021.05.002
2. Barreiro-Gomez J, Ocampo-Martinez C, Bianchi FD, Quijano N (2019) Data-driven decentralized algorithm for wind farm control with population-games assistance. Energies 12(6):1164

3. Belgioioso G, Nedić A, Grammatico S (2021) Distributed generalized Nash equilibrium seeking in aggregative games on time-varying networks. IEEE Trans Autom Control 66(5):2061–2075. https://doi.org/10.1109/TAC.2020.3005922
4. Bianchi M, Grammatico S (2021) Continuous-time fully distributed generalized Nash equilibrium seeking for multi-integrator agents. Automatica 129:109660. https://doi.org/10.1016/j.automatica.2021.109660
5. Bianchi M, Grammatico S (2021) Fully distributed Nash equilibrium seeking over time-varying communication networks with linear convergence rate. IEEE Control Syst Lett 5(2):499–504. https://doi.org/10.1109/LCSYS.2020.3002734
6. Chen G, Ming Y, Hong Y, Yi P (2020) Distributed algorithm for ϵ-generalized Nash equilibria with uncertain coupled constraints. Automatica 123:109313. https://doi.org/10.1016/j.automatica.2020.109313
7. De Persis C, Grammatico S (2019) Distributed averaging integral Nash equilibrium seeking on networks. Automatica 110:108548. https://doi.org/10.1016/j.automatica.2019.108548
8. Deng Z (2021) Distributed generalized Nash equilibrium seeking algorithm for nonsmooth aggregative games. Automatica 132:109794. https://doi.org/10.1016/j.automatica.2021.109794
9. Deng Z (2022) Distributed Nash equilibrium seeking for aggregative games with second-order nonlinear players. Automatica 135:109980. https://doi.org/10.1016/j.automatica.2021.109980
10. Deng Z, Nian X (2019) Distributed generalized Nash equilibrium seeking algorithm design for aggregative games over weight-balanced digraphs. IEEE Trans Neural Netw Learn Syst 30(3):695–706. https://doi.org/10.1109/TNNLS.2018.2850763
11. Facchinei F, Kanzow C (2010) Generalized Nash equilibrium problems. Ann Oper Res 175(1):177–211
12. Fang X, Wen G, Huang T, Fu Z, Hu L (2022) Distributed Nash equilibrium seeking over markovian switching communication networks. IEEE Trans Cybern 52(6):5343–5355. https://doi.org/10.1109/TCYB.2020.3030824
13. Feijer D, Paganini F (2010) Stability of primal–dual gradient dynamics and applications to network optimization. Automatica 46(12):1974–1981
14. Forman J, Stein J, Fathy H (2013) Optimization of dynamic battery parameter characterization experiments via differential evolution. In: Proceedings of the 2013 American control conference, pp 867–874. https://doi.org/10.1109/ACC.2013.6579945
15. Franci B, Grammatico S (2021) A distributed forward–backward algorithm for stochastic generalized Nash equilibrium seeking. IEEE Trans Autom Control 66(11):5467–5473. https://doi.org/10.1109/TAC.2020.3047369
16. Franci B, Grammatico S (2022) Stochastic generalized Nash equilibrium-seeking in merely monotone games. IEEE Trans Autom Control 67(8):3905–3919. https://doi.org/10.1109/TAC.2021.3108496
17. Gadjov D, Pavel L (2019) A passivity-based approach to Nash equilibrium seeking over networks. IEEE Trans Autom Control 64(3):1077–1092. https://doi.org/10.1109/TAC.2018.2833140
18. Gadjov D, Pavel L (2021) Single-timescale distributed GNE seeking for aggregative games over networks via forward–backward operator splitting. IEEE Trans Autom Control 66(7):3259–3266. https://doi.org/10.1109/TAC.2020.3015354
19. Ghavami A, Kar K, Gupta A (2016) Decentralized charging of plug-in electric vehicles with distribution feeder overload control. IEEE Trans Autom Control 61(11):3527–3532. https://doi.org/10.1109/TAC.2016.2516240
20. Grammatico S (2016) Exponentially convergent decentralized charging control for large populations of plug-in electric vehicles. In: Proceedings of the 55th IEEE conference on decision and control (CDC), pp 5775–5780. https://doi.org/10.1109/CDC.2016.7799157
21. Guo M, De Persis C (2021) Linear quadratic network games with dynamic players: stabilization and output convergence to Nash equilibrium. Automatica 130:109711. https://doi.org/10.1016/j.automatica.2021.109711

22. Huang B, Yang C, Meng Z, Chen F, Ren W (2022) Distributed nonlinear placement for multicluster systems: a time-varying Nash equilibrium-seeking approach. IEEE Trans Cybern 52(11):11614–11623. https://doi.org/10.1109/TCYB.2021.3085583
23. Huang B, Zou Y, Meng Z (2021) Distributed-observer-based Nash equilibrium seeking algorithm for quadratic games with nonlinear dynamics. IEEE Trans Syst Man Cybern Syst 51(11):7260–7268. https://doi.org/10.1109/TSMC.2020.2968127
24. Jiang X, Ji Y, Du M, Deng W (2015) A study of driver's route choice behavior based on evolutionary game theory. Comput Intell Neurosci 2014:47–47
25. Kisacikoglu MC, Erden F, Erdogan N (2018) Distributed control of PEV charging based on energy demand forecast. IEEE Trans Industr Inf 14(1):332–341. https://doi.org/10.1109/TII.2017.2705075
26. Krilašević S, Grammatico S (2021) Learning generalized Nash equilibria in multi-agent dynamical systems via extremum seeking control. Automatica 133:109846. https://doi.org/10.1016/j.automatica.2021.109846
27. Li N, Zhao C, Chen L (2016) Connecting automatic generation control and economic dispatch from an optimization view. IEEE Trans Control Netw Syst 3(3):254–264. https://doi.org/10.1109/TCNS.2015.2459451
28. Li Z, Li Z, Ding Z (2020) Distributed generalized Nash equilibrium seeking and its application to femtocell networks. IEEE Trans Cybern 1–13. https://doi.org/10.1109/TCYB.2020.3004635
29. Liang S, Yi P, Hong Y (2017) Distributed Nash equilibrium seeking for aggregative games with coupled constraints. Automatica 85:179–185. https://doi.org/10.1016/j.automatica.2017.07.064
30. Lu K, Li G, Wang L (2021) Online distributed algorithms for seeking generalized Nash equilibria in dynamic environments. IEEE Trans Autom Control 66(5):2289–2296. https://doi.org/10.1109/TAC.2020.3002592
31. Lu K, Zhu Q (2022) Nonsmooth continuous-time distributed algorithms for seeking generalized Nash equilibria of noncooperative games via digraphs. IEEE Trans Cybern 52(7):6196–6206. https://doi.org/10.1109/TCYB.2021.3049463
32. Ma Z, Yang N, Zou S, Shao Y (2018) Charging coordination of plug-in electric vehicles in distribution networks with capacity constrained feeder lines. IEEE Trans Control Syst Technol 26(5):1917–1924. https://doi.org/10.1109/TCST.2017.2730825
33. Ma Z, Zou S, Liu X (2015) A distributed charging coordination for large-scale plug-in electric vehicles considering battery degradation cost. IEEE Trans Control Syst Technol 23(5):2044–2052. https://doi.org/10.1109/TCST.2015.2394319
34. Ma Z, Zou S, Ran L, Shi X, Hiskens IA (2016) Efficient decentralized coordination of large-scale plug-in electric vehicle charging. Automatica 69:35–47. https://doi.org/10.1016/j.automatica.2016.01.035
35. Morimoto T, Kanazawa T, Ushio T (2016) Subsidy-based control of heterogeneous multiagent systems modeled by replicator dynamics. IEEE Trans Autom Control 61(10):3158–3163. https://doi.org/10.1109/TAC.2015.2504519
36. Nash J (1951) Non-cooperative games. Ann Math 54(2):286–295
37. Pang Y, Hu G (2021) Distributed Nash equilibrium seeking with limited cost function knowledge via a consensus-based gradient-free method. IEEE Trans Autom Control 66(4):1832–1839. https://doi.org/10.1109/TAC.2020.2995666
38. Park S, Leonard NE (2021) Kl divergence regularized learning model for multi-agent decision making. In: Proceedings of the 2021 American control conference (ACC), pp 4509–4514. https://doi.org/10.23919/ACC50511.2021.9483414
39. Park S, Martins NC, Shamma JS (2019) From population games to payoff dynamics models: a passivity-based approach. In: Proceedings of the 58th IEEE conference on decision and control (CDC), pp 6584–6601. https://doi.org/10.1109/CDC40024.2019.9029756
40. Pashaie A, Pavel L, Damaren CJ (2017) A population game approach for dynamic resource allocation problems. Int J Control 90(9):1957–1972
41. Pavel L (2020) Distributed GNE seeking under partial-decision information over networks via a doubly-augmented operator splitting approach. IEEE Trans Autom Control 65(4):1584–1597. https://doi.org/10.1109/TAC.2019.2922953

42. Qu G, Li N (2019) On the exponential stability of primal-dual gradient dynamics. IEEE Control Syst Lett 3(1):43–48. https://doi.org/10.1109/LCSYS.2018.2851375
43. Quijano N, Ocampo-Martinez C, Barreiro-Gomez J, Obando G, Pantoja A, Mojica-Nava E (2017) The role of population games and evolutionary dynamics in distributed control systems: the advantages of evolutionary game theory. IEEE Control Syst Mag 37(1):70–97
44. Salehisadaghiani F, Pavel L (2016) Distributed Nash equilibrium seeking: a gossip-based algorithm. Automatica 72:209–216. https://doi.org/10.1016/j.automatica.2016.06.004
45. Salehisadaghiani F, Shi W, Pavel L (2019) Distributed Nash equilibrium seeking under partial-decision information via the alternating direction method of multipliers. Automatica 103:27–35. https://doi.org/10.1016/j.automatica.2019.01.012
46. Srikantha P, Kundur D (2017) Resilient distributed real-time demand response via population games. IEEE Trans Smart Grid 8(6):2532–2543. https://doi.org/10.1109/TSG.2016.2526651
47. Stankovic MS, Johansson KH, Stipanovic DM (2012) Distributed seeking of Nash equilibria with applications to mobile sensor networks. IEEE Trans Autom Control 57(4):904–919. https://doi.org/10.1109/TAC.2011.2174678
48. Sun C, Hu G (2021) Continuous-time penalty methods for Nash equilibrium seeking of a nonsmooth generalized noncooperative game. IEEE Trans Autom Control 66(10):4895–4902. https://doi.org/10.1109/TAC.2020.3040377
49. Sun C, Hu G (2021) Distributed generalized Nash equilibrium seeking for monotone generalized noncooperative games by a regularized penalized dynamical system. IEEE Trans Cybern 51(11):5532–5545. https://doi.org/10.1109/TCYB.2021.3087663
50. Tajeddini MA, Kebriaei H (2019) A mean-field game method for decentralized charging coordination of a large population of plug-in electric vehicles. IEEE Syst J 13(1):854–863. https://doi.org/10.1109/JSYST.2018.2855971
51. Tatarenko T, Kamgarpour M (2019) Learning generalized Nash equilibria in a class of convex games. IEEE Trans Autom Control 64(4):1426–1439. https://doi.org/10.1109/TAC.2018.2841319
52. Tatarenko T, Shi W, Nedić A (2021) Geometric convergence of gradient play algorithms for distributed Nash equilibrium seeking. IEEE Trans Autom Control 66(11):5342–5353. https://doi.org/10.1109/TAC.2020.3046232
53. Tembine H, Altman E, El-Azouzi R, Hayel Y (2010) Evolutionary games in wireless networks. IEEE Trans Syst Man Cybern Part B (Cybernetics) 40(3):634–646. https://doi.org/10.1109/TSMCB.2009.2034631
54. Tushar MHK, Zeineddine AW, Assi C (2018) Demand-side management by regulating charging and discharging of the EV, ESS, and utilizing renewable energy. IEEE Trans Industr Inf 14(1):117–126. https://doi.org/10.1109/TII.2017.2755465
55. Wang Z, Liu F, Ma Z, Chen Y, Jia M, Wei W, Wu Q (2021) Distributed generalized Nash equilibrium seeking for energy sharing games in prosumers. IEEE Trans Power Syst 36(5):3973–3986. https://doi.org/10.1109/TPWRS.2021.3058675
56. Ye M (2021) Distributed Nash equilibrium seeking for games in systems with bounded control inputs. IEEE Trans Autom Control 66(8):3833–3839. https://doi.org/10.1109/TAC.2020.3027795
57. Ye M (2022) Distributed robust seeking of Nash equilibrium for networked games: an extended state observer-based approach. IEEE Trans Cybern 52(3):1527–1538. https://doi.org/10.1109/TCYB.2020.2989755
58. Ye M, Hu G (2015) Distributed seeking of time-varying Nash equilibrium for non-cooperative games. IEEE Trans Autom Control 60(11):3000–3005. https://doi.org/10.1109/TAC.2015.2414817
59. Ye M, Hu G (2017) Distributed Nash equilibrium seeking by a consensus based approach. IEEE Trans Autom Control 62(9):4811–4818. https://doi.org/10.1109/TAC.2017.2688452
60. Ye M, Hu G (2018) Distributed Nash equilibrium seeking in multiagent games under switching communication topologies. IEEE Trans Cybern 48(11):3208–3217. https://doi.org/10.1109/TCYB.2017.2764141

61. Ye M, Hu G (2021) Adaptive approaches for fully distributed Nash equilibrium seeking in networked games. Automatica 129:109661. https://doi.org/10.1016/j.automatica.2021.109661
62. Ye M, Hu G, Lewis FL (2018) Nash equilibrium seeking for n-coalition noncooperative games. Automatica 95:266–272. https://doi.org/10.1016/j.automatica.2018.05.020
63. Ye M, Hu G, Xie L, Xu S (2022) Differentially private distributed Nash equilibrium seeking for aggregative games. IEEE Trans Autom Control 67(5):2451–2458. https://doi.org/10.1109/TAC.2021.3075183
64. Ye M, Hu G, Xu S (2020) An extremum seeking-based approach for Nash equilibrium seeking in N-cluster noncooperative games. Automatica 114:108815. https://doi.org/10.1016/j.automatica.2020.108815
65. Yi P, Pavel L (2019) Distributed generalized Nash equilibria computation of monotone games via double-layer preconditioned proximal-point algorithms. IEEE Trans Control Netw Syst 6(1):299–311. https://doi.org/10.1109/TCNS.2018.2813928
66. Yi P, Pavel L (2019) An operator splitting approach for distributed generalized Nash equilibria computation. Automatica 102:111–121. https://doi.org/10.1016/j.automatica.2019.01.008
67. Yi P, Pavel L (2020) Asynchronous distributed algorithms for seeking generalized Nash equilibria under full and partial-decision information. IEEE Trans Cybern 50(6):2514–2526. https://doi.org/10.1109/TCYB.2019.2908091
68. Zeng X, Chen J, Liang S, Hong Y (2019) Generalized Nash equilibrium seeking strategy for distributed nonsmooth multi-cluster game. Automatica 103:20–26. https://doi.org/10.1016/j.automatica.2019.01.025
69. Zhang Y, Liang S, Wang X, Ji H (2020) Distributed Nash equilibrium seeking for aggregative games with nonlinear dynamics under external disturbances. IEEE Trans Cybern 50(12):4876–4885. https://doi.org/10.1109/TCYB.2019.2929394
70. Zhao C, Topcu U, Li N, Low S (2014) Design and stability of load-side primary frequency control in power systems. IEEE Trans Autom Control 59(5):1177–1189. https://doi.org/10.1109/TAC.2014.2298140
71. Zhu Y, Yu W, Wen G, Chen G (2021) Distributed Nash equilibrium seeking in an aggregative game on a directed graph. IEEE Trans Autom Control 66(6):2746–2753. https://doi.org/10.1109/TAC.2020.3008113
72. Zou Y, Huang B, Meng Z, Ren W (2021) Continuous-time distributed Nash equilibrium seeking algorithms for non-cooperative constrained games. Automatica 127:109535. https://doi.org/10.1016/j.automatica.2021.109535

Appendix A
The Karush–Kuhn–Tucker Conditions

This appendix provides a comprehensive explanation of the Karush–Kuhn–Tucker (KKT) conditions, which play a central role in the formulation, analysis, and solution of constrained optimization problems. Specifically, the focus is on problems characterized by continuous decision variables, convex feasible sets, and the presence of both equality and inequality constraints. These type of problems arise within the modeling framework of large-scale multi-agent systems, as developed in the main text of this book.

The exposition throughout this appendix adopts the notation and conventions previously established to ensure clarity while maintaining consistency with earlier chapters. The KKT conditions are introduced not only as necessary conditions for optimality in convex problems but also as a gateway to understanding the structure of optimal solutions, duality principles, and the development of distributed optimization algorithms that are crucial for scalable control and coordination in multi-agent environments.

As a starting point, an optimization problem in its general form is considered as follows:

$$\max_{\mathbf{x}\in\mathbb{R}^n} \quad \varphi(\mathbf{x}) \tag{A.1a}$$

$$\text{s.t.} \quad \mathbf{A}\mathbf{x} - \mathbf{b} = \mathbf{0}_{c_=} \tag{A.1b}$$

$$\mathbf{g}(\mathbf{x}) \preceq \mathbf{0}_{c_\leq}, \tag{A.1c}$$

where

- $\mathbf{x} \in \mathbb{R}^n$ is the vector of decision variables;
- $\varphi(\mathbf{x})$ is the concave and continuously differentiable objective function;
- $\mathbf{A} \in \mathbb{R}^{c_= \times n}$ and $\mathbf{b} \in \mathbb{R}^{c_=}$ are the matrices and vectors defining equality constraints; and

J. Martinez-Piazuelo et al., *Generalized Nash Equilibrium Seeking in Population Games*, Communications and Control Engineering,
https://doi.org/10.1007/978-3-032-06081-5

- $\mathbf{g} : \mathbb{R}^n \to \mathbb{R}^{c_\leq}$ is the vector of convex, continuously differentiable inequality constraint functions.

In order to analyze such an optimization problem, the following dual-Lagrangian function is introduced:

$$\mathcal{L}(\mathbf{x}, \boldsymbol{\mu}, \boldsymbol{\lambda}) = \varphi(\mathbf{x}) + \boldsymbol{\mu}^\top(\mathbf{b} - \mathbf{A}\mathbf{x}) + \boldsymbol{\lambda}^\top(-\mathbf{g}(\mathbf{x})), \tag{A.2}$$

where

- $\boldsymbol{\mu} \in \mathbb{R}^{c_=}$ are the Lagrange multipliers for equality constraints, and
- $\boldsymbol{\lambda} \in \mathbb{R}^{c_\leq}$ are the Lagrange multipliers for inequality constraints ($\boldsymbol{\lambda} \succeq \mathbf{0}_{c_\leq}$).

The Karush–Kuhn–Tucker (KKT) conditions are a set of mathematical criteria used to identify optimal solutions in nonlinear programming problems, particularly those involving both equality and inequality constraints (assumed to satisfy appropriate constraint qualifications). These conditions extend the classical Lagrange multiplier method and form a cornerstone of constrained optimization theory, with broad applications in economics, engineering, operations research, and game theory [1, 2]. First introduced by William Karush in his 1939 master's thesis [5], they remained largely unknown until Harold W. Kuhn and Albert W. Tucker rediscovered and formalized them in 1951, presenting their work in the context of nonlinear programming at Princeton [6]. Modern developments continue to expand their theoretical and practical reach; notably, [4] provide a comprehensive tutorial covering first- and second-order KKT conditions and their application in distributed optimization frameworks, highlighting their relevance in large-scale and multi-agent systems. Since their formalization, KKT conditions have underpinned the development of advanced numerical methods, such as interior-point techniques, sequential quadratic programming, and dual decomposition, and they remain essential in designing scalable algorithms for complex real-world problems.

Coming back to the formulation of the KKT conditions, for a point $\mathbf{x}^*$ to be optimal, there must exist multipliers $\boldsymbol{\mu}^*$ and $\boldsymbol{\lambda}^*$ such that

1. The gradient of the dual-Lagrangian function in (A.2) must be zero at the optimum, i.e.,
$$\nabla_{\mathbf{x}}\varphi(\mathbf{x}^*) - \mathbf{A}^\top\boldsymbol{\mu}^* - \mathbf{D}_{\mathbf{x}}\mathbf{g}(\mathbf{x}^*)^\top\boldsymbol{\lambda}^* = \mathbf{0}_n,$$
where $\mathbf{D}_{\mathbf{x}}\mathbf{g}(\mathbf{x}^*)$ is the Jacobian of $\mathbf{g}$ at $\mathbf{x}^*$. This statement is known as the *Stationarity Condition*.
2. All the constraints in (A.1) must be satisfied, i.e.,
$$\mathbf{A}\mathbf{x}^* - \mathbf{b} = \mathbf{0}_{c_=}$$
$$\mathbf{g}(\mathbf{x}^*) \preceq \mathbf{0}_{c_\leq}.$$
This statement is known as the *Primal Feasibility Condition*.

3. The Lagrange multipliers for inequality constraints in (A.1c) must be non-negative, i.e.,
$$\boldsymbol{\lambda}^* \succeq \mathbf{0}_{c_{\leq}}.$$
This statement is known as the *Dual Feasibility Condition.*
4. For each inequality constraint in (A.1c), either the constraint is active and the corresponding multiplier may be positive, or the constraint is inactive and the multiplier is zero, i.e.,
$$\boldsymbol{\lambda}^* \odot \mathbf{g}(\mathbf{x}^*) = \mathbf{0}.$$
This statement is known as the *Complementary Slackness Condition.*

In the context of population games, the KKT conditions are used to characterize equilibrium points such as generalized Nash equilibria. Hence, following Definition 2.12, the set of generalized Nash equilibrium (GNE) points can be defined as

$$\text{GNE}(\mathbf{f}) = \left\{ \mathbf{x}^* \in \mathcal{X} : \mathbf{x}^* \in \arg\max_{\mathbf{x}\in\mathcal{X}} \mathbf{x}^\top \mathbf{f}(\mathbf{x}^*) \right\},$$

where $\mathcal{X} = \Delta \cap \Omega$ is the feasible set, $\mathbf{f}(\cdot)$ is the fitness vector, and Ω encodes coupling constraints. Then, the KKT conditions for the GNE problem (as in Lemma 2.8) are listed as follows[1]:

- Stationarity:
$$\mathbf{f}(\mathbf{x}^*) - \mathbf{A}_\Omega^\top \boldsymbol{\mu}_\Omega^* - \mathbf{D}_\mathbf{x}\mathbf{g}_\Omega(\mathbf{x}^*)^\top \boldsymbol{\lambda}_\Omega^* - \mathbf{A}_\Delta^\top \boldsymbol{\mu}_\Delta^* + \boldsymbol{\lambda}_\Delta^* = \mathbf{0}_n,$$
where $\mathbf{A}_\Omega, \mathbf{g}_\Omega, \boldsymbol{\mu}_\Omega^*$, and $\boldsymbol{\lambda}_\Omega^*$ are associated to Ω, while $\mathbf{A}_\Delta$ and $\boldsymbol{\mu}_\Delta^*, \boldsymbol{\lambda}_\Delta^*$ are associated to the simplex constraints of Δ.
- Primal feasibility: $\mathbf{x}^* \in \mathcal{X}$.
- Dual feasibility: $\boldsymbol{\lambda}_\Omega^* \succeq \mathbf{0}$, $\boldsymbol{\lambda}_\Delta^* \succeq \mathbf{0}$.
- Complementary slackness: $\boldsymbol{\lambda}_\Omega^* \odot \mathbf{g}_\Omega(\mathbf{x}^*) = \mathbf{0}$, $\boldsymbol{\lambda}_\Delta^* \odot \mathbf{g}_\Omega(\mathbf{x}^*) = \mathbf{0}$.

Finally, to study such problems, the next procedure is often followed:

1. Formulate the Lagrangian function as in (A.2) with all constraints.
2. Write the KKT system (stationarity, primal/dual feasibility, complementary slackness).
3. Analyze and (if possible) solve the KKT conditions for $\mathbf{x}^*$, $\boldsymbol{\mu}^*$, $\boldsymbol{\lambda}^*$. For convex problems, any solution is globally optimal.
4. Interpret multipliers: the Lagrange multipliers can be linked to economic or game-theoretic interpretations (e.g., shadow prices, constraint incentives).

[1] Refer to [3, Sect. 1.3.2] for a detailed discussion on this topic.

Notes

See Sect. 2.5 for the formal GNE definition and KKT application. Lemma 2.8 and (2.17) provide explicit KKT conditions for the population games context. The role of Lagrange multipliers as dual variables is discussed in the context of payoff mechanisms in Sect. 2.2, Example 2.5.

References

1. Bertsekas DP (1999) Nonlinear programming, 2nd edn. Athena Scientific
2. Boyd S, Vandenberghe L (2004) Convex optimization. Cambridge University Press
3. Facchinei F, Pang JS (2003) Finite-dimensional variational inequalities and complementarity problems, vol 1. Springer
4. Ghojogh B, Ghodsi A, Karray F, Crowley M (2021) KKT conditions, first-order and second-order optimization, and distributed optimization: tutorial and survey. arXiv:2110.01858
5. Karush W (1939) Minima of functions of several variables with inequalities as side constraints. Master's thesis, University of Chicago
6. Kuhn HW, Tucker AW (1951) Nonlinear programming. In: Proceedings of the second Berkeley symposium on mathematical statistics and probability, pp 481–492. University of California Press

Appendix B
Lyapunov Stability Analysis

In this appendix, a detailed exposition of Lyapunov-based stability analysis is presented, as it is considered a foundational tool in the study of nonlinear dynamical systems. The material is intended to complement the main body of the manuscript, in which Lyapunov methods are employed to characterize stability properties. Given its relevance across various chapters, the core concepts and formal results associated with Lyapunov theory are consolidated here in a unified and self-contained manner. However, it should be noted that this material is derived from classical literature, as referenced in the *Notes* at the end of the appendix.

After the historical origins of the theory have been introduced, the direct method is examined in detail: the mathematical formalism for candidate Lyapunov functions, their time derivatives along system trajectories, and the interpretation of level sets in the state space are developed. Conditions for several notions of stability, including stability in the sense of Lyapunov (SISL), asymptotic stability (AS), and exponential stability (ES), both local and global are then presented. These conditions are formalized through theorems and are illustrated by intuitive figures that reveal the underlying dynamical behavior.

Finally, LaSalle's invariance principle is introduced, whereby the Lyapunov framework is generalized to situations in which the derivative of a candidate function is only negative semi-definite rather than strictly negative.

By the end of this appendix, a basic conceptual and mathematical foundation should have been acquired by the reader for applying Lyapunov-based techniques to the analysis and design of nonlinear systems, and the main references for in-depth study are provided in the *Notes* at the end of the appendix.

Introduction: Aleksandr Mikhailovich Lyapunov

Aleksandr Mikhailovich Lyapunov (1857–1918) was a Russian mathematician born in Yaroslavl, son of the renowned astronomer Mikhail Vasilyevich Lyapunov. After completing secondary school in Nizhny Novgorod, he studied mathematics and physics at the University of St. Petersburg, where he graduated in 1880. Influenced

J. Martinez-Piazuelo et al., *Generalized Nash Equilibrium Seeking in Population Games*, Communications and Control Engineering,
https://doi.org/10.1007/978-3-032-06081-5

by Chebyshev, he pursued a research career in mechanics, addressing the stability of a self-gravitating rotating fluid—a question central to cosmogony and long studied by Newton, Maclaurin, Jacobi, and Poincaré. His 1884 master's thesis, titled *On the Stability of Ellipsoidal Forms in the Equilibrium of a Rotating Fluid*, partially answered this foundational problem.

Lyapunov continued his academic work at the University of Kharkov, where he produced his most celebrated contribution, *The General Problem of the Stability of Motion*, published in 1892. This treatise introduced the direct and indirect methods now central to stability theory. Initially written in Russian, it gained international prominence after its 1907 publication in French. His career included appointments at Kharkov and later at St. Petersburg, where he became a full member of the Russian Academy of Sciences by 1901.

Beyond dynamical systems, Lyapunov made influential contributions to probability theory, mathematical physics, and theoretical mechanics. His legacy was further advanced by Barbashin, Krasovskii, and LaSalle, who extended the applicability of his theory via the principle of invariance and the construction of energy functions. In the context of control systems, the 1960 paper by Kalman and Bertram—*Control System Analysis and Design via the Second Method of Lyapunov*—firmly established Lyapunov's direct method as a core analytical tool in nonlinear system stability and controller design.

Definitions

When Lyapunov stability is discussed, two methods are typically considered: (i) the indirect method; and (ii) the direct method. The indirect method is regarded as a particular case in which the local stability of an equilibrium point is analyzed through the linearization of the dynamical system. In contrast, the direct method is characterized by the construction of a function that behaves like an energy measure, along with its associated derivative. In what follows, attention will be focused on the case of the direct method, where systems of the form

$$\dot{\mathbf{x}} = \mathbf{f}(t, \mathbf{x})$$

are considered, with x representing the state. For the purposes of the discussion, it will be assumed that the equilibrium point is $\mathbf{x}^* = \mathbf{0}$ (otherwise, a coordinate transformation can be applied).

Time-invariant systems[2] are now considered, i.e.,

$$\dot{\mathbf{x}} = \mathbf{f}(\mathbf{x}), \tag{B.1}$$

where $\mathbf{f} : D \mapsto \mathbb{R}^n$ is defined as a locally Lipschitz map from a domain $D \subset \mathbb{R}^n$ to $\mathbb{R}^n$, with the equilibrium point $\mathbf{x}^* = \mathbf{0}$ included in the domain. By assuming local Lipschitz continuity, local existence and uniqueness of solutions are guaranteed.

[2] In some books, these systems are referred to as autonomous systems; however, the term time-invariant is used here to avoid confusion with the term autonomous, which appears in other contexts.

Fig. B.1 Selection of $B_\epsilon(0)$

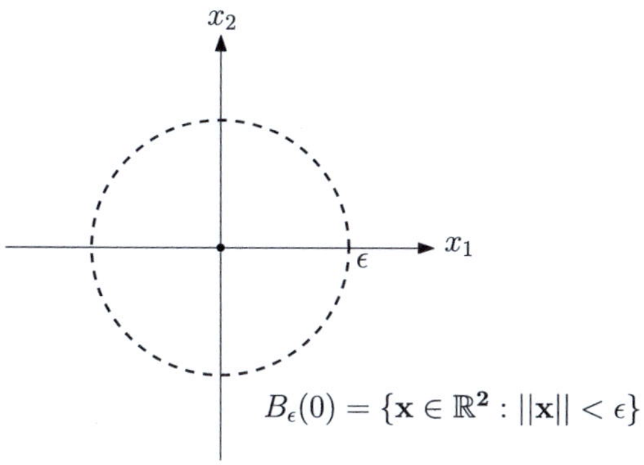

Fig. B.2 Selection of $B_\delta(\epsilon)$

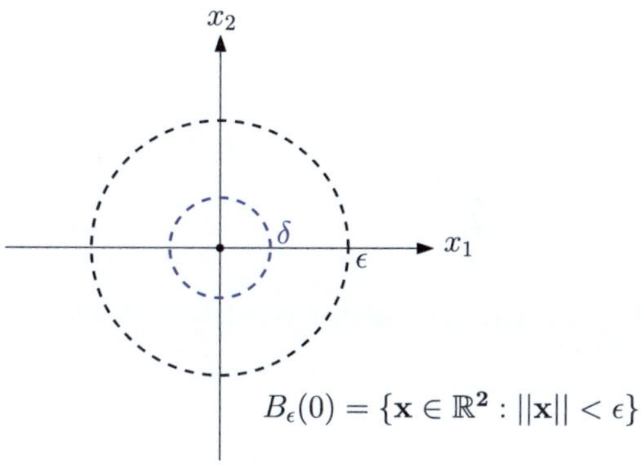

Definition B.1 (*Lyapunov Stability (SISL)*) The equilibrium point $\mathbf{x}^* = \mathbf{0}$ is Lyapunov stable (SISL) if and only if

$$\forall \epsilon > 0,\ \exists \delta(\epsilon) > 0 \text{ such that } \|\mathbf{x}(0)\| < \delta \Rightarrow \|\mathbf{x}(t)\| < \epsilon,\ \forall t \geq 0.$$

Figures B.1, B.2, and B.3 illustrate the SISL definition through geometric intuition in $\mathbb{R}^2$.

Example B.1 To illustrate this, consider the frictionless pendulum (Fig. B.4) with dynamics:

$$\begin{aligned} \dot{x}_1 &= x_2 \\ \dot{x}_2 &= -\tfrac{g}{l} \sin x_1. \end{aligned}$$

Figure B.5 shows the phase plane of this system with $l = 1$. In this case, it is clear that two balls B_ϵ and B_δ can be constructed around the equilibrium point, from which it follows that $\mathbf{x}^* = (0, 0)$ is stable in the sense of Lyapunov (SISL). Here, the maximum value of δ that can be chosen is constrained by the smallest axis of the largest periodic orbit contained within the ball of radius ϵ.

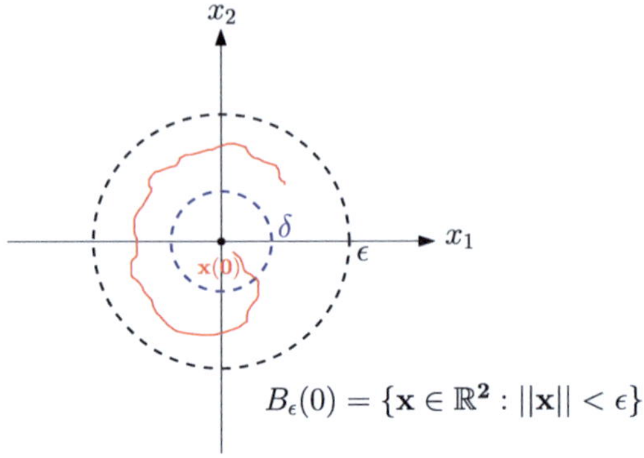

Fig. B.3 A trajectory with initial condition $\|\mathbf{x}(0)\| < \delta$ remains within $B_\epsilon(0)$

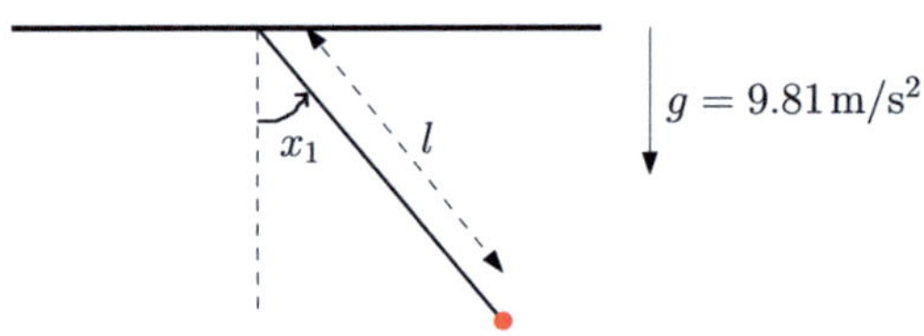

Fig. B.4 Inverted pendulum with length l, angle x_1 from the vertical, mass m, and gravity $g = 9.81\,\mathrm{m/s^2}$

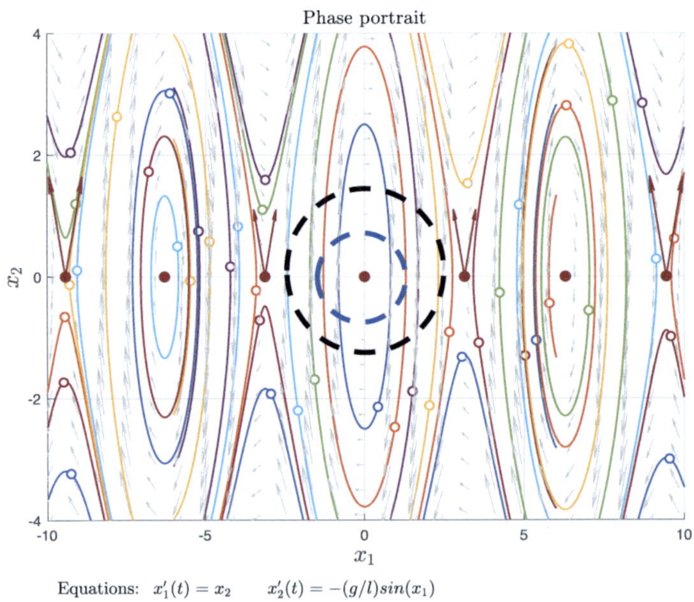

Fig. B.5 Phase portrait showing $\mathbf{x}^* = 0$ is SISL

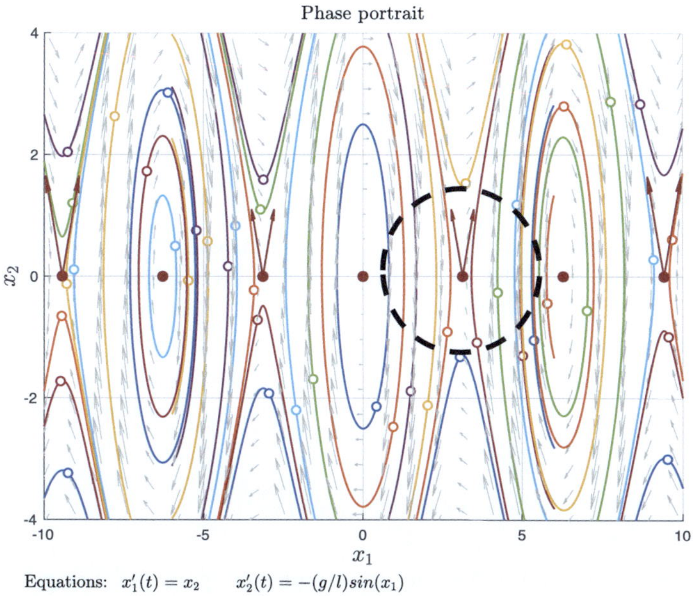

Fig. B.6 Phase portrait at $x^* = \pi$ showing instability

However, if attention is shifted to another equilibrium point, namely $\mathbf{x}^* = (\pi, 0)$, the phase plane shown in Fig. B.6 is considered. In this case, when a ball B_ϵ is drawn, it becomes evident that, for a given initial condition, the trajectories are observed to move away from the equilibrium point, leading to the conclusion that the point is unstable. ◊

Definition B.2 (*Asymptotic Stability (AS)*) The equilibrium point $\mathbf{x}^* = \mathbf{0}$ is (locally) asymptotically stable if it is SISL and

$$\forall r > 0 \text{ such that } \|\mathbf{x}(0)\| < r \Rightarrow \lim_{t\to\infty} \mathbf{x}(t) = 0.$$

This means that if a trajectory is initiated sufficiently close to the equilibrium point, it will be observed to converge to the equilibrium as time progresses. In other words, not only is the system's behavior kept bounded near the equilibrium (as in SISL), but convergence to the equilibrium is also ensured. The *region of attraction* is defined as the set of all initial conditions from which convergence to the equilibrium is guaranteed, and its importance lies in characterizing such conditions, as further discussed in the following remark.

Remark B.1 (*Region of Attraction*) In this case, the region of attraction R_A would be given by

$$B_r(\mathbf{0}) = \left\{\mathbf{x} \in \mathbb{R}^n : \|\mathbf{x}\| < r\right\}.$$

In this case $\|\cdot\|$ denotes the distance to the origin.

For a set to be the region of attraction of an equilibrium point means that if the system's initial state lies within that set, then the state will converge to the equilibrium. In this case, $B_r(0)$ is *a region of attraction* but it is **not necessarily the** region of attraction.

The region of attraction R_A is the set of all initial states such that the state converges to the origin. Thus, R_A is the union of all such regions of attraction.

Now, if $R_A = \mathbb{R}^n$, then the entire state space becomes the region of attraction, which means the result is global.

Definition B.3 (*Global Asymptotic Stability (GAS)*) The equilibrium point $\mathbf{x}^* = \mathbf{0}$ is globally asymptotically stable (GAS) if and only if it is stable in the sense of Lyapunov (SISL) and

$$\forall \mathbf{x}(0) \quad \lim_{t\to\infty} \mathbf{x}(t) = 0.$$

Clearly, if the system is GAS, the only equilibrium point is the origin.

Example B.2 To better illustrate this concept, consider the dynamic model of a pendulum with friction, given by

$$\begin{aligned} \dot{x}_1 &= x_2 \\ \dot{x}_2 &= -\frac{g}{l}\sin x_1 - \frac{k}{m}x_2. \end{aligned}$$

In Fig. B.7, the phase plane is shown for the case where $k = l = m = 1$. The convergence of trajectories toward the origin is observed when initial conditions are taken within the ball B_δ, constructed along with B_ϵ; as a result, the equilibrium point is characterized as asymptotically stable (AS).

In Fig. B.8, the region of attraction is illustrated based on the phase plane. However, in general, the determination of the region of attraction is considered a difficult (and, in some cases, intractable) task.

The natural question one might ask is whether it is necessary to verify both conditions in the definition of asymptotic stability (AS), since for linear systems convergence alone implied stability. The convergence condition would be

Convergence:

$$\forall \mathbf{x}(0) \quad \lim_{t\to\infty} \mathbf{x}(t) = 0.$$

However, in nonlinear systems, convergence does not necessarily imply stability. One classical example that shows this is the Nemytskii–Vinograd example[3]. In that case, although trajectories start near the origin, they must leave the ball $B_{r=1}$ before

[3] It is not described here, as it involves a time-varying system.

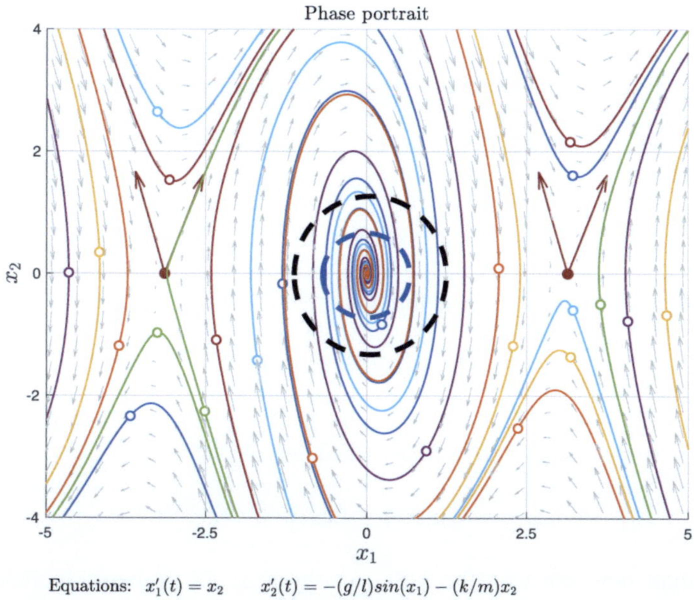

Fig. B.7 Phase portrait for the damped inverted pendulum showing AS for the case where $k = l = m = 1$

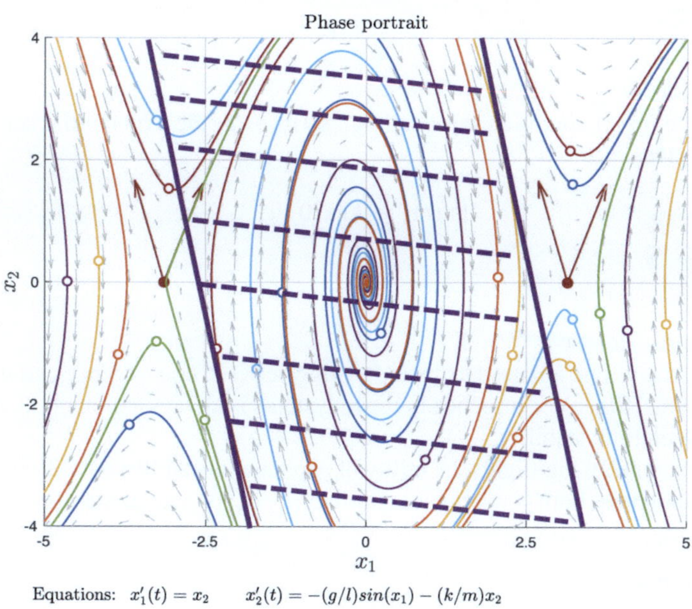

Fig. B.8 Region of attraction illustrated in phase space

eventually returning. The trajectories do converge to the equilibrium, but the point is not stable in the sense of Lyapunov (SISL).

A stronger condition than AS is exponential stability.

Definition B.4 (*Exponential Stability (ES)*) The equilibrium point $\mathbf{x}^* = \mathbf{0}$ is (locally) exponentially stable (ES) if and only if there exist $r, k, \lambda > 0$ such that

$$\|\mathbf{x}(0)\| < r \quad \Rightarrow \quad \|\mathbf{x}(t)\| \leq k\|\mathbf{x}(0)\|e^{-\lambda t}, \quad \forall t \geq 0.$$

In this case, the region of attraction would be

$$B_r(\mathbf{0}) = \left\{\mathbf{x} \in \mathbb{R}^n : \|\mathbf{x}\| < r\right\}.$$

Definition B.5 (*Global Exponential Stability (GES)*) The equilibrium point $\mathbf{x}^* = 0$ is globally exponentially stable (GES) if and only if there exist $k, \lambda > 0$ such that

$$\forall \mathbf{x}(0) \quad \Rightarrow \quad \|\mathbf{x}(t)\| \leq k\|\mathbf{x}(0)\|e^{-\lambda t}, \quad \forall t \geq 0.$$

One might now ask why the SISL condition is not explicitly included in the definition. The answer is that exponential convergence implies SISL. To show this, note that for any $\epsilon > 0$, one can choose a δ such that SISL is satisfied. In this case, let:

$$\delta < \frac{\epsilon}{k}.$$

Since the initial condition must satisfy $\|\mathbf{x}(0)\| < \delta$, the exponential convergence condition gives

$$\begin{aligned}\|\mathbf{x}(t)\| &\leq k\|\mathbf{x}(0)\|e^{-\lambda t}\\ &< k\delta\\ &< \epsilon.\end{aligned}$$

This holds because $e^{-\lambda t} < 1$.

This is why we are not used to distinguishing between convergence and stability in linear systems. Since convergence in those systems is always exponential, it directly implies asymptotic stability.

This relationship highlights the utility of exponential stability as a stronger condition that guarantees both convergence and stability (Exponential Stability $\Rightarrow$ Asymptotic Stability).

Direct Method

The motivation for using the direct method lies in the fact that if energy is continuously dissipated in an electrical or mechanical system, then the system will stabilize. That is, the system state will tend toward an equilibrium point. Consider a ball moving inside a bowl, as shown in Fig. B.9. The ball will roll down, then up, then down

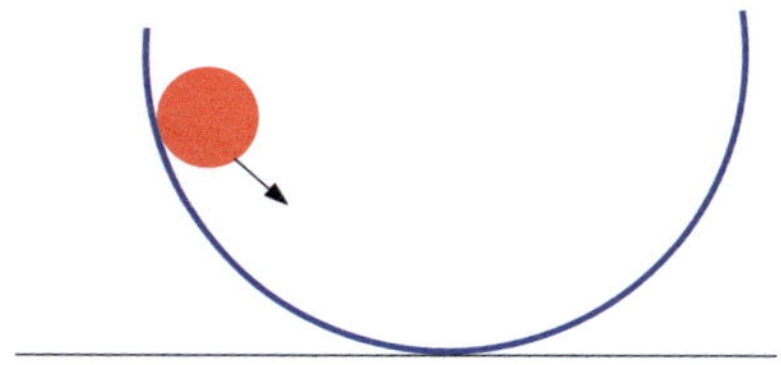

Fig. B.9 Ball rolling inside a bowl illustrating energy dissipation

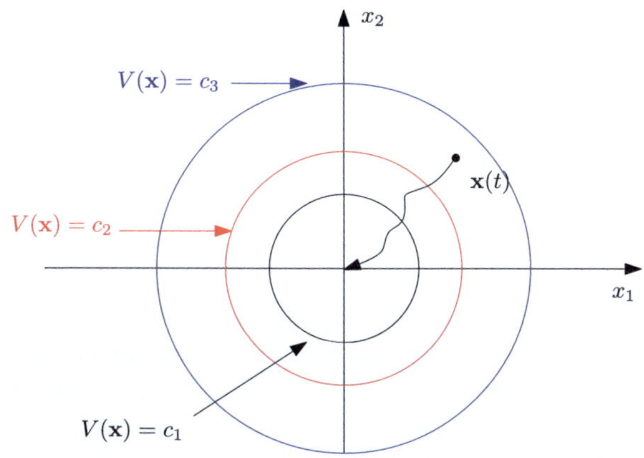

Fig. B.10 Level curve $V(\mathbf{x}) = c_1$

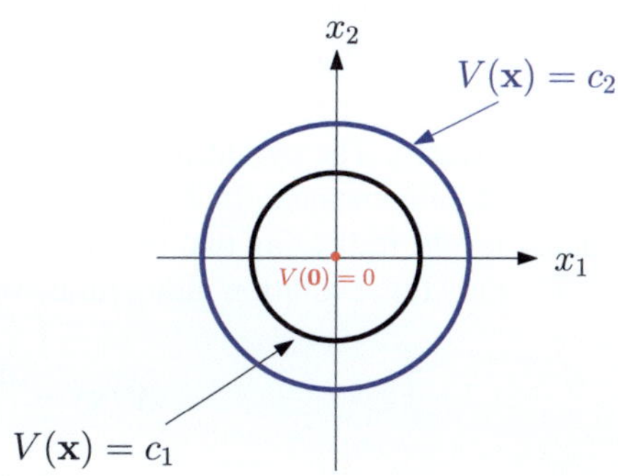

Fig. B.11 Two level curves $V(\mathbf{x}) = c_1$ and $V(\mathbf{x}) = c_2$

again, but due to friction, energy will be dissipated, and eventually the ball will come to rest at the bottom of the bowl (i.e., at the equilibrium).

To facilitate a better understanding of the concept, a two-dimensional system is considered, and the behavior of the equilibrium point $\mathbf{x}^* = 0$ is analyzed. The energy of the system is described by a scalar function of the state, denoted as $V(\mathbf{x})$. This energy is defined to be zero at the equilibrium, i.e., $V(\mathbf{x}^* = 0) = 0$. A set of lines in the state space along which the energy remains constant is then drawn. For instance, by selecting a constant value $c_1 > 0$, the configuration illustrated in Fig. B.10 is obtained.

Now consider another curve where the energy is constant at value $c_2 > c_1 > 0$, as shown in Fig. B.11.

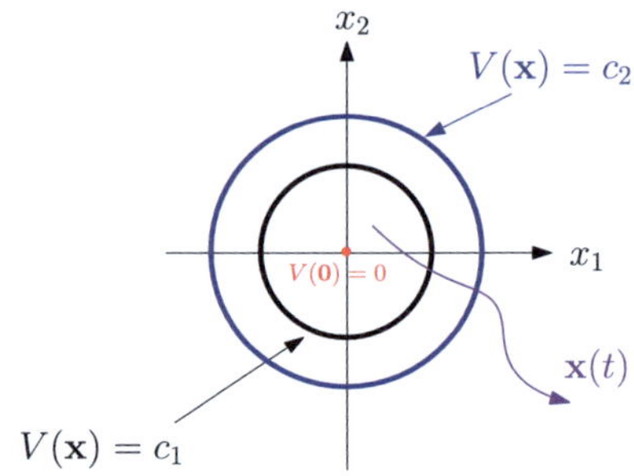

Fig. B.12 Trajectory moving toward higher energy level curves, hence moving away from the equilibrium

These curves are called level curves or surfaces, which are given by

$$V(\mathbf{x}) = c_i, \quad 0 < c_1 < c_2 < c_3 < \ldots,$$

and represent different constant energy levels in the phase plane. For all states on the curve $V(\mathbf{x}) = c_i$, the energy remains constant and equal to c_i.

How the properties of the equilibrium at $\mathbf{x}^* = \mathbf{0}$ can be analyzed using this energy function?

Time-invariant systems of the form

$$\dot{\mathbf{x}} = \mathbf{f}(\mathbf{x}), \quad \mathbf{x}^* = \mathbf{0}$$

are considered. The evolution of energy over time along the system trajectories is analyzed. For instance, in Fig. B.12, a trajectory is shown moving toward higher energy levels, indicating that the equilibrium is being moved away from.

To study this evolution along trajectories, the following expression is computed:

$$\begin{aligned}\dot{V}(\mathbf{x}) &= \frac{dV}{d\mathbf{x}}\mathbf{f}(\mathbf{x}) \\ &= \nabla V(\mathbf{x})\mathbf{f}(\mathbf{x}) \\ &= \sum_{i=1}^{n} \frac{\partial V}{\partial x_i} f_i(\mathbf{x}).\end{aligned}$$

This quantity is referred to as the directional derivative of $V(\mathbf{x})$ along the vector field $f(\mathbf{x})$. In other words, the change in the value of the energy function along the system trajectories is described.

Accordingly,

1. If $\dot{V}(\mathbf{x}) > 0$, the energy increases along $\mathbf{x}$.
2. If $\dot{V}(\mathbf{x}) < 0$, the energy decreases along $\mathbf{x}$, meaning trajectories intersect level curves corresponding to lower energy until the energy reaches zero. Thus, the equilibrium is AS. This convergence is illustrated in Fig. B.13.

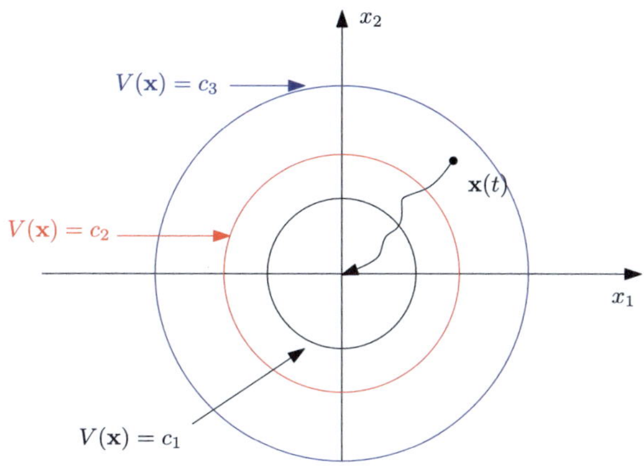

Fig. B.13 Intuitively, the condition $\dot{V} \leq 0$ implies that once a trajectory crosses the surface $V(\mathbf{x}) = c$, it remains in the set $\Omega_c = \{\mathbf{x} \in \mathbb{R}^n | V(\mathbf{x}) \leq c\}$. If $\dot{V} < 0$, it moves to lower energy surfaces approaching the origin

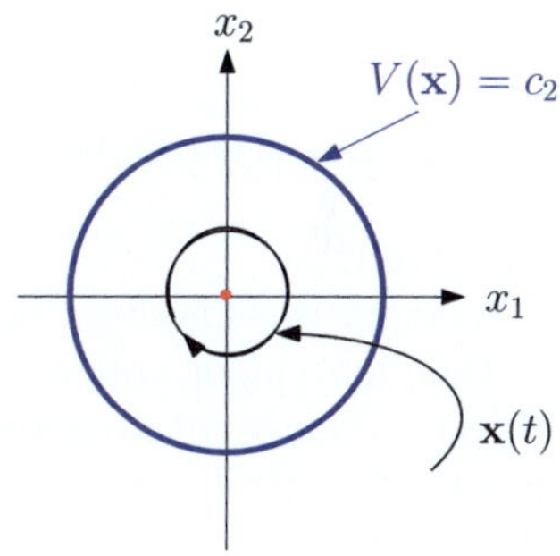

Fig. B.14 Trajectory moving along level curves and converging to a limit cycle. This fact occurs when $\dot{V}(\mathbf{x}) \leq 0$

Clearly, this analysis does not require solving the system equations; the directional derivative of $V(\mathbf{x})$ provides the necessary information.

What happens if the energy does not decrease but remains constant? In this case, $\dot{V}(\mathbf{x}) \leq 0$, and when $\dot{V} = 0$, the energy is constant, meaning trajectories move along the level curves. This scenario may not guarantee convergence, as shown in Fig. B.14, but the behavior resembles that of a stable equilibrium.

These energy-based concepts are directly applied to electrical and mechanical systems. However, the generalization to arbitrary dynamical systems was achieved in Lyapunov's 1892 doctoral thesis, *The General Problem of the Stability of Motion*, which marked a key contribution to stability theory.

Before this idea is generalized, some preliminary concepts related to continuity and differentiability are introduced. In this context, vector-valued functions mapping from an n-dimensional space to an m-dimensional space are considered, i.e., $\mathbf{f} : \mathbb{R}^n \mapsto \mathbb{R}^m$. In other words, functions of the form

$$\mathbf{f}(\mathbf{x}) = \left[f_1 \begin{pmatrix} x_1 \\ x_2 \\ \vdots \\ x_n \end{pmatrix} f_2 \begin{pmatrix} x_1 \\ x_2 \\ \vdots \\ x_n \end{pmatrix} \cdots f_m \begin{pmatrix} x_1 \\ x_2 \\ \vdots \\ x_n \end{pmatrix} \right]^\mathsf{T}$$

are considered.

If each of the functions $f_i(\mathbf{x})$, for $i = 1, \ldots, m$, is continuous at a point x_0, then the function f is said to be continuous at that point. By convention, the vector-valued function is referred to as C^0 at x_0. If each component function is continuous throughout a set $\mathcal{S}$, then continuity is said to hold on $\mathcal{S}$ for all f_i, $i = 1, \ldots, m$.

Moreover, if the function is continuous and its first derivative exists and is also continuous, the function is said to be *continuously differentiable*. This can be summarized as follows:

If

- $f_i(\mathbf{x})$ is continuous for all $i = 1, \ldots, m$, and
- $\frac{\partial f_i}{\partial x_j}$ exists and is continuous for all $i = 1, \ldots, m$ and $j = 1, \ldots, n$,

then, $f \in C^1$.

As in the previous case, this property can hold at a point x_0 or over a set $\mathcal{X}$.

This concept of continuity of higher-order derivatives can be extended. For instance, if the second-order partial derivatives of all components exist and are continuous, then the function f is said to be twice continuously differentiable, or C^2. This can be extended further to functions that are C^∞, i.e., differentiable infinitely many times. Such functions are referred to as smooth.

For general cases, an *energy-like functions* for general stability analysis is introduced.

Definition B.6 (*Lyapunov Candidate Function*) Let $V(\mathbf{x}) : D \to \mathbb{R}$ be a C^1 function defined on a set $D \subset \mathbb{R}^n$ containing the origin. The derivative along the trajectories of (B.1) is given by

$$\dot{V}(\mathbf{x}) = \sum_{i=1}^{n} \frac{\partial V}{\partial x_i} \dot{x}_i = \sum_{i=1}^{n} \frac{\partial V}{\partial x_i} f_i(\mathbf{x})$$

$$\dot{V}(\mathbf{x}) = \left[\frac{\partial V}{\partial x_1}, \ldots, \frac{\partial V}{\partial x_n} \right] \begin{bmatrix} f_1(\mathbf{x}) \\ \vdots \\ f_n(\mathbf{x}) \end{bmatrix} = \frac{\partial V}{\partial \mathbf{x}} f(\mathbf{x}).$$

This function $V(\mathbf{x})$ is called a Lyapunov candidate function.

Definition B.7 (*Lyapunov Function*) A function $V(\mathbf{x})$ is a Lyapunov function for $\mathbf{x}^* = 0$ if and only if:

1. $V(\mathbf{x}) \in C^1$.
2. $V(\mathbf{0}) = 0$ and $V(\mathbf{x}) > 0$ in $D \setminus \{\mathbf{0}\}$.

3. $\dot{V}(\mathbf{0}) = 0$ and $\dot{V}(\mathbf{x}) \leq 0$ in $D \setminus \{\mathbf{0}\}$.

If, in addition,

4. $\dot{V}(\mathbf{0}) = 0$ and $\dot{V}(\mathbf{x}) < 0$ in $D \setminus \{0\}$, then $V(\mathbf{x})$ is a strict Lyapunov function.

Remark B.2 Condition 2 implies that $V(\mathbf{x})$ is positive definite on the set D, so it has a minimum at the origin. This means the level surfaces approach the origin.

Condition 3 implies that $\dot{V}(\mathbf{x})$ is negative semi-definite in D, ensuring that V either decreases or remains constant along the system trajectories. This is the only condition involving the system dynamics.

In condition 4, the difference is that $\dot{V}(\mathbf{x})$ is strictly negative in $D \setminus \{0\}$.

In the following theorem, **sufficient** conditions are provided for the stability of the equilibrium point located at the origin. It should be noted that this theorem does not allow conclusions to be drawn regarding the instability of the equilibrium point. Furthermore, if multiple equilibrium points are present, the analysis must be carried out for each of them individually, with the caveat that a coordinate transformation should be performed beforehand in order to apply the following results.

Theorem B.1 *Let $\mathbf{x}^* = \mathbf{0}$ be an equilibrium point of system (B.1), and let $D \subset \mathbb{R}^n$ be a domain containing $\mathbf{x}^*$.*

- *If a Lyapunov function $V(\mathbf{x})$ is found, then the equilibrium point is stable in the sense of Lyapunov (SISL).*
- *If a strict Lyapunov function $V(\mathbf{x})$ is found, then the equilibrium point is asymptotically stable (AS).*

LaSalle's Invariance Principle

In many cases, it is only ensured that trajectories do not diverge when $\dot{V}(\mathbf{x}) \leq 0$. To establish convergence, the LaSalle Invariance Principle is applied.

Theorem B.2 (LaSalle's Invariance Principle) *Let $V(\mathbf{x})$ be a Lyapunov function for $\dot{\mathbf{x}} = f(\mathbf{x})$, and suppose that $\dot{V}(\mathbf{x}) \leq 0$ in a compact, positively invariant set Ω. Then, every trajectory starting in Ω approaches the largest invariant set contained in*

$$E = \{\mathbf{x} \in \Omega : \dot{V}(\mathbf{x}) = 0\}.$$

If $E = \{\mathbf{0}\}$, then $\mathbf{x} = \mathbf{0}$ is asymptotically stable.

Remark B.3 This principle is essential when $\dot{V}$ is not strictly negative. It identifies the asymptotic behavior of trajectories by their convergence to invariant sets within $\dot{V} = 0$.

Many additional results could be presented in this appendix, but only the main ideas have been chosen for inclusion. Concepts that connect passivity and Lyapunov stability have also been developed and presented in seminal works, such as those by Willems. In the following *Notes* section, some fundamental references are provided for readers interested in exploring the topic further.

Notes

The material presented in this appendix is drawn from both classical and contemporary sources. The foundational reference is Lyapunov's original monograph, recently reprinted in [5], which laid the theoretical groundwork for modern stability analysis. The primary references are the textbooks by Khalil [3] and by Haddad and Chellaboina [1], both of which offer a rigorous and comprehensive treatment of nonlinear systems and Lyapunov-based stability methods. Additional sources include standard and widely cited texts such as [2, 4, 6–9], which have informed the development of the theoretical material, history, examples, and interpretations presented throughout this appendix. Complementary visual explanations can be found in a series of illustrative videos available at @KYPTeachTech, where key concepts related to Lyapunov stability and nonlinear systems are presented.

References

1. Haddad WM, Chellaboina V (2008) Nonlinear dynamical systems and control: a Lyapunov-based approach. Princeton University Press
2. Isidori A (1995) Nonlinear control systems. Springer, NYC
3. Khalil H (2002) Nonlinear systems. Pearson Education. Prentice Hall
4. Kokotović P, Arcak M (2001) Constructive nonlinear control: a historical perspective. Automatica 37(5):637–662
5. Lyapunov AM (1992) The general problem of the stability of motion. Int J Control 55(3):531–534
6. Marquez H (2003) Nonlinear control systems: analysis and design. Wiley-Interscience
7. Michel A (1996) Stability: the common thread in the evolution of feedback control. IEEE Control Syst Mag 16(3):50–60
8. Mitropolskii Y, Borne P, Martynyuk A (2007) Alexander Mikhaylovich Liapunov. Nonlinear Dyn Syst Theory 7(2):113–120
9. Sastry S (1999) Nonlinear systems: analysis, stability and control. Springer

Index

J. Martinez-Piazuelo et al., *Generalized Nash Equilibrium Seeking in Population Games*, Communications and Control Engineering,
https://doi.org/10.1007/978-3-032-06081-5

FSC
www.fsc.org

MIX
Papier aus verantwortungsvollen Quellen
Paper from responsible sources
FSC® C105338

If you have any concerns about our products,
you can contact us on
ProductSafety@springernature.com

In case Publisher is established outside the EU,
the EU authorized representative is:
Springer Nature Customer Service Center GmbH
Europaplatz 3, 69115 Heidelberg, Germany

Printed by Libri Plureos GmbH
in Hamburg, Germany